Mathematics and Visualization

Springer
Berlin
Heidelberg
New York
Barcelona
Hong Kong
London
Milan
Paris
Tokyo

Claude P. Bruter
Editor

Mathematics and Art

Mathematical Visualization in Art and Education

With 284 Figures, 127 in Color

Springer

Editor

Claude P. Bruter

Université Paris XII
Mathématiques
UER Sciences
61 Avenue du General de Gaulle
94010 Creteil Cedex
e-mail: bruter@univ-paris12.fr

Cataloging-in-Publication Data applied for

Die Deutsche Bibliothek - CIP-Einheitsaufnahme

Mathematics and art : mathematical visualization in art and education / Claude P. Bruter ed.. -
Berlin ; Heidelberg ; New York ; Barcelona ; Hong Kong ; London ; Milan ; Paris ; Tokyo :
Springer, 2002
(Mathematics and visualization)
ISBN 3-540-43422-4

The cover figure reproduces a classical Kleinian tessellation of the hyperbolic plane by triangles (Klein, 1878-1879). In the present case, the angles of each triangle are $(\pi/2, \pi/3, \pi/7)$. All the triangles have the same area, $\pi/42$: they are the smallest triangles with which the hyperbolic plane can be tiled.

Mathematics Subject Classification (2000): 97D20, 97C50, 97D30, 97U99, 00B10

ISBN 3-540-43422-4 Springer-Verlag Berlin Heidelberg New York

Springer-Verlag Berlin Heidelberg New York
a member of BertelsmannSpringer Science+Business Media GmbH

http://www.springer.de

Typeset by the authors. Edited by Kurt Mattes, Heidelberg, using a Springer TEX macro package.
Cover design: *design & production* GmbH, Heidelberg

SPIN 10856665 46/3142LK - 5 4 3 2 1 0 – Printed on acid-free paper

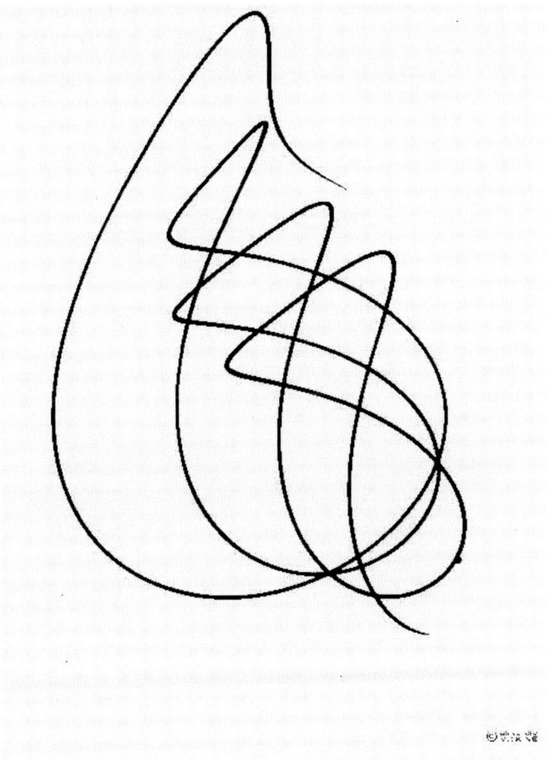

Born in 1910, Alexandre VITKINE, vitkine@wanadoo.fr, became a photographer and a graphic artist after a career in the industry. He is now a sculptor.

His drawings, photos (obtained through electronic equipments made by himself), and his sculptures produced by computer-controlled machines (info-sculptures), are based on mathematical forms, mainly those of Lissajous curves.

The drawing above was chosen as the basis for the poster announcing the Colloquium.

Preface

I am convinced that the work of the artists
is to create order from chaos
Fred Uhlman[1]

A Colloquium on *Mathematics and Art* was hold in the French city of Maubeuge in September 2000. The scientific committee included Jacek Bochnak (Amsterdam), Ronald Brown (Bangor), Claude-Paul Bruter (Paris 12), Manuel Chaves (Porto), Michele Emmer (Roma), Tzee-Char Kuo (Sydney), Richard Palais (Brandeis) and Valentin Poenaru (Paris 11). We would like to warmly thank Francis Trincaretto and his team who arranged to have the meeting in such an agreable venue: the "Théâtre du Manège".

Placed at the transition of the second and the third millennium, this Colloquium presented original ideas related to the development of new forms of civilization based on the many recent and rapid technological advances in communication and computation. With the strong encouragement of the local organizer, Francis Trincaretto, the Colloquium was – unlike more formal mathematical conferences – videoed and could be attended on the web. The speakers were true artists and mathematicians of rather unusual standard: while the artists were partly inspired by advanced mathematics, or even were sometimes ahead of mathematics, the mathematicians intended to show the beauty of their work and to share their feeling with the greater part of the population. They used all the old and new means of static and dynamic visualizations. Their works may be understood as symbolic and iconic representations of our environment and as essential tools for the understanding of our world, and the development of mankind.

Indeed, this Colloquium can be related to a renewal in the ways of diffusion and of teaching of mathematics. While schools of plastic or musical art are beginning to ask for some mathematics, mathematicians are seriously thinking of setting forth the artistic qualities of their work to attract the mind, and to support and facilitate the learning of their discipline.

We hope that readers may find in these proceedings ideas, projects and realizations which can contribute towards the inspiration and promotion of new cultural developments in Society.

The order of the articles follows that of the talks. Pictures and images appear in grey in the articles. Images appear also in colour in the Appendix. Such an image is labelled [XY]k where XY are the main initials of the author, while k numbers the image.

[1] (1901–1985) German advocate, then British painter and writer: his booklet *Reunion* is a true masterpiece.

To conclude, we would like to thank Mike Field who accepted to help the translations into English of most French written texts, Bill Mac Callum who did that work for the second part of my first contribution, and the Springer team who accepted to publish these Proceedings, and provided their help to the editor.

Claude-Paul Bruter

Table of Contents

Presentation of the Colloquium. The ARPAM Project

Claude-Paul Bruter

Mathématiques, Université Paris 12, Av. du Général de Gaulle, 94010 Créteil, France <bruter@univ-paris12>

1 The Colloquium

1.1 Introduction

My intervention has two parts. The first one is devoted to a general presentation of the Colloquium, through an evocation of the works of the artists who are present among us. Thus that presentation does not address mathematicians in a particular way. It reveals some of the reasons which have directed the scientific organisation of the Colloquium. As its architecture shows, it turns over the art of visualisation of mathematics, either for the general public, or for the one of mathematicians. The second part is devoted to a succinct description of the ARPAM project.

1.2 Presentation of the Colloquium

On the foundations of the relations between Mathematics and Arts
As a preliminary comment, it is fitting to say a few words on the relations which tie Mathematics and the Arts: they are so tight that sometimes Mathematics is compared with one of the Fine Arts.

One of the reasons, the main one to my eyes, which solders the arts to mathematics is probably the following: the tangible object, the living being, are not only present in space, and are evolving in space, but are moreover highly elaborated constructions, obtained from the unfolding of the properties of the primordial space.

In other respects, the existence of the object, that is its inward properties of stability, are themselves dependent on the stability of its constituents, of their internal arrangement according the various levels of integration. This existence also depends upon the capabilities of the object to resist against shocks of any kind, of internal or external origin, created by all that makes its environment, close or distant, into space and time. Thus, knowledge of this environment, in all its modalities, is the essential means whereby the being can guarantee its spatio-temporal stability. So we are always brought back to the fundamental problem of the knowledge of the space, of all the richness of its manifestations.

Indeed, mathematicians as much as artists are preoccupied with deepening this knowledge. They succeed by using representations, primarily abstract and scriptural for the mathematicians, more physical for the artists. As both of them are sometimes representing the same objects, one cannot but wonder about the common points of these processes of representation, and that each borrows subjects of representation, resources, discoveries from the other, in such a way that art and mathematics progress together by mutual enrichment.

The recent films by Michele Emmer, one of the pioneers in the study of the connection between mathematics and art, put in the limelight the discreet but significant role of the development of mathematics on contemporary art. The weight of this influence will also be revealed by an inspection of some of the themes on which our artist friends have been working.

Regarding the plastic arts, six topics will be introduced during this Colloquium: the first, titled "perspective and geometry", starts with the artistic technique; the three following, "polyhedra", "curves", "surfaces", concern mathematical objects of stiff appearance, very classical, which hold an important position in the mathematical universe; the fifth theme, recurrence and dynamical systems, is more recent, and the development of the computer has given it an important boost; the sixth and last theme, the "sphere eversion", has likewise the interest of novelty, not only from the mathematical point of view *per se*, but also from a pedagogical point of view. To this end, a video was made both for mathematicians and also to popularize mathematics, a consequential video by its cost and by the size of the team which has produced it. Finally, to conclude this first part, homage will be paid to those who make us share joys among the purest, those of music.

First theme: perspective and geometry The choice of this theme is partly directed by historical reasons. For at least 32 000 years, when they used to ornament the walls of the caves, the artists have painted on plane or curved surfaces, using sometimes the rules of perspective in a spontaneous manner. We are indebted to the painters for having founded a rational theory of perspective. According to the roman architect Vitruve, the painter Agatharchus, from the creation of sceneries for the Eschyleus theatre, would have been the forerunner of the theory. Anaxagoras and Democritus would have begun to develop it, but all their works are lost. The Renaissance artists, like for instance Brunelleschi around 1415, in their turn, brought into focus the first known rudiments of this theory for the practice of their art. These elements lead the architect from Lyon, Gérard Desargues, around 1639, to base projective geometry: that is a classical example of the phenomenon of symbiosis between art and mathematics.

Projective geometry holds an important position in mathematics because its interest is much more general than the one of classical Euclidean geometry. Indeed, in Euclidean geometry, the bright source which illuminates the objects is located at infinity. In projective geometry, the bright source lies at any point of space, at a finite distance or not: thus, in this respect, projective geometry contains classical Euclidean geometry.

Many painters are working on canvas which, from the point of view of mathematical representation, are understood as pieces of plane surfaces. The point of their canvas towards which the eyes of the observers seem to converge is called the vanishing point. It plays an important role in the construction of the pictures.

From the mathematical point of view, a surface is an ideal, infinitely thin skin. The diversity of the surfaces is infinite. We are going to stick to perfectly smooth surfaces, without any asperity, as for instance the plane surfaces or the spheres. A plane surface is a very singular surface characterized by the fact that its curvature is null at any point. Apparently, a few things distinguish the plane from the sphere: an important difference (which implies others) lies into the value of the curvature, which is constant at any point in both case, but null in the case of the plane, and non null in the case of the sphere.

The curvature is a local data: the one at a point of a more or less elastic thread is tied to the effect of the internal tensions, to the ability to resist to a stretch at that point. If no resisting force is present, the thread seems to be able to strech out indefinitely, there is no natural curvature, the physical and mathematical curvatures are null. Let us take now an elastic and smooth surface, as the canvas is a little bit. It is a kind of fabric whose stitches are infinitely fine and close. At each point, two elastic and perpendicular threads cross, each one having a local curvature at the point.

From these data, one defines two notions of curvature, first the *Gaussian curvature* at the point under consideration, which is the product of the local curvatures of each of the two threads crossing over the point. This notion of curvature allows us to classify the smooth surfaces within three categories: the spherical or elliptic ones with a positive curvature, the hyperbolic ones with a negative curvature, the parabolic ones with a null curvature. Among these, are the plane surfaces whose local curvature is null in all the directions. Indeed, when the local curvatures of the threads are for instance positive, their product which is the Gaussian curvature is also positive, as it happens in the case of the sphere. When the local curvature of a thread is positive while the local curvature of the other thread is negative, the Gaussian curvature which is their product is negative, as in the case of the surfaces of some water-towers, and which are generated by the rotation of a piece of hyperbola around one of its symmetry axis.

Let us come back to painting. We are accustomed to look at paintings which mainly are painted on plane surfaces. But why to stick at that? Would not it be possible to paint on a spherical or hyperbolic surface? But then what could be the reasons which would lead a painter to put forth his genius on such or such type of surface? There are the natural data indeed: the painter of caves will practise on the spherical shape of a stone, on those, plane, spherical or hyperbolic of the walls of his cave. But there might be other reasons, as those that the painter Dick Termes will show in detail.

His wish was to represent the whole space, not only what meets our eyes, before us, but also what is on our sides, at the right, at the left, above us, underneath, and behind. He has then taken six different canvas, stretched over a cube, with which he can represent all that space. By blowing inside the cube, not too much strongly in order not to burst the canvas, the cube becomes a sphere, that the topologists like to cover with six curved disks, equivalent to the six faces of the cube.

On each canvas, Dick Termes chooses a vanishing point and represents the part of space that faces it. He will explain us how he chooses his vanishing points in such a way that the partial images join together harmoniously. He proceeds quite as the geometers who construct local representations, then, by using analytical techniques, fit them together to obtain coherent wholes. This point of view is the one of the theorist. Dick Termes' work is interesting, not only because of his remarkable artistic qualities, but also because he solves a concrete problem of reconciling images. The artist enriches the corpus of problems brought up to the mathematician, suggesting to undertake a fine study of the junctures between local geometries by changes of the vanishing points on the sphere, and more generally on smooth surfaces of any curvature. One can elsewhere set up the question: given a representation on a sphere, of which types of spaces is it the image?

Second theme: polyhedra Here is an other example of interaction between art and mathematics. We have just met, for the needs of the complete representation of the usual space, a first polyhedron, the cube. Let us notice that the cube has this marvellous properties to be able to be easily mass-produced, and that the stacking of infinitely many cubes in all the directions allows us to fill up the space. This property makes the felicity of one of my friends. One should maybe find him as ill-seeing, if not a little bit foolish, but for him, all the objects of Nature have the shape of a cube. He is so very happy, because he is one of the rare people to be able to answer one the fundamental questions, how does Nature fill up the space?

Mathematicians have found other polyhedra which, joined, permit to fill up the space. One for instance will quote this result by Poincaré used in the film *Not Knot* according to which one can make a tessellation of an hyperbolic space, i.e. with negative curvature, using hyperbolic dodecahedra, which are polyhedra with 12 curved faces. Many mathematicians and artists are captivated by polyhedra. Their study is the starting-point of an important part of the works by George Hart and Charles Perry. Starting with known polyhedra, they proceed to learnedly controlled deformations to get objects which are full of power, of dynamism, and of novelty. George Hart among other things deals with partial but regular simplicial subdivisions on the edges of nested polyhedra. He thus implicitly makes new local groups of symmetry, and contributes to enlarge the theory of the 230 classical crystallographic groups by introducing, over that basis, kinds of algebraic fibers. The proceeding can be widely generalised, leading on the limits to fractal structures, first passing through combinatorial structures of a giddy extent.

Let us suppose that the space is regularly tiled so that all the tiles have the same shape, the same dimension. Let us cut our space by a surface, for instance a plane surface. What is the trace of this tiling on the plane: a regular tiling? That may happen. Antonio Costa will show us the famous tilings that the Arabian artists have produced in the Spanish city of Grenada, on the walls of the Alhambra palace. Five centuries before us, they discovered the fact that there exist 17 really distinct ways to tile a plane, each different type of tiling being characterized by a particular family of internal symmetries. We are in presence of motives which, on the plane or in the space, are infinitely repeated. By using a learned play of mirrors constituting a pedagogical tool of great interest, Maria Dedo will explain how she makes these motives discover as well as the set of symmetries which characterise the polyhedra. Such a play of mirrors is used in the film *Not Knot*, the Poincaré's tessellation of the hyperbolic space appears.

Whereas George Hart enriches the motives of the spatial tilings by the use of methods belonging to static mathematics, Michael Field beautifies the motives of the plane in a rich and elegant manner, by calling upon techniques which are used in dynamics. Indeed, the study of dynamical systems shows us this remarkable phenomenon, the birth of new morphologies at singular moments, for singular values of the parameters. These creative bifurcations can put in light hidden internal symmetries, new shapes of trajectories. The artist and mathematician work these phenomena to create new remarkable ornamental pictures.

Third theme: knot shaped curves The curves, the trajectories, the threads without thickness which close on themselves are called knots of topological dimension 1. Their diversity, their interweaving, their infinite variations of shapes immerse the mind into reverie, or on the contrary fix it on perfection. They have inspired among the most impressive works of the sculptors Nat Friedman, Charles Perry and John Robinson.

Knots play an important role in physics and in mathematics. Ronald Brown and Nathaniel Friedmann will show all their artistic and pedagogical value. As one will see in the film *Not Knot*, there exists very tight connections between polyhedra and knots. Let us take a knot: prick it in some points we call vertices, then stretch the portion of curve located between any couple of vertices up to get a rectilinear segment called an edge; we have set up a sequence of edges which are closed on itself which has the same topological properties as the initial knot. The skeleton of dimension 1 of a polyhedron made with its edges is then an combination of knots having common parts, and that elsewhere we can separate one from each other in many ways.

Forth theme: the surfaces The theory of knots belongs to topology, i.e. to the study of the properties of the space, independently from considerations of distance, the physical meaning of the distance being that of the energetic cost of the transfer from a point to an other. Topology becomes geometry when these supplementary metrical considerations are taken into account in the study of space. Three talks will show us very various shapes of geometrical surfaces, conceived from sometimes very different motivations.

On a technical point of view, Konrad Polthier's talk addresses means of studies and of representation of minimal surfaces. From the mathematical point of view, they are defined from a notion of local curvature that is different from the Gaussian curvature. The Gaussian curvature at a point is the product of the curvatures of the perpendicular threads that cross at that point. The other curvature, called the *mean curvature* is simply the half sum of the local curvatures of the previous curves. When the mean curvature of a surface is everywhere null, one says that the surface is minimal because then the value of some energy tied with the surface and depending of the local curvature of the threads is minimal. A noteworthy case is that of soap bubbles, which were deeply studied by the Belgium physicist Plateau in the 18th century. Konrad, one of the best scientist involved in mathematical visualization (see for instance http://www.eg-models.de), will discuss this subject in detail. He will project films he did on such surfaces, which sometimes have inspired architects to make important roofings.

The two last talks of this Colloquium will be devoted to other geometrical surfaces. Bruce Hunt will guide us along a rich and beautiful gallery devoted to algebraic surfaces which are not necessarily smooth. The video by the sculptor Helaman Ferguson will show some of them, modelled in stone or in metal.

The ruled sculptures by Philippe Charbonneau, generated by moves of lines, are inspired from some of these surfaces of order 3. Some of theses sculptures have the property to be movable around privileged axis, and introduce symmetries with respect cylindrical or conic elements, which make us think of those one can sometimes see in the gears. François Apéry proposes a generalisation of that constructing process of ruled surfaces by replacing lines by conics. Through the creation of these elements of sometimes branched and with various symmetries algebraic varieties, these sculptures are maybe the indication of an enrichment of the mathematical museum.

Richard Palais is trying to set forth such a museum. He will show us some of the precious objects it contains. Dick Palais' project goes far beyond of a simple virtual museum. In fact, it will be used as an intelligent interactive library of a new style, that will be a very useful tool to inform and to discover, in as much for the mathematician as for the neophyte. I hope you will be charmed by discovering this project, of a high elevation in its conception, to the realisation of which we shall be doubtless numerous to contribute.

Sixth theme: recurrence and dynamical systems The study of recurrence and of discrete dynamics has more especially led to the discovery of the fractal world. This one has inspired the realisation of a fascinating video, *The Mandelbloom*, a marvellous ballet of roses, dancing on a choreography suggested by the events proper to this universe. The fractal phenomenon is tied to the presence of iterative procedures together with scales reductions: one finds it in the methods for solving polynomial equations used by Scott Crass, and in the study of a great number of dynamical systems, such as those studied by Mike Field and John Hubbard. The first development of the fractal theory concerns recurrence equations defined on the space **C** of Chuquet'numbers, the so-called complex numbers, representing similarities on the plane. Such equations have equally been studied using the Hamilton numbers, the so-called quaternions representing similarities in the usual space. Is it not surprising to find among some sculptures by Charles Perry an anticipation of the representation of some quaternionic Julia sets, discovered more lately, and that we shall find in the Jean-François Colonna virtual exhibition?

Last theme: the sphere eversion Let us imagine Dick Termes located at the centre of a sphere, painting it. Obviously Dick wishes to show his work to the largest possible public. But this public stands outside the sphere. What to do? Does not the solution simply consist in eversing the sphere? In a mail, Stewart Dickson informed me of his present preoccupation, to materialize objects in process of transformation, as for instance the sphere during the steps of its eversion. The precise problem set up by this eversion is to deform the sphere so that its first interior, red and hidden, become *in fine* the exterior face of the sphere, without by no means injuring the sphere, without causing any traumatism to this infinitely thin skin.

The president of the Association Idem + Arts, Francis Trincaretto, is a surgeon. You will understand that he asked me the origin of this problem, since the work of surgeons would be greatly made easier if they could turn us over like gloves. There is nothing less sure that the mathematicians have much take care of helping the surgeons. Usually, it is rather the contrary that happens.

In a pictorial way, the mathematical origin of the problem is the following. From the "practical" point of view, mathematicians are interested in problems of removal, of transportation. For instance take one of the painted Dick Termes sphere. As a sphere, it is initially a totally smooth surface, without any asperity to the contrary for instance of a mountain, and whose realisation has required a great care. The artist does not look upon it that his work be injured during its transportation from Searfish to Maubeuge. Mathematicians call a carriage a mapping, and to qualify this transport without any accident, use the aquatic term of immersion.

In 1957, the mathematician Steven Smale questioned whether, two transports without accident being given, it would not be possible to pass from the first to the second along a sequence of such transports. In the case of the carriage of the sphere, his answer has been positive, which implies that one can mathematically turn over the sphere without subjecting it to any traumatism. This transport can have anyhow a particularity which sometimes partakes of the science fiction. In the same way that the famous hero of a French story, by the humorist Marcel Aymé, could cross over the walls, it is admitted that the surface may go across itself.

From these data, mathematicians have searched to create procedures to everse the sphere. Bernard Morin, François Apéry and John Sullivan have been important actors of that saga. In order to get for themselves a better understanding of the techniques they had to work out, and also to give a better understanding to their colleagues, the mathematicians have first been led to prepare drawings and iron-wired or papered models. In their workshop, Bernard Morin and Richard Denner will show some of these models inspired by the first methods of eversion developed by Bernard Morin.

John Sullivan has worked on another procedure of eversion using a minimal energetic cost constraint. Stewart Dickson will show some main steps with the help of physical realisations, while John's film, *the Optiverse*, will allow us to follow the continuous unfolding of the eversion.

This unfolding has inspired the creation by François Apéry of a flexible, luminous steel-wired sculpture, that can be taken into pieces: then one looks at the unfolding of a double covering of the sphere. The object, the demonstration are fascinating.

De la musique avant toute chose ... "De la musique avant toute chose", in the "old times", pupils in high schools used to learn this verse from the French poet Verlaine. It is possible that, among all the arts, the music be the closest to mathematics by the sensations aesthetic it brings. One can for instance find them again in the ethereal practice of arithmetic. And maybe the Pythagoreans would have not refuted this proportion: the geometry is with respect to plastic art what the number is with respect to musical art. From the past mathematicians who worked on the foundation of the musical art, we can remember the names of Pythagoreas, those of Archytas and Euler. It is on the study of the same foundations or on their expressions that today the musician Michel Deneuve, the mathematicians Carlota Simoes and Yves Hellegouarch, equally a professional cellist, are working. Hellegouarch's mathematical theory of scales precisely describes the most diverse musical situations. We should not thank enough these musicians for having accepted to make us share these sometimes divine suggestive sensations that their art affords to us.

The objectives of the Colloquium I would like to conclude this beginning upon the subject by a return to the objectives of the Colloquium. It is not merely a matter of showing the wonderful objects that the artists and the mathematicians, each one on his side, are able to produce. It is not only the question of promoting the fructuous exchanges between these two communities. The Colloquium has not only the simple purpose of serving as a prefiguration to the realisation of the park of geometries, whose main outlines I shall describe in a moment, and to better validate this project in the opinion of the scientific community.

The central objective of the Colloquium is of pedagogical nature. In some subtle way, it would like to concern the formation of the mind. In a more apparent manner, it turns around the transmission and the acquisition of knowledge through visual representations, be they static like painting, sculpture or architecture, or more dynamic like the modern audio-visual animations, the qualities of these representations being supported by the resources of the musical art and of the literary art. And so, the important objectives of the Colloquium are to encourage the formation of contacts, the reflection on these animations and their creation in the midst of community structures.

2 The ARPAM project

2.1 The pedagogical data

Two essential reasons, of more or less biological and psychological nature, justify the interest of the project:

1) the first takes as a starting point the saying that "the animated precedes the inanimate one", and the aphorism of Charles Darwin, "any construction is genealogical". From there results that the acquisition of the knowledge and the formation of the spirit, which have a phylogenesis, deserve to be conceived according to a process of ontogenesis which respects this phylogenesis. One will remember that the first developments of science were accomplished within the framework of societies at a pace very slow, appearing even almost fixed to the local temporal glance. The first discoveries were established by the patient observation of structures or objects which appeared immutable. A fixed object has the advantage of being able to be contemplated for as a long time as it is wished so that our senses have the possibility of taking a detailed image of it which can, with much more facility that the fleeting image, print itself deeply in the memory. Therefore is it natural to support any process of training by repeatedly presenting of objects illustrative of the subject matter, in particular, if possible, objects motionless or subjected to slow and repeated movements.

Project ARPAM is a pedagogical project which, as one of its bases, rests on the paramount presentation of fixed objects.

2) Any object which can be seen, touched with the eye, has a semantic field much larger than that of its idealisation, of the abstract object. It has one signifier, one meaning, more or less hidden, and through it, has an emotional potential. As any process of attractive training can only be based on the use of attractive objects, it is advisable to choose the latter with understanding,

by taking account at particular emotional effects that they can induce. The sight of certain objects can contract the spirit, to cause in it violent tensions. Others, on the contrary, can produce a relaxation, release the thought of tensions and concern locked up it in a kind of reducing vice, until it is prevented from being receptive on other subjects, in particular the innovations. Many artistic objects have this physical and psychological virtue. Inserted in the process of training, they bring a supplement of heart to science.

Project ARPAM aims at including, in a way more conscious than usual, an artistic dimension in pedagogy, in particular within disciplines considered difficult, like mathematics. The examples, the illustrative objects, through the complex work of our senses, must strike the spirit, and allure it. This is a characteristic of works of art. They can, in thousand ways, present a character of strangeness by their dimension, their unexpected forms, the successful layout of their decoration and their colours, and by their unusual sound effects.

It is on these bases which the project that I will briefly present was drawn up. Although their cost does not allow their multiplication in great number, it goes without saying that other projects of the same nature can be and undoubtedly will be conceived, for, we hope for it, the benefit of the human community. Allow me to recall here that the practice of mathematics tends to abolish the borders of any nature which can separate societies. This practice reveals the universality of the steps of thought. It is a factor of exchange and unity between men. UNESCO carries from there testimony, which made year 2000 not only one year of peace between the people, a wish alas still pious, but also that of mathematics, a wish finally fulfilled.

2.2 General characteristic of the project

The fixed objects chosen are of modest size, on the one hand to respect a certain intimacy, on the other hand not to burden the cost with the realization. They are however of sufficient size to impress the senses with a measured strength.

These objects are either gardens, or small buildings called "folies", or sculptures. Although their basic structure is fixed, decorations which they carry are likely to be modified with time. Their principal function is to illustrate concepts and facts belonging to the greatest possible extent of the mathematical field, while evoking the history of their discovery.

In order to reinforce the effect of surprise, the majority of the objects should a priori be hidden. One will discover them at the turning of a path, behind a curtain of trees, by reaching a saddle, or the top of an hillock.

In addition, the objects will not be crowded together as one finds in museums. Relaxing paths will separate them; this relaxation can also be created by pleasures of the eye, as by physical effort which constrains the work of the thought. The time spent in walking between two objects should be sufficient to make it possible to fix in memory and to assimilate the previous object, and to make the spirit sufficiently available to discover the following curiosity with interest.

These constraints, to which it is naturally necessary to add that of the accessibility, weigh on the choice of a site for the project. There is undoubtedly no ideal site. That which the Town of Maubeuge, offers, in the fortifications created at the 16th century by the geometer Sebastien Vauban, has, by the weight of its famous past, the advantage of the originality.

An introductory panel will be placed at the entry of each garden and each folie. Folders of various types established according to the mathematical level of knowledge of the reader, will give additional information. To the interior of each folie will be placed one or more data-processing consoles for which interactive tools for visualization will be designed. Young researchers present in the folies to answer the possible questions of the visitors will in addition be charged partly to develop these tools, which will have to meet teaching standards.

2.3 A word on the gardens

The gardens, like the folies, bear names. In the preliminary draft, five gardens were envisaged:

A. the garden of symmetries, close to the Seventh Temple
B. the projective park, close to the Cap of Apollonius
C. the phyllotaxic clearing, close to the Horn of Plenty
D. the knotted forest, close to the Knotted Stained Glass
E. the Eulerian gardens, close to Euler's Bridges

The garden of symmetries, the projective park, and the Eulerian garden are typically French gardens. Their creation should make happy a renewed family of gardeners. Free course will be given to their imagination within the geometrical data which we will be able to propose to them. Topiary art, of the size of suitable shrubs, will be renewed. It will not be a question any more of simply cutting parallelepipeds, spheres and cones, but also of reproducing the various shapes of the objects which one meets in the geometry of surfaces.

Here is a very simple example of a flower-bed, illustrating the configuration of Pappus of Alexandria (IVth century after J.C.), the last of the large Greek geometers. This configuration is composed of a polygon with 6 sides or hexagon, whose alternate nodes are aligned on two lines represented on the ground by alignments of red flowers. The sides of the hexagon are represented by alternate alignments of yellow and blue flowers. It is remarkable that the points of intersection of the opposite sides of this hexagon are also aligned. The poets will observe in this configuration the presence of two M intertwined. The modern gardeners could try to execute this flower-bed with a new flower, recently discovered if one believes the scientist newspaper "Nova Biologica". This flower is very interesting: one could show indeed that it could carry only four colours at most, and that it had a symmetry of a new nature, like the number of nodes of the configuration. This is purely anecdoctal. The scientific name of this flower is: *Septembris Malbodiensis* (or *Maubeugensis*) *Calunaris*.

We will see now why the garden of symmetries, whose flower-bed will be able to comprise friezes and plane tilings made up of quite selected flowers, deserves to be placed not far from this folie called "the Seventh Temple".

2.4 Succinct presentation of some folies

Introduction Each one is free to conceive as many folies as he wishes. In reaction to the sheep-like and anaesthetizing shape of uniformly cubic constructions which populate our cities, a common point between these projects would be the concern of building folies whose uncommon form stimulates, on the contrary, the spirit and imagination. I have not yet completely defined all those of which I have thought, in particular the three following ones:

"The Knotted Stained glass", entirely made with transparent material, illuminated from inside, intended to illustrate differential topology.

"The Poincaré Surprises", folie whose roof comprises two parts between which circulates a fluorescent fluid. The lower part of the roof is fixed, whereas the other part is mobile and transparent, which makes it possible to follow the evolutions of the fluid.

"The Luminous Torus", a building in the shape of a torus, made partly out of glass. It allows one, from the interior of the torus, to observe the effects, and in particular the caustics, produced by lightbeams projected on a suitable transparent object of variable index of refraction.

Here is the description of six computed folies. The drawings which here accompany them reproduce the original drafts made ten years ago on an old computer.

The Cap of Apollonius The steps of thought, in particular that of analytical thought, are progressive. After the study of linear properties of objects, of degree 1, comes that of objects of degree 2, the first of the nonlinear ones. These objects are also known as quadratic. They played and continue to play an essential role in geometry, the theory of numbers, and in the applications of mathematics in the physical world, in particular in mechanics. This part of mathematics is partially illustrated by the Cap of Apollonius (born in Pergamon in 262, died in Alexandria into 190). This folie makes it possible to display some basic results of Euclidean geometry in usual space. The shape of the building resembles the cap which the noble ladies wore in the Middle Ages. It is about a truncated cone of revolution, divided by a vertical plane which contains the way out. At the interior, one shows the circles of Dandelin and the principal traditional properties of the conical sections. The higher part of cone is out of transparent material to benefit from the natural light. The remainder of the truncated cone is out of clear metal. Dick Mee will show on one of his beautiful CD-Roms how one can decorate the outside of this truncated cone, isometric with a portion of the Eucliden plane.

The Seventh Temple Encouraged by the remarks of epistemologists, the mathematicians of the 18th century revealed some structures in the families of objects that they handled, in particular that of group, initially in connection with the numbers, then in geometry. A set of objects has the structure of group, this structure is known as an algebraic structure, if, in particular, these objects can combine between them, and if any object admits a symmetric one. This theory thus makes it possible in particular to study the manner of filling space using standard tiles, where symmetry plays an essential role. The form of the Seventh Temple resembles a little that of certain small temples of Antiquity, and intends to illustrate this significant algebraic theory, in particular, from the visual point of view, through the presentation of tilings of Euclidean and hyperbolic spaces. One of the interests of plane hyperbolic space is to be able to play the role of universal covering of unspecified kinds of smooth surfaces.

Mathematicians established that there are 7 types of possible friezes having internal symmetries. This is why the vertical interior of the folie has 7 faces; on each face, a removable panel illustrates one of the 7 elements of the types of possible friezes.

17 is a number known as a Fermat number, and Gauss showed that one can build, with the rule and the compass, a regular polygon of which the number on sides is such a Fermat number. But 17 is also the number of elements of the types of tessellations of the Euclidean plane. This is why the vertical outside of the temple has 17 faces. On each face, a removable panel shows one of the 17 tilings. Two adjacent faces are separated by a column in form of braid. The top of these columns carries coloured material polyhedrons, sometimes mobile around an axis of symmetry. Every two years, one will try to launch competitions, at various levels, for the creation of friezes and pavings.

In the interior, the ground carries an example of a paving of the hyperbolic plane. The fact that the interior of the building comprises seven faces invites to pave the plane with hyperbolic Klein triangles of interior angles ($\pi/2$, $\pi/3$, $\pi/7$). This hyperbolic paving is raised on the spherical cupola which is used as the roof of the building. It is a spherical stained glass the colour of which can be either the same as, or complementary to the colours of the paving of the ground. The position of the dome is such that at the zenith of summer, the rays of the sun project exactly the tiling of the stained glass on that of the ground.

Several color fillings of such a tessellation are possible, corresponding to distinct surface coverings. Thanks to a mechanical device, it will be possible to substitute one paving by another. Lastly, from the conceptual point of view, the invariance of properties with respect to a group of transformations makes it possible to introduce the fundamental concept of stability.

The Horn of Plenty The preceding folie is relates to a paramount state of the universe which fills space with particles and identical forms. The Horn of Plenty corresponds to a second phase of the evolution of the universe, where the descendants of the first generation are identical to their forebears only scaled down by a constant factor which is set at the beginning. The Horn of Plenty thus ad infinitum will illustrate the concepts of recurrence, those of sequence and series, fractals, and thus the premises of analysis.

The building is composed of coupled cells which converge towards a point. The scale factor is naturally the famous golden section $\frac{\sqrt{5}+1}{2} = 1,618\ldots$ which comes from the 12th century Fibonacci sequence. One finds this number in the regular pentagon: the length of a diagonal is equal to that of the side multiplied by the golden section. This is why one chose for cross section of the Horn such a pentagon. While varying with constant step the dimension of the pentagon and the rotation angle between two consecutive pentagons, one obtains an unfolding in the space of curves of pursuit giving contours of the Horn. On the walls of the interior rooms, will be drawn or projected the creations made using recurrence algorithms, and thus in particular those of the fractal world. At the interior point of convergence of the rooms will be placed a source of light which will make it possible to irradiate the small final rooms.

I would like to thank Mr. Vitkine for having carried out by infosculpture a first outline of the model of this building.

The Gauss' Observatory This folie is devoted to differential geometry, notably to that of smooth surfaces in usual space. It is composed of a cylindrical trunk surmounted by a spherical cupola. At a point of this cupola, one makes appear a tangent plane, similar to the cap which English-speaking students sometimes wear, and which will be mobile with the point of contact.

The base of the building is a piece of a Scherk surface, one of the minimal surfaces discovered at the 18th century. One enters the building by one of the cells of this surface. A piece of an other minimal surface, the helicoid, is used as staircase to reach the level of the principal room, whose ground is out of transparent material.

Here still, from the conceptual point of view, one will insist on the concept of stability, whose extremality is one of its metamorphosis, or sometimes one of the substitutes.

The Whitney's Umbrella Located at the crossway of various branches of mathematics, the Whitney's umbrella was selected to represent algebraic surfaces, as much for historical reasons and mathematics as for the richness of the fundamental concepts which it can illustrate: stability, singularity, stratification, bifurcation.

Being a ruled surface, it is in addition easily constructible. The central part which surrounds the handle will be made of transparent material to better ensure the interior luminosity of the building. Note that the frontal parts of this folie are pieces of swallowtails, which are homeomorphic with the umbrella.

The House of Number The forms, either the partly circular base one, or the logarithmic curve of the roof, are selected to evoke of course the essential numbers π and e. The input for drawing the base is a piece of the right strophoid with polar equation $r = 4\frac{\cos 2\mathrm{b}}{\cos \mathrm{b}}$, which one will compare with that of the Scherk surface previously evoked of the standard form $z = \mathrm{Log}\frac{\cos \mathrm{ky}}{\cos \mathrm{kx}}$. Elements of right helicoids are erected above the towers on the sides of the House.

Arithmetic always inspired a double feeling to me, of great brightness on the one hand, great mystery and darkness on the other hand. Therefore I would have wished to translate this double feeling outside by constructing a building coated with metallized and thus luminous glass plates, but quite dark in its interior, except for some luminescent hearths.

2.5 The future of the project

Thanks to the support of all the participants mentioned in the introduction to this talk, project ARPAM might develop in the years which come, within a few meters from here. It will be necessary to refine it on the technical level to hold account inter alia particular constraints which weigh on the site. In collaboration with the mathematicians, the structural engineers will get busy working on it.

The choice of materials, decorations interior and external of the folies are not always defined. For that purpose, a close cooperation between mathematicians, artists, and perhaps industrials, will be necessary. The project makes sense, and will be perennial, only if we make it our objective to create objects that are beautiful in themselves, jewels which, by their aesthetic qualities, profit from a very strong capacity of attraction, and thus allow everyone to perceive the beauty inherent in mathematics, thus helping to break down some psychological barriers which prevent to enter the mathematical world.

Just like this work of completing the project, the design and realization of animations, meant either for the general public, or for the community of mathematicians themselves, will also require the cooperation and goodwill of professionals coming from all the walks of life, academic or not. Community structures of preparation will have to be set up. Let us all take part in the realization of this project in the service of the works of the spirit, the Beautiful, and the Good.

References

1. C.-P. Bruter (1992) Le Parc Mathématique: Eléments pour l'étude de faisabilité architecturale et muséographique, Arpam.

Solid-Segment Sculptures

George W. Hart

Department of Computer Science, Stony Brook University, Stony Brook, NY 11794, USA
<george@georgehart.com>, <http://www.georgehart.com>

Abstract. Several sculptures and designs illustrate an algorithmic technique for creating solid three-dimensional structures from an arrangement of line segments in space. Given a set of line segments, specified as a position in 3-dimensional space for each endpoint, a novel algorithm creates a volume-enclosing solid model of the segments. In this solid model, a prismatoid-like strut represents each segment. The method is very efficient with polygons and produces attractive lucid models in which the sides of the "prismatoids" are oriented in directions relevant to the structure. The algorithm is applicable to a wide range of structures to be realized by 3D printing techniques.

1 Sculpture by 3D Printing

As an artist of constructive geometric sculpture, I often visualize forms and then need to develop new techniques which enable me to create them.[5]–[10] This paper describes a new method for creating geometric structures which correspond to a given arrangement of line segments. The procedure is an essential step in my design of several recent sculptures.

Figure 1 and [GH]1 shows a 10 cm diameter sculpture titled *Deep Structure*, consisting of five nested concentric orbs. Each of the five has the same structure as the outer, most visible, orb: there are 30 large 12-sided oval openings, 12 smaller 10-sided openings, 80 irregular hexagonal openings, and 120 small rectangular openings. Oval "corkscrew spirals" in the 12-sided openings connect the layers with each other. The concept is based on familiar concentric ivory spheres which are traditionally turned on a lathe and hand carved, with holes in each layer providing access to the inner layers. However, Fig. 1 [GH]1 is created in plaster by an automated 3D printing process, without any human hand [2].

After I design such a sculpture as a computer file, it is fabricated in a machine which scinters, laminates, or solidifies thousands of very thin layers. This piece and the next were printed by Zcorp [16].

Figure 2 shows a second 3D printing sculpture, about 8 cm in diameter. This is based on a projection of the regular 4-dimensional polytope called the "120-cell," but adapted into a more textured, chiseled form.

There are many 5-sided openings and passages throughout its volume. The title *Five-Legged-Bee Hive* physically refers to the fact that bees with six legs

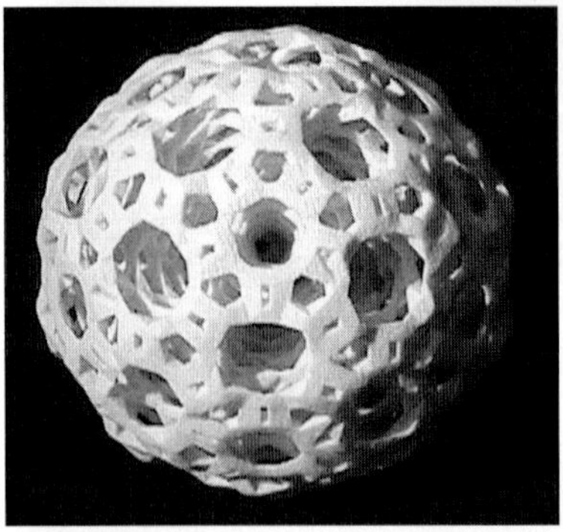

Fig. 1. [GH]1: *Deep Structure, George W. Hart, 2000.*

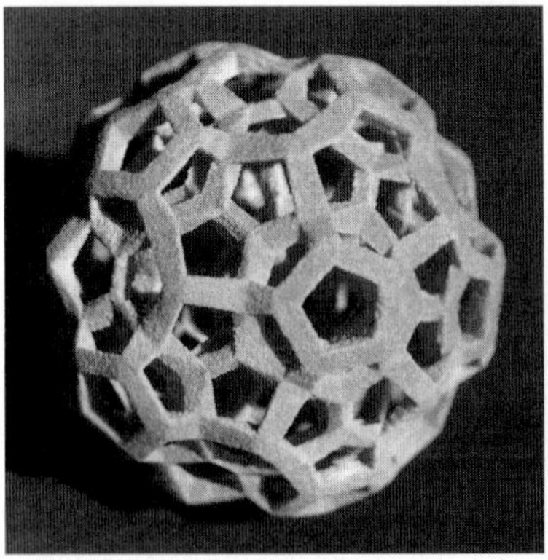

Fig. 2. [GH]2: *Five-Legged-Bee Hive, George W. Hart, 2000.*

make six-sided hives. (Thank you to Sandor Fekete for the suggestion). Both these sculptures can be idealized as a set of line segments connecting points in three-dimensional space. However, the lines here are not simple cylindrical rods. I designed angular, faceted struts to give these works more sculptural presence.

2 Visual presentation of segment structure

An ideal mathematical line has zero thickness. For rendering lines on two-dimensional computer displays, simple standard algorithms have been developed which produce lines one pixel in thickness, to be as close as possible to the mathematical ideal, yet clearly visible. For producing lines in three-dimensional space, other techniques are necessary.

Leonardo da Vinci invented a *solid segment* representation of polyhedra for the illustrations to Luca Pacioli's 1509 book, *De Divine Proportione.* Prior

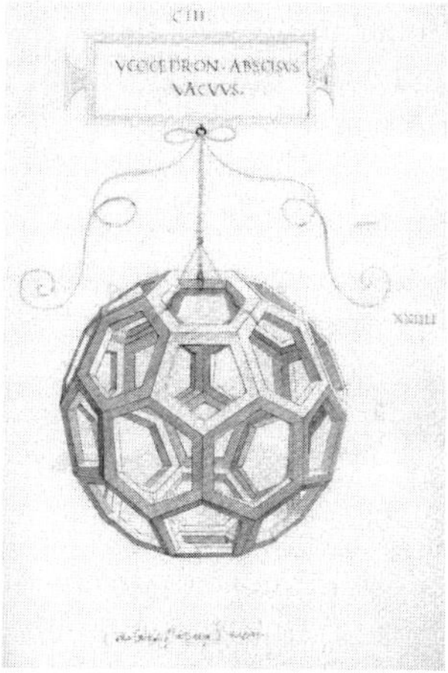

Fig. 3. [GH]3: *Leonardo "solid segment" presentation of truncated isocahedron.*

to this, polyhedra were illustrated as either opaque solids or transparent line drawings. Opaque representations have the disadvantage of hiding the rear structure. Simple line drawings can produce the "Necker cube" illusion, in which it is not clear whether a line belongs to the front or rear surface. In Leonardo's new form of representation, thick segments and open faces show the front and back structure, avoiding both problems.

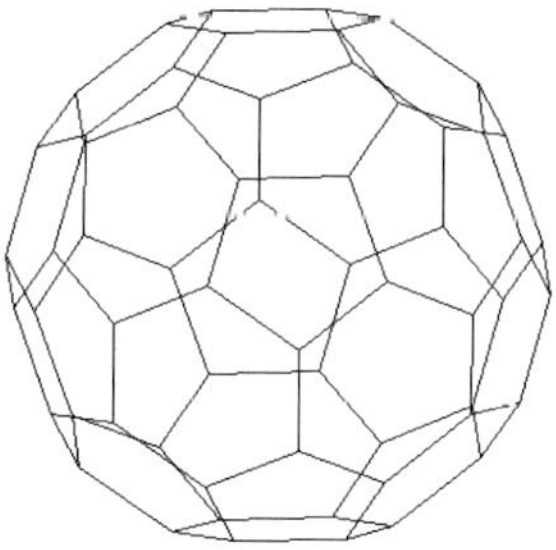

Fig. 4. *Line drawing of truncated icosahedron displays "Necker cube" ambiguity between front and rear surface.*

Many styles for solid-segment models are possible. Consider, for example, molecular models used by chemists, with cylindrical struts and spheres at each vertex. Leonardo used consistent, implicit rules in his 3D designs such as Fig. 3: (1) outer faces of struts are made in the planes of the polyhedron faces, (2) the inner faces are parallel to the outer faces, and (3) the "side faces" are perpendicular to the inner and outer faces. This makes for a very lucid presentation of convex polyhedra, but it does not generalize to an arbitrary network of vertices and segments, in which the structure need not determine planes tangent to the surface of a sphere. In the design of geometric sculpture, I faced the problem of connecting a set of points in space with faceted struts of some form, and needing an algorithmic procedure to design interesting struts.

3 Statement of problem

The abstract problem considered here is that we are given a set of (x, y, z) points in space, plus the information that various pairs of these points are to be connected via line segments. It is not difficult to produce the necessary vertex coordinates and edge lists for polyhedra [11]–[13] or polytope projections [3]. However, to produce 3D printings, line segments are not sufficient; one must have a set of polygons which enclose volume. This is an open-ended problem that can be approached in many ways. One tradeoff is between the number of polygons and how closely the result surrounds the segments. There are also many approaches to choosing the struts' shapes, sizes, and orientations. Selecting among these choices will depend on the application. A different approach with different properties is described in [1].

As background, consider two approaches to this problem that were found unsuitable. The *Union* operator of constructive solid geometry is a standard tool for combining 3D objects into more complex objects [2]. It is straightforward to produce prisms, antiprisms, or other approximations to cylinders for each segment and then call a *Union* subroutine to combine them. However, this approach fails in practice for large data sets, because of the instability of the union operator when using floating point data, even in large, expensive, commercial software packages. Plus, a serious drawback to this approach is that if it succeeds, it results in a large number of polygons.

Another approach is to create cylindrical approximations for each strut as above, but not to union them as 3D objects. Instead the effect of the union operation may be achieved on a voxel scale in the cross-sectioning software which creates the layer data that control the 3D printing. Again, there is a likelihood that the software will fail when given a complex structure with many overlapping components. But more fundamentally, these structures will not be visually lucid, because the orientation of the faces of the cylindrical approximations are designed individually, so are not relevant to the overall structure.

4 New approach

The algorithm is presented in reverse here, starting with the final step.

Step 3. Make Struts. Given two polygons that surround the ends of a line segment, take their convex hull to construct a *"prismatoid"* that surrounds the body of the segment. A prismatoid is the convex hull of two parallel polygons; its sides are triangles and/or quadrilaterals; special cases include the prism and the antiprism. Our "prismatoids" might not be formally prismatoids because the two bases may not be exactly parallel. This "prismatoid" step is done for every given segment after constructing appropriate polygons at each end.

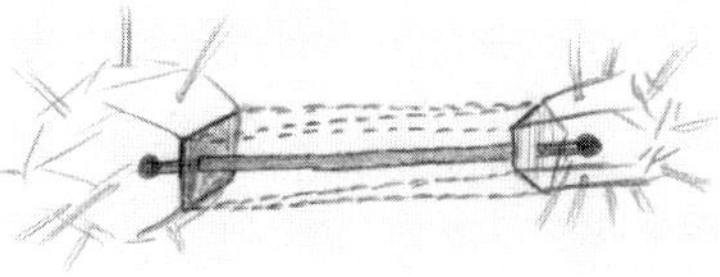

Fig. 5. *In step 3, the convex hull of two polygones (shaded) gives a "prismatoid" (dotted) around each segment.*

Step 2. Make Polygons around Endpoints. In order to form these segment-surrounding polygons, we construct a simple convex polyhedron around each endpoint in such a way that its vertices lie between the outgoing segments. This is simply the dual to the convex hull of the points where the segments intersect a small sphere. For example, if six orthogonal segments meet at a given endpoint (Fig. 6a) they define the six vertices of a small octahedron surrounding that point (Fig. 6b). The dual to the octahedron is a cube (Fig. 6c) which has six square faces, one surrounding each segment.

Notice that this dual polyhedron can be scaled for thicker or thinner struts, within certain limits. This is called the *scale factor* parameter, and can be different for each vertex.

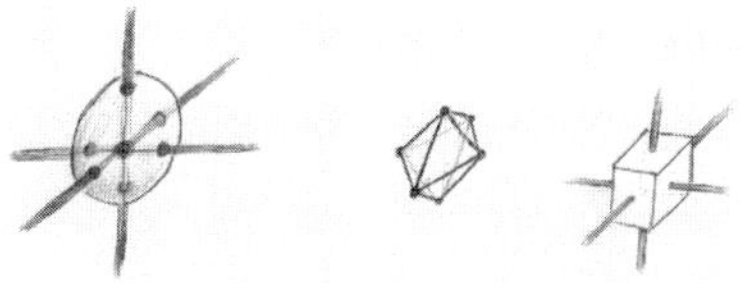

Fig. 6. *In step 2, (a) construct points around a vertex where segments meet small sphere, (b) form their convex hull, and (c) contruct dual polyhedron.*

Step 1. Ensure Endpoints are Enveloped. The step 2 procedure would fail if there were not enough points to define a polyhedron. So if there are fewer than four segments at a vertex, we first add "pseudo segments" to make a tetrahedral arrangement. Even with more than four, the arrangement may fail to form a polyhedron that surrounds the given endpoint, e.g., if the segments all lie in a plane or half-space. There is a short list of troublesome configurations to search for, illustrated in Fig. 7.

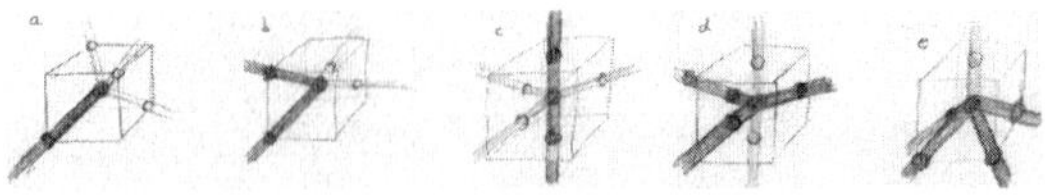

Fig. 7. *In Step 1, if the real segments (dark) do not produce an arrangement which surrounds the vertex, add "pseudo segments" (light) which do.*

The following list of cases, in this sequence, seems to be sufficient (other methods could be employed):

a) If there is only one segment, add three more as a regular tetrahedron.
b) If there are two segments not opposed, add two more as a tetrahedron.
c) If there are two opposed segments, add three on their "equator" as a triangular dipyramid.
d) If all the segments lie in a plane, add two "poles" orthogonal to the plane.
e) If all the segments lie in a half-space, add one in the direction opposite their mean.

Cases (d) and (e) are not exclusive; consider a small fan of segments. The nature of duality guarantees that every face of the dual surrounds either a segment or a pseudo-segment. The faces which surround the real segments are used as the bases of the prismatoids. They are not output as part of the result. Only the sides of the prismatoids and the faces which surround pseudo-segments are output.

All the above operations are straightforwardly implemented with standard vector operations. A suitable 3D convex hull subroutine is required, but this is a well-studied problem and many implementations exist.

5 Examples

For experimental purposes, the algorithm has been implemented in *Mathematica* using floating-point values. When efficiently implemented, the computational complexity of the entire process will be that of the many small convex hull operations, one about each vertex and one about each segment. The results are attractive and lucid, producing well-designed facets, with a low number of polygons. It is found to function well even in complex structures with several thousand segments to enclose. Three finished pieces and a new design are discussed below.

5.1 Deep Structure

Shown in Fig. 1 above, there are 720 struts in each of the five layers plus 480 struts which connect adjacent layers. The 4080 segments are wrapped with 27214 triangles altogether, which bound the struts. The close-up image in Fig. 8 shows some of the diagonal inter-layer struts, and makes clear the slight degree of randomness introduced because they start at a random phase in each oval.

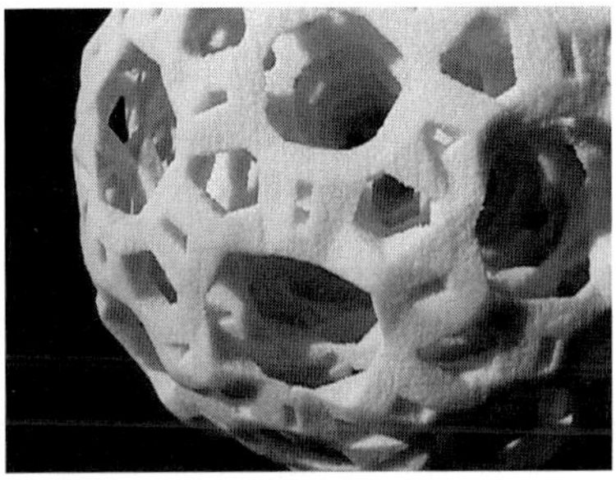

Fig. 8. *Close up view of Figure 1 shows texture and layering.*

For the structure of each layer, I chose the edges of a 31-zone zonohedron, illustrated in Fig. 9 [GH]9, which has edges in the 31 directions of the icosahedron's axes of symmetry. It is a 242-sided polyhedron bounded by thirty 12-gons, twelve 10-gons, eighty 6-gons, and one hundred twenty squares [11]. It is particularly easy to build a physical model of this polyhedron with the Zometool modeling set [14].

Figure 9 [GH]9 is generated by a program I wrote which presents polyhedra in the solid-segment style of Leonardo da Vinci, discussed above, which helps to make clear the elegant structural aspects of this polyhedron. Although it is straightforward to make a 3D printing of the form in Fig. 9 [GH]9, I would consider that a mathematical model, not a sculpture. Comparing it to Fig. 1 [GH]1 and Fig. 8 should make clear the type of sculpted, chiseled look I was after.

5.2 Five-Legged-Bee Hive

Shown in Fig. 2 above, this sculpture is based on the 120-cell. The underlying 4-dimensional structure consists of 120 regular dodecahedra, arranged so three meet around each edge and four meet at each vertex.

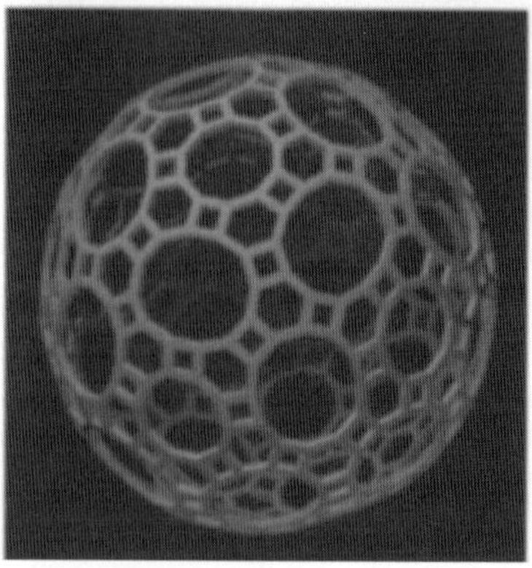

Fig. 9. [GH]9: *31-zonohedron that undelies each layer of Figs. 1 and 8.*

In 3D, regular dodecahedra do not pack like this without gaps, but in 4D there is freedom to rotate them slightly to pack snugly. A shadow of the 4D edge arrangement onto 3D space gives the mathematical form underlying this sculpture. If the edges were straightforwardly rendered by the above algorithm, the structure in Fig. 10 [GH]10 could result, but for 3D printing the struts would probably be made thicker. However, as a sculptor, I am interested in modifications which produce visually interesting effects, so I made a small variation to the algorithm, by which the faces corresponding to pseudo-struts are placed further from their vertex than other faces. This is actually accomplished by making the pseudo-strut points closer to the vertex before taking the dual in step 2 above. Doing so produces the sculptural form of Fig. 2 rather than the mathematical model in Fig. 10 and [GH]10.

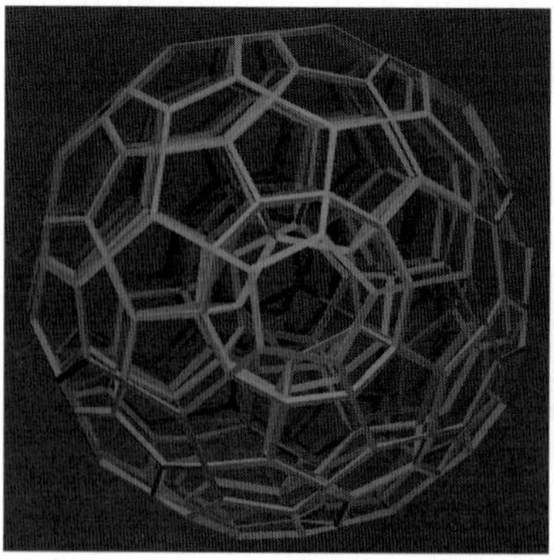

Fig. 10. [GH]10: *Edges of 120-cell orthogonally projected from 4D to 3D, along the direction to the center of a cell.*

5.3 120-cell projected as "Schlegel Diagram"

Figure 11 [GH]11 shows the design for a perspective projection of the 120-cell into 3D space, taken from very close up in 4D space. The effect is analogous to a 2D "Schlegel diagram" [3] of a 3D polyhedron. There are 119 dodecahedral cells packed inside the 120th outer cell. In general, elegant effects can result if the strut scale factor varies as a function (either increrasing or decreasing) of the distance of a vertex from the polytope center. In this case, the small inner cells are made with a strut thickness proportional to their smaller distance from the center. This design was worked out in discussions with Carlo Sequin. The 3D printing in Fig. 12 is 5 cm across, executed and photographed by Bathsheba Grossman [4] with a Sanders prototype machine [17].

5.4 Truncated 120-cell

An as-yet-unrealized form is analogous to Fig. 13 [GH]13, but truncated (in 4D) at each vertex. This is a cell-first orthogonal projection of the "truncated

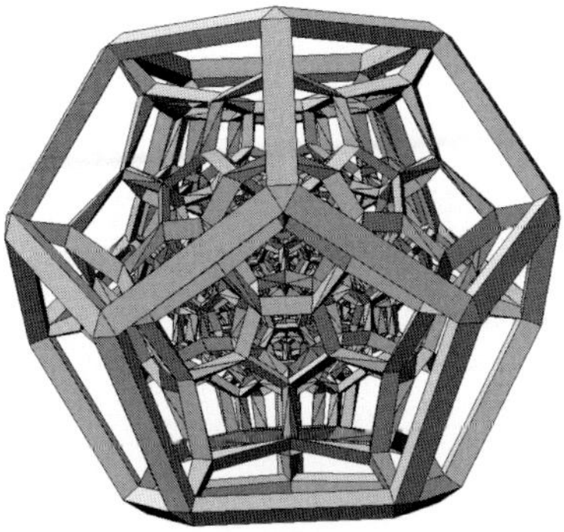

Fig. 11. [GH]11: *120-cell projected in perspective, like Schegel Diagram*

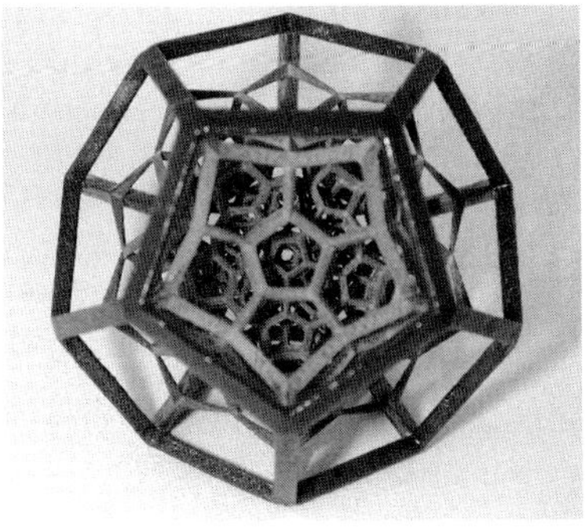

Fig. 12. [GH]12: *Wax realization of Fig. 11, 5 cm diameter.*

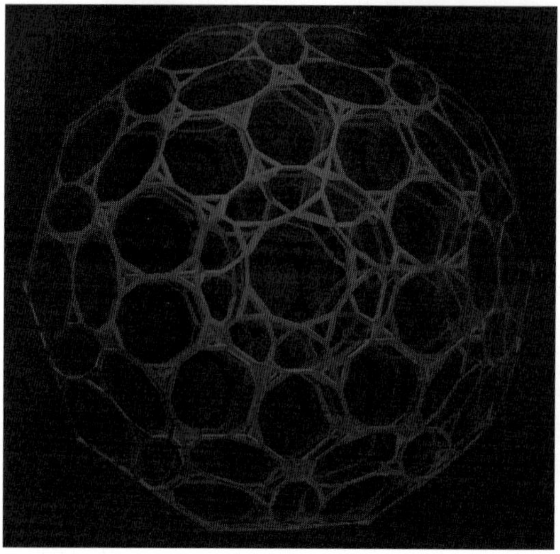

Fig. 13. [GH]13: *Projection of edges of truncated 120-cell.*

120-cell," a 4D polytope consisting of 120 truncated dodecahedra and 600 regular tetrahedra, first described in a 1910 paper written by Alicia Boole Stott (a daughter of George Boole) [3], [5]. In this projection it produces a stunningly beautiful 3D form with 10-sided passageways piercing it in six different directions. A large and very rewarding physical model can be made using Zometools; instructions are given in [14].

6 Conclusions

A simple, efficient algorithm has been presented which can be applied to problems in 3D design where segments need to be rendered as struts. Its use as a tool in sculpture and the design of mathematical models has been demonstrated. Beautiful forms have been constructed by this means, including several examples of sculpture and models which I hope to realize soon. For additional information about my sculpture, see my web pages at [5].

References

1. Boylan J., (1989) "Ugworm," in Carlo Sequin, Procedural Generation of Geometric Objects, U.C. Berkeley Comp. Sci Division Report #UCB/CSD89/518
2. Burns M. (1993) Automated Fabrication: Improving Productivity in Manufacturing, Prentice Hall
3. Coxeter H.S.M., (1963) Regular Polytopes, Dover reprint, New York
4. Bathsheba Grossman, `http://www.protoshape.com/`
5. Hart G.W., Geometric Sculpture, `http://www.georgehart.com/`

6. Hart G.W., (2000) "Sculpture based on Propellorized Polyhedra," Proceedings of MOSAIC 2000, Seattle
7. Hart G.W., (2000) "The Millennium Bookball," Proceedings of Bridges 2000: Mathematical Connections in Art, Music and Science, Southwestern College, Winfield, Kansas, July 28–30
8. Hart G.W., "Reticulated Geodesic Constructions," Computers and Graphics, to appear
9. Hart G.W., "Loopy," to appear in Humanistic Mathematics.
10. Hart G.W., (1998) "Icosahedral Constructions," in Proceedings of Bridges: Mathematical Connections in Art, Music and Science, Southwestern College, Winfield, Kansas, July 28–30, 195–202
11. Hart G.W., (1999) "Zonohedrification," The Mathematica Journal, vol. 7 no. 3
12. Hart G.W., (1998) "Zonish Polyhedra," Proceedings of Mathematics and Design '98, San Sebastian, Spain, June 1–4
13. Hart G.W., (1997) "Calculating Canonical Polyhedra," Mathematica in Research and Education, Vol. 6 No. 3, 5–10
14. Hart G.W. and Picciotto H., (2001) Zome Geometry: Hands-on Learning with Zome Models, Key Curriculum Press
15. Boole Stott A., (1910) "Geometrical deduction of semiregular from regular polytopes and space fillings," Verhandelingen der Koninklijke Akademie van Wetenschappen te Amsterdam, (eerste sectie), Vol. 11, No. 1, 1–24 plus 3 plates
16. Zcorp, `http://www.zcorp.com/`
17. Sanders, `http://www.sanders-prototype.com/`

Visualizing Mathematics – Online

Konrad Polthier

Technische Universität Berlin, Institut für Mathematik, Straße des 17. Juni 136, 10263 Berlin, Germany
<polthier@math.tu-berlin.de>

Abstract. Modern mathematical visualization has always been related with special graphics workstation although visualization was always part of mathematics. Here we start from historical roots, bring interactive visualization into the classrooms and create online mathematical publications. The topics include Java applets and online videos, a new electronic journal for geometry models, an interactive mathematical dissertation and online experiments.

1 *Multimedia* of the Old Masters

Over the recent years, mathematical visualization driven by computer graphics has proven to be a successful tool for mathematicians to investigate difficult mathematical problems which seem to be inaccessible by standard mathematical tools.

But visualization is not a new word in mathematics. Mathematical history contains a rich set of visualizations. For example, the famous copper plate engravings of Hermann Amandus Schwarz. The image appeared in his research publication on the Gergonne problem in 1865. He discovered new minimal surfaces which solved long-standing questions in geometry and analysis on the existence of solutions of elliptic boundary value problems. Although his proof was of theoretical nature he found it worthwhile to invest a lot of energy to include visual images of the new surfaces into his research publications. This becomes even more remarkable if his effort is compared to the much less effort one needs today for the generation of computer images. The head of Apollonius shows lines connecting parabolic points on the surface. Felix Klein wanted to find characteristic lines describing the nature of a shape.

One of the most thorough approaches in mathematics to use physical models and experimental instruments in education and research is the famous collection of mathematical models in Göttingen. This model collection already had a long history when Hermann Amandus Schwarz and Felix Klein overtook the direction of the collection. Especially under the direction of Klein the model collection was systematically modernized and completed for the education in geometry and geodesy. This collection was considered so important that Klein exhibited the models on the occasion of the World's Columbian Exposition 1893 in Chicago [2]. The models were produced among others by the publisher Martin Schilling in Halle a.S., see his catalog of mathematical

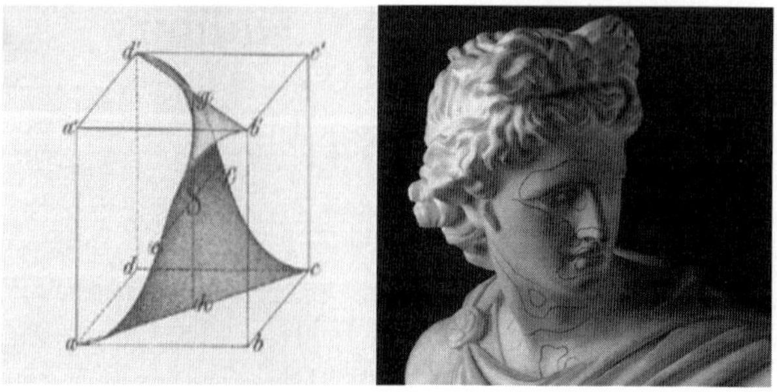

Fig. 1. Left image: Copper plate engraving of the Gergonne surface by Hermann Amandus Schwarz. Image taken from his Collected Works, Springer Verlag 1890. Right image: Head of Apollonius. Image: Hilbert/Cohn-Vossen Anschauliche Geometrie, Springer Verlag.

models [7]. The price of approximately $250 per models was relatively high, and therefore the large size of the collection of more than 500 plaster models is even more impressing. The collection can still be seen in the mathematical department in Göttingen, and a description including photos of many models is given in Fischer's book [3].

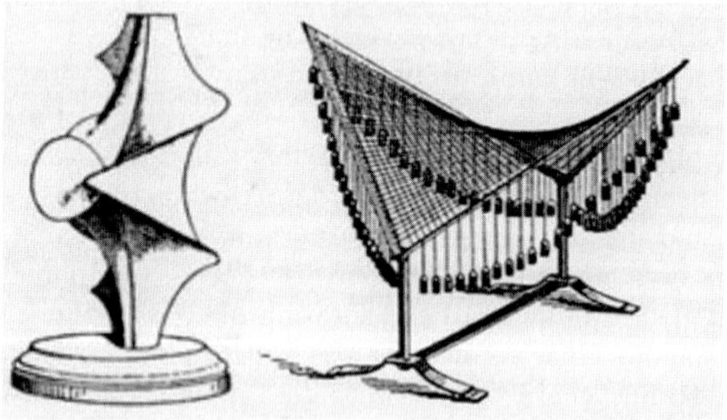

Fig. 2. Left image: Plaster model by Martin Schilling in Halle a.S. Large collection of models collected in Göttingen by Hermann Amandus Schwarz and Felix Klein since 1870. Right image: Steel model of hyperbolic paraboloid from Göttingen Collection. Used in education in geometry and geodesy. Production stopped after 1930. Focus on more abstract ideas in Mathematics, books of van der Waerden and Bourkaki appeared. Images: G. Fischer Mathematical Models, Vieweg Verlag.

2 Gaining Mathematics from Visualization

Among the first breakthroughs of mathematical visualization was the proof of embeddedness of a minimal surface. Celsoe Costa discovered the mathematical formulae of a genus 1 minimal surface which was a candidate to solve a 200-year long question: whether there exists a third embedded and complete minimal surfaces with finite total curvature beside the trivial examples, the flat plane and the rotational symmetric catenoid. David Hoffman and William Meeks [5] developed computer programs to visualize Costa's surfaces 3 and watch surface properties which they could later successfully prove after having enough insight into the complex shape of the surface.

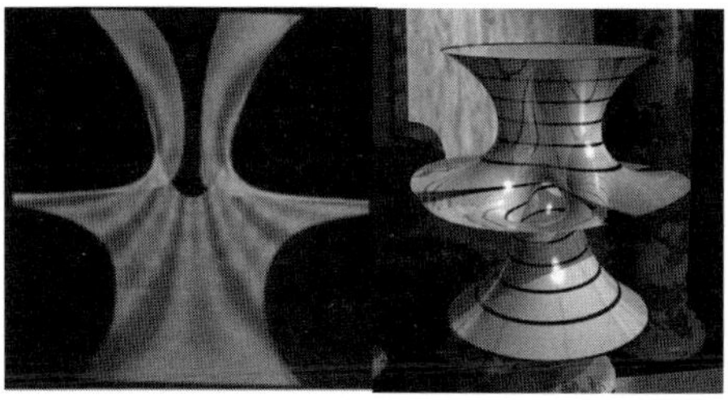

Fig. 3. [KP]1 : Costa-Hoffman-Meeks minimal surface (1985). Costa gave formulas of the surface, but how does surface look like, i.e. what geometric properties? For example: does it have self-intersections? The left image shows the very early computer pictures by Hoffman and Meeks (left). Years later, we see a raytraced image ot the Costa-Hoffman-Meeks minimal surface (right).

Mathematical visualization has proven to be an efficient tool for analyzing complex mathematical phenomena, and it has given decisive hints leading to rigorous mathematical proofs of long-standing problems. Visualization is not only a tool to visualize complex objects but in combination with modern numerical methods allows to perform mathematical experiments and simulations in an artificially clean environment. For example, the unveiling of the Costa-Hoffman-Meeks surface, or the first numerical examples of compact constant mean curvature surfaces with genus greater than two by Große-Brauckmann and Polthier [4] are among the most prominent results of the fruitful interaction of mathematics with the new toolkit mathematical visualization.

3 Discrete Minimal Surfaces – Online

Minimal surfaces are among the most prominent shapes in mathematics. For an introductory overview and their relation to physical soap films have a look

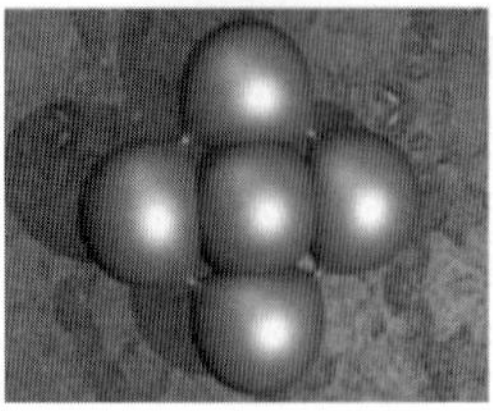
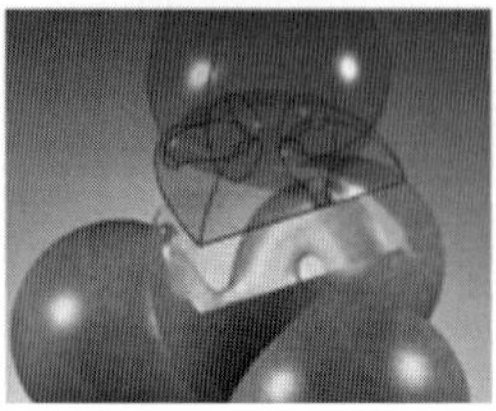

Fig. 4. [KP]2: Compact Constant Mean Curvature Surfaces (1995, Große-Brauckmann – P., Oberknapp – P.). Free-boundary value problem for an elliptic partial differential equation. Numerical method based on the notion of discrete differential geometry and on triangulations. Compact soap bubble with genus four (left) and tetrahedral symmetry (right).

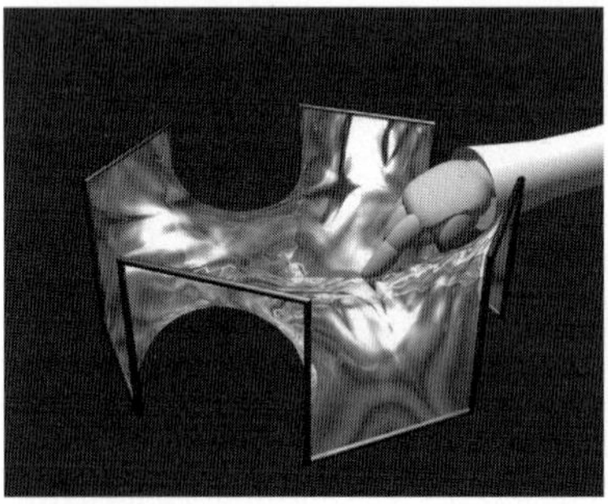

Fig. 5. [KP]3: Snapshot of a minimal soap film taken from the video Touching Soap Films.

at the online version of a booklet which accompanies the english version of the video Touching Soap Films.

Discrete surfaces are made up of a set of triangles which are especially convenient when doing numerics and visualization with a computer. In fact, in finite element mathematics and discrete geometry these simplicial complexes have become objects of their own interest. For example, the interactive applets on the online pages of this article work with triangle meshes which are minimal in a discrete sense.

In the following we give several demonstrations of combining interactive experiments, publications and research.

Making Textbooks Interactive A classical text book may easily be enhanced with interactivity. The following page (not shown in the printed book) is taken from the recent introductory textbook on differential geometry "Einführung in die Differentialgeometrie" by Wolfgang Kühnel. Here the static image of a helicoid was replaced with a JavaView applet. Note, the smooth transition from a static book to an interactive book. There is no need for a revolution on the reader's side, for example a changed reading behaviour. The revolution is of technical nature and resides inside the online software hidden to the reader.

Discrete Minimal Catenoid This discrete catenoid is a critical point of the discrete area functional. Discrete minimal surfaces are minimizers of a discrete area functional. The authors (K. Polthier and W. Rossman) have found an explicit representation of discrete catenoids whose set of vertices is dihedrally symmetric. These catenoids are the first explicitly known discrete minimal surfaces for the discrete area functional beside the trivial plane. The explicit representation allows to generate exact discrete minimal surfaces without numerical errors which are especially useful for index computations.

Theorem 1. *There exists a four-parameter family of embedded and complete discrete minimal catenoids $C = C(\theta, \delta, r, z_0)$ with dihedral rotational symmetry and planar meridians. If we assume that the dihedral symmetry axis is the z-axis and that a meridian lies in the xz-plane, then, up to vertical translation, the catenoid is completely described by the following properties:*

1. *The dihedral angle is $\theta = \frac{2\pi}{k}$, $k \in \mathbb{N}$, $k \geq 3$.*
2. *The vertices of the meridian in the xz-plane interpolate the smooth* cosh *curve*
$$x(z) = r \cosh\left(\frac{1}{r} az\right),$$
with
$$a = \frac{r}{\delta} \operatorname{arccosh}\left(1 + \frac{1}{r^2}\frac{\delta^2}{1+\cos\theta}\right),$$
where the parameter $r > 0$ is the waist radius of the interpolated cosh *curve, and $\delta > 0$ is the constant vertical distance between adjacent vertices of the meridian.*
3. *For any given arbitrary initial value $z_0 \in \mathbb{R}$, the profile curve has vertices of the form $(x_j, 0, z_j)$ with*
$$\begin{aligned} z_j &= z_0 + j\delta \\ x_j &= x(z_j) \end{aligned}$$
where $x(z)$ is the meridian in item 2 above.
4. *The planar trapezoids of the catenoid may be triangulated independently of each other.*

The applet below allows to modify all parameters of the catenoid and study their geometric meaning. Such a web-based applet allows to simulateneously work on this example while one authors is in Germany and the other in Japan.

Further references on the discrete catenoid may be found on the Electronic Geometry Models server at http://www.eg-models.de/ model number 2000.05.002.

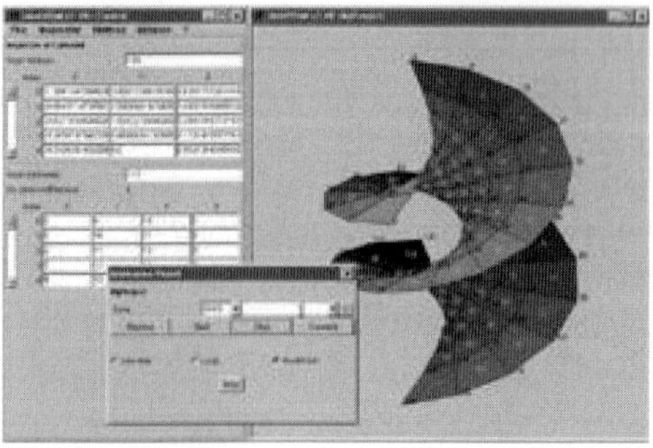

Discrete Harmonic Maps The research applet on this page minimizes a surface energy. For example, when selecting the Dirichlet energy then the following problem is solved on an arbitrary initial surface:

Find: A minimal surface bounded by a given curve G

Algorithm: Use initial surface M_0 and construct sequence of surfaces M_{i+1} by finding (Laplace-Beltrami) harmonic maps F_i with

$$F_i : M_i \to M_{i+1} \text{ with } \partial M_{i+1} = G.$$

The limit surface is a minimal surfaces under certain conditions.

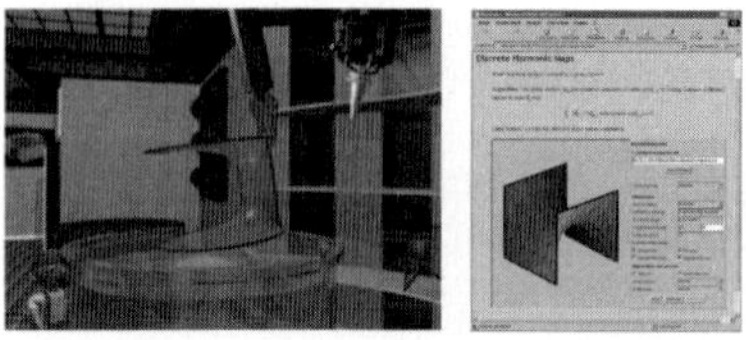

Fig. 6. Soap film machine from video *Touching Soap Films* [6] (left). Computing minimal surfaces in an interactive applet on the JavaView web site [1] (right).

During minimization boundary vertices are retained. Pick and drag vertices with the left mouse button while holding key "p" pressed to modify vertices of the surface. Set the number of iteration loops for the minimization algorithm in the textfield "Num Loops".

The checkboxes "Tangential" and "Normal" define in which directions the minimizer is allowed to move vertices. The checkboxes "Update Normals" and "Update Domain" determine if surface normals and domain surface are recomputed in every minimization step.

The button "Step" invokes one minimization step, the button "Minimize" starts as many minimizing iterations as are specified in the "Num Loops" textfield. By the "Resume" button you can stop and continue the iteration.

4 Mathematical Videos

There exists already quite a remarkable set of mathematical videos. The common drawbacks of videos like being non-interactive, requiring special equipment and projection facilities, are nowadays no longer true.

Sequences of videos may be inserted in online documents in the same way as Java applets find their way into mathematical publications. Still, there is a lot of effort required for the production of videos. Have a look at the following sites for online MPEG versions of sequences of the videos.

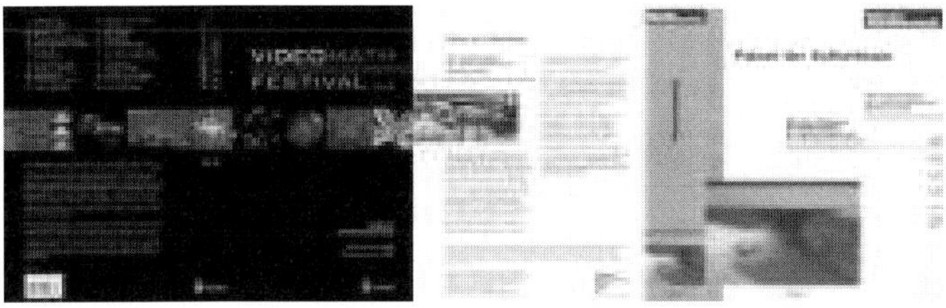

The VideoMath Festival Cassette is a collection of juried mathematical videos presented at ICM'98, the International Congress of Mathematicians in Berlin. The videos are winners of a worldwide competition. The mathematical themes include problems in topology and geometry and their recent solutions, visualizations of classical ideas of Archimedes, Eratosthenes, Pythagoras, and Fibonacci, topics in high school mathematics, and applications of modern numerical methods to real world simulations. Length 76 min. (1998). Look at the VideoMath home page for links to many other longer videos.

Touching Soap Films is a popular scientific video on minimal surfaces and soap films. All animations are computer generated. The english edition is distributed by Springer Verlag since 1999. The german edition Palast der Seifenhäute is available through Bild der Wissenschaft and Komplett-Media. Length 41 min. (1995). The Portuguese version was distributed to all schools in Portugal by CMAF/Lisbon in the year 2000.

The video Geodesics and Waves documents recent work on discrete geodesics and their application to numerical problems on curved surfaces and visualization algorithms. It premiered at Siggraph'97 Electronic Theater, received the Science Award of the Multimedia Grand Prix (Nicograph'97) in Japan, and was awarded at other festivals. Length 4:50 min. (1997).

A new mathematical video series was launched by Springer Verlag in June 1999. Among the first films are "Touching Soap Films" by Andreas Arnez, Konrad Polthier, Martin Steffens and Christian Teitzel; "N is A Number – A Portrait of Paul Erdös" by George Paul Csicsery; "Arabesques and Geometry" by Antonio F Costa; and "The New Shepherd's Lamp" by Francois Tisseyre, Claire Weingarten, and Jean-Pierre Bourguignon.

The award-winning computer animations like Outside In, Shape of Space and Not Knot. Videos are available through AK Peters.

Under the direction of Tom Apostol and Jim Blinn Project MATHEMATICS! has produced highly awarded video tapes that explore basic topics in high school mathematics in ways that cannot be done at the chalkboard or in a textbook. The tapes use live action, music, special effects, and imaginative computer animation.

5 EG-Models Journal: An Archive for Electronic Geometry Models

The electronic geometry models archive (at http://www.eg-models.de) is a new journal for the publication of digital geometry models. Among the main characteristics are:

- Refereeing of the digital geometry models ensures highest quality of the data sets.
- Each model has a full description and a unique identification number which ensures the intellectual property of the author and allows for a unique citation.
- Possibility to validate geometric experiments by making the data available to 3rd parties through this archive.
- Review of best models in Zentralblatt in a similar way as research articles.

The first collection of reviewed electronic geometry models is open for any geometer to publish new geometric models, or to browse this site for material to be used in education and research. Access to the server is free of charge.

The geometry models in this archive cover a broad range of mathematical topics from geometry, topology, and, to some extent, from numerics. Examples are geometric surfaces, algebraic surfaces, topological knots, simplicial complexes, vector fields, curves on surfaces, convex polytopes, and, in some cases, experimental data from finite element simulations.

All models of this archive are refereed by an international team of editors. The criteria for acceptance follow the basic rules of mathematical journals and are based on the formal correctness of the data set, the technical quality, and the mathematical relevance. This strict reviewing process ensures that users of the EG-Models archive obtain reliable and enduring geometry models. For example, the availability of certified geometry models allows for the validation of numerical experiments by third parties. All models are accompanied by a suitable mathematical description. The most important models will be reviewed by the Zentralblatt für Mathematik.

We are advocating the construction and submission of digital geometric models from various areas of mathematics. The advantages of these digital models go beyond those of the classical plaster shapes and dynamic steel models of earlier days. At the end of the 19th century several mathematicians felt the need to physically handle the geometric objects they thought about. In particular, Felix Klein and Hermann Amandus Schwarz in Göttingen built many models of curves, surfaces and mechanical devices for teaching and other educational purposes.

Using the digital models, interested mathematicians can verify the claims on their own, using appropriate software of their choice. Moreover, once there is a model available, it is possible to perform one's own computational experiments on this data set. This could be a numerical evaluation as well as a search for another property yet to be analyzed for this model.

Each model comes with a detailed description which identifies the author, explains the mathematical purpose, and includes references to other sources of information. Each model has a unique identification number for unambiguous citation. Each model is equipped with qualified meta data information; therefore, the archive can be searched via specialized search engines such as those from EMIS http://www.emis.de and MathNet http://www.math-net.de/search/germany/. Each model itself is represented by a master file which is from a fixed set of file formats, including XML formats specified by DTDs. By restricting the data formats we want to ensure that the server's information can be kept up to date on a technical level. Additional files in arbitrary formats are welcome for explanatory purposes.

6 Dissertation Online

Publication of dissertations is a good playground for testing various technical issues related with online documents. JavaView was used to enhance a sample dissertation in mathematics with multimedia elements.

The project Dissertation Online funded by the Deutsche Forschungsgemeinschaft (DFG) is a co-operation of the Humbolt-Universität Berlin, Gerhard-Mercator-University Duisburg, Universität Erlangen-Nürnberg and Carl von Ossietzky-Universität Oldenburg and of the main libraries in Germany. The project defines criteria and guidelines for the digital publishing of dissertations in Germany.

Among the many open problems are:

- Different document structures for different scientific disciplines.
- Designated meta informations for retrieval.
- Copyright and other legal aspects.
- Authentication and unchangeability must be guaranteed.
- Exploit the multimedia capabilities of digital publications and many more issues ...

JavaView has been selected to produce an online version of a mathematical dissertation that demonstrates the possibilities to valorize a thesis with interactive components. The sample dissertation by Tim Hoffmann enhanced with online experiments by Eike Preuß and himself is available at the JavaView homepage.

7 JavaView

JavaView [6] is a 3d geometry viewer and geometry library to publish interactive geometry in any HTML document, and to perform interactive numerical experiments on the internet. A stand-alone version of JavaView also runs as application from a Unix or Dos command prompt, and it can be attached as 3D viewer to other programs.

JavaView's software library focusses on problems in differential geometry and mathematical visualization. Its class library can be used and extended for own numerical experiments in Java, while always profiting from the advanced 3D visualization capabilities and the web integration.

- Tight Integration of Mathematical Experiments, Visualization and Online Publications
- Exchange and Communication of Experiments for research and educational purpose

Since its first public release in November 1999 JavaView was awarded several prizes. As a proof of usability, JavaView was selected by the project "Dissertation Online" of the German science foundation DFG to produce a reference online dissertation in mathematics.

Selected Features of JavaView

- 3D Scientific Viewer in Web Pages
- Software Library for Differential Geometry and Numerics
- Import and Export of Geometries in Multiple Data Formats,
- PostScript PS+EPS; Images GIF, PPM; XML Formats
- Tight integration with Mathematica and Maple.

Images and geometries exported with JavaView may easily be included in TeX publications as well as online documents. Additionally, JavaView is a class library for advanced numerics in differential geometry including tools for finite element numerics. A standalone version of JavaView runs in a Unix or Windows shell from the command prompt, and can be attached as 3D viewer to other programs like *Mathematica*. The JavaView homepage at http://www.javaview.de contains several applications and tutorials.

In the following subsections we discuss some properties of JavaView and give a number of sample experiments. Since this book is paper-based we refer to an online version of this article at the web site http://www-sfb288.math.tu-berlin.de/vgp/ which contains interactive versions of the presented applets.

How to Publish a JavaView Enhanced Web Page? There are different possibilities to add interactive visualizations to a web page using JavaView. Here we assume that a 3d geometry has been produced with any software tool such as Maple, Mathematica, JavaView or others and which was saved in a geometry file star.mgs. For an interactive web page we need:

1.	The geometry model file	models/star.mgs
2.	Write a web page including the applet below	myPage.html
3.	Get the JavaView library	jars/javaview.jar
4.	Upload all three files to a web server	

The sample applet tag somewhere inside the document *myPage.html* looks as follows:

```
<applet
      code=javaview.class
      archive=''jars/javaview.jar''
      width=200 height=200>
      <param name=model value=''models/stars.mgs''>
</applet>
```

Note, although this example is functional, it is just a demonstration of the required inputs. This applet visualizes the geometry model inside a small window of 200*200 pixels on the web page. The model need not be a geometry file on a local computer but the model parameter may be any internet address referring to a model on an arbitrary web server.

This example stresses the fact that the installation of the JavaView software is no longer an issue compared to the installation process of other software. The browser keeps care to download the required Java archive when it encounters the archive parameter inside the applet tag. The browser also ensures that the archive is downloaded only during first usage, and later reuses the version it has stored in the browser cache.

The easy download mechanism is especially useful for library servers offering Java enhanced electronic publications. The digital article and the JavaView archive are both stored, for example, in the same directory on the library server. The files must be uploaded by the author as described above, and are automatically downloaded by a browser when a user accesses the web page. Therefore, the librarian has no additional duties related with software installation. The library must only offer the usual upload mechanism for documents which it has already installed.

Responsibilities (for Online Publications) Electronic libraries require self-contained (online) publications. Especially, external links must be kept to a minimum in order to allow the libary to maintain and ensure the functionality of a document. For example, an interactive dissertation cannot be allowed to access a database outside the library in order to be functional.

Java enabled publications have all required software bundled with the publication. This frees the library from any maintenance task and requires no special support from the provider beside storing the publication which has been uploaded from an author.

Further, on the client side, the reader needs no special software installation to be able to read a Java enhanced publication. The base Java system is installed with custom web browsers and the special modules used in a publication are downloaded automatically by the web browser when a user accesses the document for the first time.

Author	Create content and upload
Online Library	no special activity, just store the data
Reader	Java enabled browser

For example, the mathematical visualization software JavaView just requires Java 1.1 installation which is automatically available within all modern web browsers.

8 Hardware for 3D Visualization

For a long time scientific visualization was beyond the budget of many mathematical departments. Large research institutes, military organizations, and commercial companies were among the first who could afford specialized graphics hardware. In science, specially funded research groups where able to afford high-end graphics workstations including the necessary staff to manage the machines and simultaneously do the scientific experiment. In the meantime, the computational power of personal computers with relatively cheap graphics card suffice to perform most of the scientific visualization tasks found in research. Nevertheless, one still encounters the following drawbacks of the current software running on specialized workstations and mainframes:

Workstation Based Specialized and expensive graphics hardware.

– Large program size since operating system just supports basic functionality.
– Usually only the programmer is able to run the experiments.
– Installation at other sites requires experts, and does not allow regular update.
– Advantage: extremely fast execution speed.

These drawbacks are in strong contrast to the situation we have encountered during the development and usage of the software JavaView. JavaView is a scientific visualization software completely written in the programming language Java. Java is an object-oriented programming language similar to the language C and C++ but different in the sense that Java is designed to run on any computer. Further, Java programs may run inside web browsers. Both properties are the reason that Java has become the major programming language for interactive web applications since its first presentation in 1995. A program written in Java has the following advantages:

Online Visualization

- Runs on Standard PC and Workstation.
- Tiny program size because Java base classes are already installed.
- Each application has a user interface per default since it runs in a browser.
- No installation beside a browser with Java since browser performs the data transfer.
- Speed: depends.

These advantages have the following reasons:

(1.) Java is automatically installed on a computer if a web browser is installed. Therefore, the popularity of web browsers helped to install Java on nearly any computer world-wide.

(2.) The size of Java programs is usually very small compared to classical stand-alone application software since the Java base classes, which are comparable to software libraries, are already installed. Therefore, an application must only deliver its additional functionality, and not system routines.

(3.) An application inside a web page must have a well-designed graphical user interface since it is by default used by some other people than the programmer. This is in contrast to classical experimentation software, and leads to a great benefit in the design of better products.

(4.) The installation of classical software systems has often been a pain. The customer often needed to compile the package again on his machine, or make special adjustments depending on his specialized hardware. The author was in an even worse situation. He needed to offer and maintain different versions for different platforms. When using Java then there exists only one version independent of the hardware platform and operating system. This is possible since the Java virtual machine must cope with system differences, so the responsibility is transferred from the author of applications to the supplier of the Java virtual machine. Therefore, the installation process of a Java application such as JavaView is reduced to downloading an archive, i.e. one or more library files, which is done automatically through a web browser. This allows the author to concentrate on the development of the software without keeping to much care on the destination platform, and it frees him from providing installation mechanisms. The user is freed from any installation task, he just starts his browser and selects a Java enhance web page.

(5.) The speed of Java applications not only depends on the hardware but to a large extend on the quality of the installed Java virtual machine (JVM). A Java application consists of machine independent byte code which is interpreted by a JVM and executed on a local computer. JVMs differ largely in quality, for example, when loading a Java application some JVMs compile the byte code into machine dependent code, which leads to a drastic increase in execution speed.

9 Conclusion

The internet will dramatically change the classical way of communicating and publishing mathematics. We have given some ideas on possible changes to expect, and the benefits which mathematics may gain from these new developments. The interactive, exploratory component of mathematics, which has been removed from mathematical publications for a too long time, is now available in the form of Java enabled software. We have given several examples of multimedia enhanced experiments which allow to imagine the possibilities waiting at the horizon.

For more information and interactive versions of the experiments described in this paper we refer to the JavaView home page. These pages also include tutorial material how to include interactive geometries into own web pages.

A Few Links

JavaView	http://www.javaview.de
Mathematical Videos	http://www.springer.de/math/livingmath/
Dissertation Online	http://www-sfb288.math.tu-berlin.de/vgp/dol/
Electronic Geometry Models	http://www.eg-models.de
Mathematics Archives	http://archives.math.utk.edu/
Mathematical Atlas	http://www.math-atlas.org/welcome.html
Scientific Graphics Project	http://www.msri.org/publications/sgp/SGP/

References

1. Arnez A., Polthier K., Steñens M., Teitzel C., (1999) Touching Soap Films, Springer Video-MATH Series
2. Chicago Historical Society. http://www.chicagohistory.net/history/expo.html.
3. Fischer G. (1986) Mathematische Modelle/Mathematical Models, Vieweg, Braunschweig
4. Große-Brauckmann K., Polthier K., (1997) Compact constant mean curvature surfaces with low genus, Experimental Mathematics, 6(1) 13–32
5. Hoffman D., MeeksIII W. H., (1985) A complete embedded minimal surfaces with genus one, three ends and finite total curvature, J. of Diff. Geom., 21, 109–127
6. Polthier K., Khadem-Al-Charieh S., Preuß E., Reitebuch U., (1999) http://www-sfb288.math.tu-berlin.de/vgp/javaview/
7. Schilling M., (1911) Catalog Mathematischer Modelle Für Den Höheren Mathematischen Unterricht, Leipzig

The Design of 2-Colour Wallpaper Patterns Using Methods Based on Chaotic Dynamics and Symmetry

Michael Field

University of Houston, Department of Mathematics, Houston, TX 77204-3008, USA
<mf@uh.edu>

Abstract. We describe the theoretical basis for the design of symmetric patterns using dynamics, chaos and symmetry. We show examples of some of the one- and two-colour wallpaper patterns that we have created using these ideas.

1 Introduction

In this article, I describe the design and colouring of planar symmetric patterns – in particular, two-colour wallpaper patterns – using techniques based on dynamics and chaos. Apart from showing examples, some coloured, of the pictures created using these ideas, my main aim in writing this article was to provide an overview of the theory that underpins the techniques. In a companion article [13], I discuss some of the ways in which aesthetics and mathematics become intertwined in attempts to create an art based on symmetry and chaos.

The images shown in this article were all designed and coloured (or 'grey scaled') using software that I started to develop about twelve years ago. This software, called *prism* (an acronym for 'PRograms for the Interactive Study of Maps'), allows the interactive design and colouring of planar figures with non-trivial discrete symmetry. Some of the early pictures produced using *prism* can be found in the 1992 book *Symmetry in Chaos* [14], written jointly with Marty Golubitsky. (Many of the iterative algorithms used in *prism* were developed in collaboration with Marty Golubitsky.) We refer to [11] for a relatively up-to-date description of *prism* and the way real (as opposed to virtual) pictures are obtained.

Prism can generate a wide range of symmetric patterns including bounded symmetric patterns in the plane and all of the frieze and wallpaper patterns, including the 46 two-colour wallpaper patterns. In practice, the development of *prism* has been strongly influenced by my interest in using the program to create artistically satisfying designs. This consideration has particularly influenced the choice of algorithms I use for colouring two-colour designs (see [10]). It may at first sight seem strange to use relatively sophisticated methods based on symmetry and chaos to construct symmetric designs. Indeed, there are many combinatorial techniques, and computer programs, that can produce symmetric designs (for example, see [3,18,17]). However, the typical image designed using *prism* exhibits an unusual and striking global coherence as well as a wealth of rich and complex detail. These features result from the image being a visual representation of an attractor of a globally defined dynamical systems.

It turns out that *prism* is a useful tool in teaching both geometry and design [12]. In recent years, I have used *prism* as the basis of a course on 'Patterns, Designs and Symmetry' that I have given in the Department of Art at the University of Houston (see the *URL:* nothung.math.uh.edu/~patterns/ for some of the designs produced by the Fall 1998 class). In another direction, I have used *prism* in a seminar on 'Symmetry, Patterns and Designs' that I gave in 1999 for the *Houston Teachers Institute* (see the *URL:* www.uh.edu/hti/).

We now describe the contents of the article by section. In section 2, we survey the mathematical theory underlying the design of images using methods based on symmetry and chaos. After a brief review of planar symmetry, we discuss the concept of an attractor and give examples of the construction of bounded symmetric attractors using both deterministic and non-deterministic (or 'random') dynamical systems. We conclude the section with some comments about the numerical implementation of these ideas and colouring ('coloured measures'). Section 3 is devoted to the topic of one-colour wallpaper patterns. In section 4, we provide a definition of two-colouring that applies to dynamically generated patterns. We also describe two different ways in which we can create two-colour patterns using dynamics. We conclude by showing some coloured examples of two-coloured wallpaper patterns created using dynamics.

2 The Theory of Designer Chaos

In this section, we describe how we can use methods based on symmetry and chaos to construct symmetric designs[1]. We start with a brief review of some foundational results on planar symmetry groups.

2.1 Planar symmetry groups

Let $\mathbb{R}^2$ denote the Euclidean plane and $E(2)$ denote the Euclidean group of rigid motions of $\mathbb{R}^2$. We recall that if $T \in E(2)$ is not the identity, then T is either a reflection, or a rotation or a translation or a glide reflection. We denote the subgroup of $E(2)$ consisting of translations by $E_T(2)$ and note that $E_T(2)$ is naturally identified with $\mathbb{R}^2$.

[1] We refer the reader to the book *Symmetry in Chaos* [14] for a more leisurely and elementary introduction to these ideas and to [8,11] for more recent developments.

Suppose that X is a closed subset of $\mathbb{R}^2$. A transformation $T \in E(2)$ is a *symmetry* of X if $T(X) = X$. We let $\Sigma = \Sigma(X)$ denote the group of all symmetries of X and remark that Σ defines a closed subgroup of $E(2)$. Let $\Sigma_T = \Sigma \cap E_T(2)$ denote the group of translational symmetries of X.

We shall only be interested in patterns X for which Σ is a 0-dimensional (discrete) subgroup of $E(2)$. We recall the well-known classification of discrete subgroups Σ of $E(2)$.

(B) If Σ is finite, then there exists $n \geq 1$ such that Σ is either isomorphic to $\mathbb{D}_n$ (the dihedral group of order $2n$) or $\mathbb{Z}_n$ (the cyclic group of order n).
(F) If $\Sigma_T \cong \mathbb{Z}$, then Σ is a *frieze* group. There are seven isomorphism classes of frieze groups.
(W) $\Sigma_T \cong \mathbb{Z}^2$, then Σ is a *2-dimensional periodic* or *wallpaper* group. There are seventeen isomorphism classes of wallpaper groups.

The reader may find proofs of these results, as well as examples of patterns realizing each of the discrete subgroups of $E(2)$, in [1, chapters 19,26]. In the sequel, we follow the notation for wallpaper patterns used in [1,16,20] (see also section 3).

2.2 Attractors

For the present, we restrict attention to attractors of planar dynamical systems. However, all of what we say generalizes easily to dynamical systems defined on more general spaces. Our definitions are formulated so as to cover both deterministic and random (or non-deterministic) dynamical systems. Readers unfamiliar with dynamical systems theory are advised to skim quickly through the following until they come to the paragraphs on numerical implementation.

Let $\mathcal{F}$ denote a set of n (continuous) planar maps $f_i : \mathbb{R}^2 \to \mathbb{R}^2$, $1 \leq i \leq n$. Let $S(n)$ denote the space of all infinite sequences $s = (s_j)_{j\geq 1}$ of integers such that $s_j \in \{1, \ldots, n\}$, $j \geq 1$. Given $x \in \mathbb{R}^2$, $s \in S(n)$, we define the sequence $(x^s_m)_{m\geq 0} \subset \mathbb{R}^2$ inductively by

$$x^s_0 = x,$$
$$x^s_m = f_{s_m}(x^s_{m-1}), \quad m > 0.$$

Often we drop the superscript s if the sequence s is implicit from the context or $n = 1$. Indeed, if $n = 1$, the sequence (x_m) is the result of repeated iteration of the point x by f: $x_m = f^m(x)$, $m \geq 0$. In this case, we regard f as defining a deterministic dynamical system.

If $n > 1$, the result of the first m iterations generally depends on the sequence s. In this case, it is natural to place a probability measure on $S(n)$[2]

[2] For our applications we typically fix a probability measure on $\{1, \ldots, n\}$ and take the corresponding Bernoulli product measure on $S(n)$.

and look for properties of (x_i^s) that hold for 'most' sequences s. We call $\mathcal{F}$ (together with the probability measure on $S(n)$) a *random dynamical system* or *iterated function system*. Roughly speaking, successive iterations are defined by picking maps at 'random' from the set $\mathcal{F}$.

Remark 1. The theory of iterated function systems when the maps f_i are affine linear contractions of the plane may be found in [4,7]. For the general theory of random dynamical systems, see Arnold [2].

For both deterministic and random dynamical systems the asymptotic behavior of the iteration (x_m^s) is sometimes relatively insensitive to the choice of initial point and sequence s. More precisely, let $\varpi(x,s)$ denote the *ϖ-limit set* of the iteration (x_m^s). That is, $y \in \varpi(x,s)$ if there exists an infinite, strictly increasing sequence of integers n_i such that

$$\lim_{i\to\infty} x_{n_i} = y. \tag{1}$$

Definition 2. A closed set $X \subset \mathbb{R}^2$ is an *attractor* for the dynamical system defined by $\mathcal{F}$ if we can choose an open neighbourhood U of X such that

(a) For all $x \in U$, and sequences s, $\varpi(x,s)$ is a non-empty subset of X.
(b) For (Lebesgue) almost all $x \in U$ and almost all sequences s, $\varpi(x,s) = X$.

Remark 3. There are many different definitions of attractor in the literature. Our particular definition is chosen for its generality and relevance to our intended applications. Later, when we discuss issues of colouring, we have to strengthen the definition of attractor by requiring that it supports an ergodic measure naturally defined through the iteration.

2.3 Symmetric Attractors

We are going to represent symmetric designs as *attractors* of symmetric dynamical systems. We consider both *deterministic* and *non-deterministic* dynamics. We illustrate the general methods with examples leading to patterns with *finite* symmetry. We defer to the following section the construction of wallpaper patterns

Deterministic dynamics leading to patterns with finite symmetry. Let $G \subset E(2)$ be finite. We consider (continuous) mappings $f : \mathbb{R}^2 \to \mathbb{R}^2$ which are G-symmetric (equivariant). That is, if we apply a symmetry $g \in G$ to $x \in \mathbb{R}^2$ and then apply f, the result is the same as if we had applied the symmetry g to $f(x)$. In symbols, for all $x \in \mathbb{R}^2$, $g \in G$, we have

$$f(gx) = gf(x).$$

Suppose that X is a closed and bounded subset of the plane. Define

$$\Sigma(G,X) = \{g \in G \mid gX = X\}.$$

Obviously, $\Sigma(G,X)$ is a subgroup of G. We refer to $\Sigma(G,X)$ as the *G-symmetry group* of X. If G, X are clear from the context, we set $\Sigma(G,X) = \Sigma$. For most of our applications, we will have $\Sigma(G,X) = \Sigma(X)$ – the symmetry group of X.

Parts 1 and 2 of the next result follow easily from the G-symmetry of f (see [5]).

Lemma 4. *Suppose that $X \subset \mathbb{R}^2$ is an attractor for the G-symmetric map $f : \mathbb{R}^2 \to \mathbb{R}^2$. Then*

(1) For all $g \in G$, gX is an attractor with G-symmetry group $g\Sigma(G,X)g^{-1}$.
(2) If $g \in G$ and $g \notin \Sigma(G,X)$, then $gX \cap X = \emptyset$.
(3) For generic f, $\Sigma(G,X) = \Sigma(X)$.

There are restrictions imposed by symmetry on the symmetries of attractors. For example, it is not possible to find a $\mathbb{D}_6$-symmetric planar map f which has an attractor with $\mathbb{D}_3$-symmetry, even though $\mathbb{D}_3$ is a subgroup of $\mathbb{D}_6$. Further, if f has an attractor with $\mathbb{D}_6$-symmetry, then X must be infinite. We refer to [19] for proofs and generalizations of these results (see also [15] for the case of invertible maps).

In general, a G-symmetric continuous planar mapping may not have an attractor. Even if we restrict to G-symmetric polynomial mappings – it is possible using classical invariant theory to describe these maps rather precisely – there are only a few, very few, cases where we can explicitly *prove* the existence of infinite G-symmetric attractors. However, numerical evidence suggests that G-symmetric attractors are quite common. This was first observed in the paper by Chossat & Golubitsky [5] who also proposed mechanisms of 'symmetry creation' which could lead to G-symmetric attractors. Unfortunately, numerical investigation also shows that small changes in the planar map can result in a non-finite attractor changing to a finite attractor.

Example 5. In Figure 1 we show the result of numerically iterating three maps with $\mathbb{D}_5$-symmetry. The first map is defined by

$$f_1(z) = (-2.25 + 0.9|z|^2 + 0.041\mathrm{Re}(z^5))z + 0.1\bar{z}^4, \quad (z \in \mathbb{C}),$$

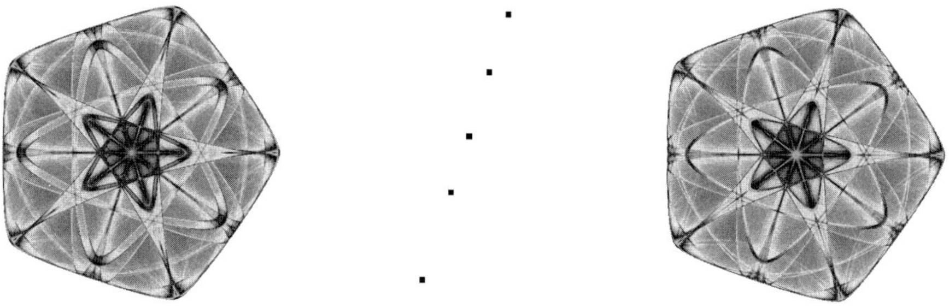

Fig. 1. Varying parameters

where we have identified the complex plane $\mathbb{C}$ with $\mathbb{R}^2$. The remaining maps f_2, f_3 are obtained by successively incrementing the coefficient of $\mathrm{Re}(z^5)$ by -0.001. The images shown, from left to right, are the results of iterating each of the maps f_1, f_2, f_3, 50 million and one thousand times and plotting the fifty million points after the first 1,000 iterations (so as to ignore transients). The initial point x_0 was chosen to be $(0.1, 0.21)$. For the first and third images, we used a grey scale colouring that depended on the number of times each pixel was hit in the iteration. For the middle image, the attractor is an attracting periodic point of period 5 which lies on an axis of symmetry of $\mathbb{D}_5$. We represented the points on the periodic orbit as (large) squares to enhance visibility of the attractor. Numerical experimentation confirms that the attractors in this sequence are robust to changes of the initial point. However, as follows from lemma 4, the attractor for f_2 can lie on any one of the five axes of symmetry of $\mathbb{D}_5$ – depending on the choice of x_0. Note also that colourings of the attractors corresponding to f_1 and f_3 are not identical.

A consequence of the behaviour shown in Example 5 is that is it likely to be extremely difficulty to find computable conditions on the coefficients of G-symmetric polynomials that imply the existence of an infinite chaotic G-symmetric attractor. Similar 'windows' of attracting periodic orbits are well known in the theory of 1-dimensional mappings.

Rather than attempt further theoretical analysis of symmetric attractors, we shall instead make the working assumption that attractors we find by numerical experimentation are attractors in the sense of Definition 2. We need also to assume some facts about the statistical properties of the iteration on and near X. To this end, let δ_x denote the Dirac probability measure supported at x. We assume that if X is a G-symmetric attractor for the G-symmetric map f, and U is an open neighbourhood of X satisfying the conditions of Definition 2, then for almost all $x \in U$, $\frac{1}{n}\sum_{i=0}^{n-1} \delta_{x_i}$ converges (weakly) to a unique G-invariant measure ρ supported on X. As we explain shortly, we use the existence of ρ as the theoretical basis for colouring the attractor X. In other words, we regard ρ as a coloured measure on X where 'colour' represents the probability of visiting a pixel (or Borel set) during the iteration.

Non-deterministic dynamics. It is straightforward to construct iterated function systems on the plane which produce symmetric attractors. Let G be a finite subgroup of $E(2)$ of order n. Let f be any affine linear contraction of the plane[3]. Define $\mathcal{F} = \{g \circ f \mid g \in G\}$. It is easy to verify that $\mathcal{F}$ consists of affine linear contractions. Define the probability measure p on $\{1, \ldots, n\}$ by $p(i) = \frac{1}{n}$, $1 \le i \le n$, and take the corresponding Bernoulli product measure μ on $S(n)$. The iterated function system defined by $\mathcal{F}$, μ has a unique G-symmetric attractor X. The attractor X supports a natural

[3] We might equally well have started with a finite set of affine linear contractions.

G-invariant ergodic measure ρ such that for almost all $x \in \mathbb{R}^2$ and sequences $s \in S(n)$, $\frac{1}{n}\sum_{i=0}^{n-1} \delta_{x_i^s}$ converges (weakly) to ρ. We refer the reader to [14,11] for details.

Example 6. In Figure 2 we show three 'symmetric fractals' constructed using an iterated function systems generated by a single affine linear contraction. From left to right, the images have $\mathbb{D}_{10}$, $\mathbb{Z}_8$ and $\mathbb{Z}_{14}$-symmetry respectively.

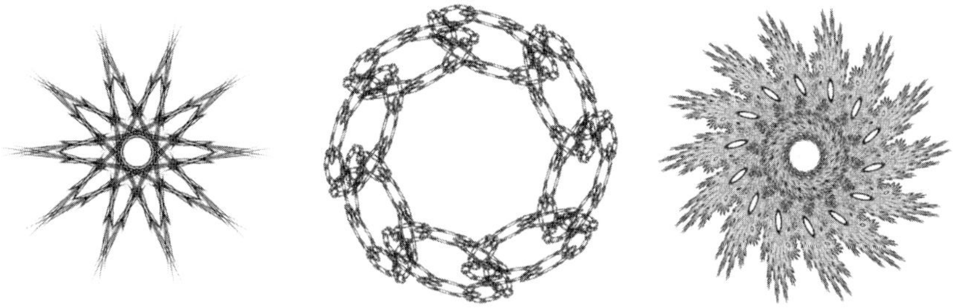

Fig. 2. Symmetric Fractals

Following example 5, we have a used a grey scale colouring that reflects the structure of the natural ergodic measures on the attractors.

2.4 Numerical implementation

We divide the numerical investigation of symmetric attractors into two components: experimentation (design) and computation.

Experimentation – Design. The aim of the experimentation or design process is to find aesthetically appealing patterns with specified symmetry. This process is implemented in *prism* in the following way.

The user initially specifies the symmetry of the pattern (any discrete subgroup of $E(2)$), and the type of iteration (deterministic or iterated function system). In the case of wallpaper patterns, there is also a choice of one- or two-colour symmetry. Depending on the type of iteration and symmetry chosen, various algorithms are available. For example, if a dihedral group $\mathbb{D}_n$, $n \geq 3$, and deterministic iteration are selected, then one possibility is to use the family of $\mathbb{D}_n$-symmetric polynomial maps

$$P(z) = (\lambda + \alpha|z|^2 + \beta\text{Re}(z^n))z + \gamma\bar{z}^{n-1}, \quad (z \in \mathbb{C}).$$

In this case the user can vary the (real) coefficients $\lambda, \alpha, \beta, \gamma$ as part of the experimentation and design process. For further details on this algorithm (and extensions), as well non-deterministic algorithms for bounded patterns, we refer the reader to [14]. The user can control all aspects of the iteration. For example, the initial point x_0, the number of iterations plotted, and the number of iterations that are not plotted during the initial part of the iteration. Further, the effect of incrementing parameters on the iteration can be shown with a sequence of plots on the screen (as in Figure 1).

Computation. Once a design has been selected, a numerical approximation to the (assumed) natural measure on the attractor is computed. This is done by making a choice of resolution (size) and then computing a large number of iterations, recording the number of hits on each pixel. For example, if a resolution of 1000×1000 is chosen for an attractor computed using a deterministic algorithm, it might be reasonable to compute 100 million iterates. In practice, the number of iterations needed depends on the specific attractor and also the type of iteration – attractors produced using non-deterministic algorithms tend to need fewer iterations because the associated measures are often relatively uniform.

2.5 Colouring

For one-colour designs, the attractor is coloured on the basis that the colour of a pixel depends only on the number of times the pixel was hit in the iteration.

More formally, suppose that X is a G-symmetric compact attractor with associated ergodic measure ρ defined as a weak limit of Dirac measures along trajectories. Since X is G-symmetric, if follows from the uniqueness of ρ, that ρ is G-symmetric. That is, if A is a measurable subset of X and $g \in G$, then $\rho(A) = \rho(gA)$ for all $g \in G$.

In order to develop a theory of coloured measures that we can apply – especially to 2-colour quilts – we need to make a further assumption about the measure ρ. In practice this assumption is harmless and, more importantly, it is consistent with the way we compute approximations to a coloured measure.

Let $d\lambda$ denote Lebesgue measure on the plane.

Definition 7 (cf [11]). We say the probability measure ρ is a *density* on X if there exists a Lebesgue measurable function $\xi : X \to \mathbb{R}$ such that

(1) ξ is G-invariant: $\xi(gx) = \xi(x)$, all $x \in X$, $g \in G$.
(2) $\xi(x) > 0$ for all $x \in X$.
(3) For all measurable subsets A of X, $\rho(A) = \int_A \xi \, d\lambda$.

We refer to ξ as a *density function* on X.

Let $\mathcal{C}$ denote a space of colours (for example, RGB- or CMYK-space). A *symmetric colouring* of the attractor X is given by a map $C : X \to \mathcal{C}$ that factors through the density function ξ.

That is, C is a colouring if there exists $c : \mathbb{R} \to \mathcal{C}$ such that

$$C(x) = c(\xi(x)), \quad (x \in X).$$

Computing densities. Simple examples, such as the Sierpiński triangle, show that the natural measure ρ on X is generally not a density. However, it is straightforward to show that we can arbitrarily approximate ρ by an (approximately) G-symmetric density. Furthermore, we can realize the approximation numerically. We briefly sketch the way we do this.

Without loss of generality, assume that X is contained in the unit square S in $\mathbb{R}^2$. Choose a large positive integer N. Let $\mathcal{D}$ denote the partition of S into N^2 squares, each of side length $1/N$. Let $\mathcal{P}(S) = \{s \in \mathcal{D} \mid s \cap X \neq \emptyset\}$. Suppose that (x_i) is a sequence of iterates such that the corresponding averages of Dirac measures along the trajectory converges to ρ. For $n > 0$, $s \in \mathcal{D}$, let $i(n)$ denote the number of times points of the finite trajectory $x_1, \ldots, x_n$ enter s. (We ignore the question of iterates lying on the boundary of s and assume that every point of (x_i) lies in exactly one square from $\mathcal{D}$.) We define a density function on X by

$$\xi_N = \lim_{n \to \infty} \frac{1}{n} \sum_{s \in \mathcal{D}} i(n) \chi_s,$$

where χ_s denotes the characteristic function of the square s. Let ρ_N denote the corresponding measure. Since ρ is ergodic, it follows easily that the limit defining ξ_N exists for almost all choices of (x_i) and depends only on the subdivision $\mathcal{D}$. As we increase N, ρ_N will converge weakly to ρ. Furthermore, it is easy to see that since ρ is G-symmetric, the density ξ_N is approximately G-symmetric. All of these statements about approximation can be made quite precise granted our original assumptions about ρ.

In the sequel, we assume that attractors always have a density. We refer the reader to [14, chapter 1], and [8,11] for more details on colouring. Matters are more complex for two-colour patterns and we discuss this later in section 4.

3 Wallpaper Patterns

We start with a rapid review of some basic facts about lattices, tori and wallpaper patterns. Most of what we say can be found in the text by Armstrong [1]. We conclude the section with some details and examples on the construction of wallpaper patterns using dynamics.

3.1 Geometry of wallpaper patterns

Let $O(2)$ denote the orthogonal group consisting of all rotations about the origin and reflections in lines through the origin of $\mathbb{R}^2$. In what follows, we identify the group of translations $E_T(2)$ with $\mathbb{R}^2$. Every transformation $g \in E(2)$ can be written uniquely as $g = r \circ t$, where $t \in \mathbb{R}^2$ and $r \in O(2)$. It follows that we have a natural homomorphism $\pi : E(2) \to O(2)$, defined by $\pi(g) = r$.

Let $\mathbb{Z}^2 = \{(m, n) \mid m, n \in \mathbb{Z}\}$ denote the integer lattice in $\mathbb{R}^2$. The quotient group $\mathbb{R}^2/\mathbb{Z}^2$ is isomorphic to the two-dimensional torus $\mathbb{T}^2$. Topologically, $\mathbb{T}^2$ is the product of two circles, or the space obtained by identifying opposite edges of the unit square in $\mathbb{R}^2$. For future use, we define the hexagonal lattice $\mathbb{Z}^2_h = \{m\mathbf{a} + n\mathbf{b} \mid m, n \in \mathbb{Z}\}$, where $\mathbf{a} = (1, 0)$, $\mathbf{b} = (\frac{1}{2}, \frac{\sqrt{3}}{2})$.

Suppose that G is a wallpaper group. Since the translational subgroup of $E(2)$ is naturally identified with $\mathbb{R}^2$, it follows that the translational subgroup G_T of G is naturally identified with a (lattice) subgroup $\mathcal{L}$ of $\mathbb{R}^2$. Necessarily $\mathcal{L}$ is isomorphic to $\mathbb{Z}^2$ and so $\mathbb{R}^2/\mathcal{L}$ is also isomorphic to $\mathbb{T}^2$ – the isomorphism will depend on the choice of isomorphism between $\mathcal{L}$ and $\mathbb{Z}^2$. In the sequel we set $\mathbb{R}^2/\mathcal{L} = \mathbb{T}^2_{\mathcal{L}}$ to emphasize the dependence of the torus on the lattice. Let $J = \pi(G)$ denote the *point group* of G. Since $\mathcal{L}$ lies in the kernel of π, it follows that J is a finite subgroup of $O(2)$ and $J(\mathcal{L}) = \mathcal{L}$ [1, Theorem 25.2].

At this point it is customary to divide the lattice subgroups into five classes (oblique, rectangular, centered rectangular (or rhombic), square and hexagonal) and then classify the wallpaper groups according to the point groups that can occur for each lattice class.

We shall proceed a little differently. Since $J(\mathcal{L}) = \mathcal{L}$, J acts as a finite group of transformations on $\mathbb{T}^2_{\mathcal{L}}$. It follows that if X is a closed J-invariant subset of $\mathbb{T}^2_{\mathcal{L}}$, then the lift of X to $\mathbb{R}^2$ will be a doubly periodic pattern with wallpaper symmetry group (at least as big as) G. We use this observation as the basis of our construction of wallpaper patterns using dynamics.

It follows straightforwardly from the J-invariance of the lattice $\mathcal{L}$ that rotations in a point group can only be of order 2,3,4 or 6. Further, order 3 or 6 rotations can only occur when the corresponding lattice is hexagonal. Five of the wallpaper patterns (**p6m, p6, p3, p31m, p3m1**) are only supported on a hexagonal lattice. The remaining twelve patterns can all occur on a square lattice. Amongst these twelve patterns, **p4m, p4, p4g** are *only* supported on a square lattice. While the remaining eight patterns all occur on a square lattice, they also occur on other lattices. For example, patterns of type **p1, p2** occur on an oblique lattice. Rather than attempt to construct wallpaper patterns for every class of lattices, we shall only work with the integer and hexagonal lattices $\mathbb{Z}^2, \mathbb{Z}^2_h$. Once we have constructed a particular pattern on a square lattice that is normally supported on a different lattice class L, we can apply linear transformations to the pattern that map $\mathbb{Z}^2$ onto a lattice in the class L. For example, if the pattern is of type **pmm** (normally supported on a rectangular lattice), composition with any non-singular diagonal matrix will lead to a **pmm** pattern supported on a general rectangular lattice.

3.2 Dynamics

Deterministic dynamics. Suppose G is a wallpaper group with associated point group J. For simplicity we suppose that $G_T = \mathbb{Z}^2$. As we indicated above every closed J invariant subset of $\mathbb{T}^2$ lifts to a wallpaper pattern with wallpaper group (at least) G. In order to construct J-invariant sets using dynamics, we search for J-invariant attractors of continuous (or smooth) J-symmetric mappings $f : \mathbb{T}^2 \to \mathbb{T}^2$. Every $f : \mathbb{T}^2 \to \mathbb{T}^2$ lifts to a doubly periodic mapping $F : \mathbb{R}^2 \to \mathbb{R}^2$. The doubly periodic mappings of $\mathbb{R}^2$ are given by Fourier series. For deterministic iterations, we work with classes of low order trigonometric polynomials and vary coefficients just as we described above for bounded attractors in the plane. In order to obtain J-invariant attractors, we need to work with J-symmetric trigonometric polynomials on $\mathbb{T}^2$. The conditions for J-symmetry are straightforward to compute using the dual lattice (see [14, Appendix D]).

In order to obtain a sensible symmetry classification of *dynamically* generated wallpaper patterns, we regard the pattern as being determined by the pair (X, ρ), where X is a J-invariant attractor and ρ is a G-invariant measure on X. In particular, we may (and often do) have $X = \mathbb{T}^2$. If the associated measure ρ has (say) only the trivial symmetry, then (X, ρ) will define a wallpaper pattern of type **p1**.

Non-deterministic dynamics. We continue to assume G is a wallpaper group with associated point group J. Let $f : \mathbb{T}^2 \to \mathbb{T}^2$ (possibly not continuous everywhere). Form the iterated function system $\mathcal{F} = \{g \circ f \mid g \in J\}$. We find numerically that for most choices of f, the iterated function has a J-symmetric attractor that is independent of the initial point of the iteration. This situation is theoretically well-understood when the elements of $\mathcal{F}$ satisfy appropriate Lipschitz conditions (see, for example, [7, chapter 9]).

3.3 Wallpaper patterns through dynamics

We conclude this section with some examples of wallpaper patterns constructed via dynamics.

In Figure 3, we show two quilts generated using an iterated function system. The quilt on the left is of type **p4** and the generating function was chosen so as to give an angular effect. As the measures produced by this algorithm are often rather flat, we have given a 3-dimensional effect to the image by using the embossing option of the *gimp* software package. The image on the right is of type **pmg** and is a grey scale version of 'Blue Columns', a picture that has been exhibited at a number of centers in Europe as part of *The Frontier between Art and Science, International Exhibition.*

In Figure 4, we show two quilts generated using a deterministic algorithm. The quilt on the left is of type **p6m**, the one on the right of type **pmm**. Both patterns are grey scale versions of pictures that were originally designed as

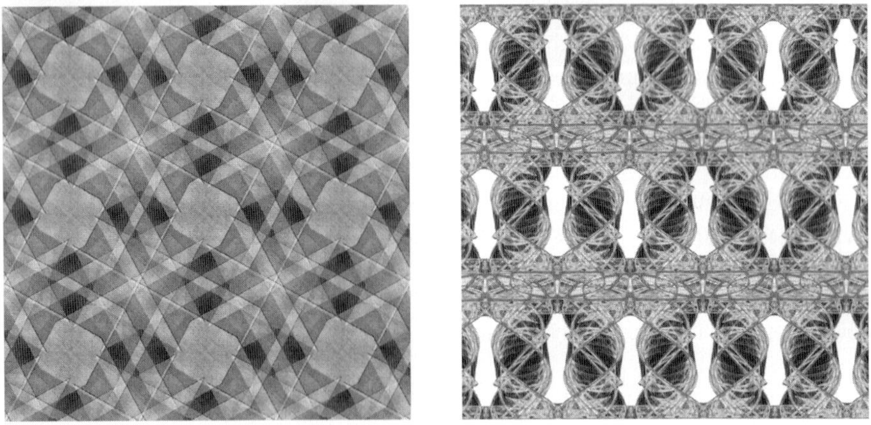

Fig. 3. Wallpaper patterns of types **pmm** and **pmg**

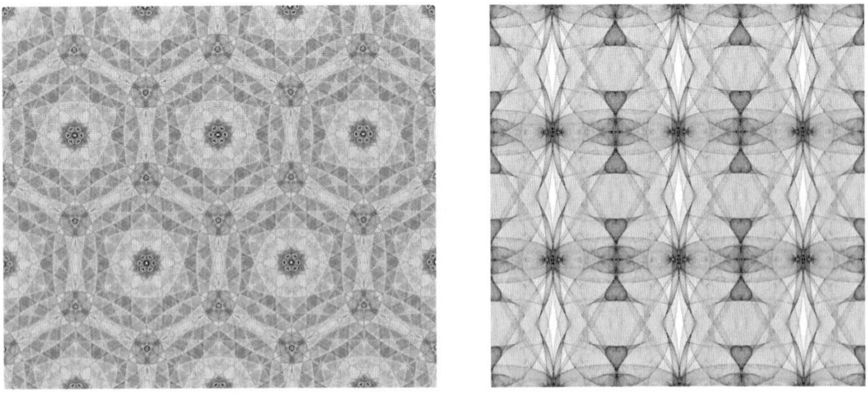

Fig. 4. Wallpaper patterns of types **p6m** and **pmm**

possible entries for an additional display during the visit of the *The Frontier between Art and Science* exhibition to Granada. Eventually, the second image was chosen for the exhibition.

4 Two-colour Patterns

We start by reviewing the definition and basic properties of two-colour wallpaper patterns. We then introduce a definition of two-colour pattern suitable for the analysis and description of dynamically generated patterns. We conclude by showing some examples of two-coloured patterns

4.1 Two-colour wallpaper patterns

The reader may find definitions of two-colour patterns in Grünbaum and Shephard [16, chapter 8] and Washburn & Crowe [20]. Roughly speaking, a 2-colour wallpaper pattern $\mathbf{P} \subset \mathbb{R}^2$ is a wallpaper pattern that uses two colours and is such that

(a) Every symmetry of the (uncoloured) pattern either reverses or preserves colour.
(b) There exist colour reversing symmetries.

Woods [21] showed that there were exactly 46 2-colour wallpaper patterns. Examples of all 46 of the 2-colour wallpaper patterns may be found in the books by Washburn & Crowe and Grünbaum and Shephard cited above (see also the article by Coxeter [6]). There are two notations for the 2-colour wallpaper patterns, one due to Belov the other to Coxeter. In what follows we shall assume some familiarity with both notations (see [20, §3.4.2]).

Example 8. In Figure 5 we show a simple example of a two-colour pattern of type $\mathbf{pg}'$ (Belov's notation). In this case, the uncoloured pattern has symmetry $\mathbf{pg}$. Translational symmetries preserve colour, while glide reflection symmetries (along either of the glide lines L or M) reverse colour. There are no other symmetries of the pattern. Note that either of the sets P_r, P_b is a wallpaper pattern of type $\mathbf{p1}$. Hence, if we use Coxeter's notion, the pattern is of type $\mathbf{pg/p1}$.

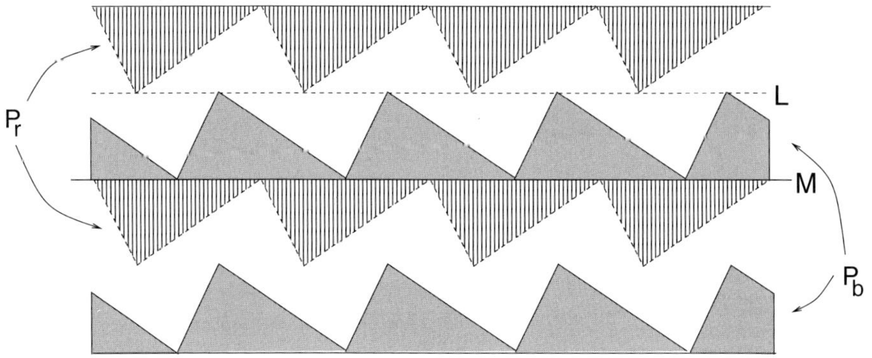

Fig. 5. A 2-colour pattern of type $\mathbf{pg}'$

Even for two-colour *tilings* of the plane, there are some subtleties in giving an unambiguous definition. Indeed, if we assume tiles are closed (or open) sets, then either some points of the plane are not covered in the tiling or some (boundary) points are covered at least twice and by tiles of different colours.

Since we intend working with dynamically generated sets which may have a very large fractal boundary we need to take some care with definitions.

As a way of motivating our definition, suppose that $\mathbf{P}$ is a two-colour wallpaper pattern. Suppose that the colours used in the pattern $\mathbf{P}$ are red and blue. Let $\mathbf{P}_r$ denote the red points, $\mathbf{P}_b$ the blue points. We shall assume that $\mathbf{P}_r$ is a proper closed subset of $\mathbb{R}^2$. Since there exist colour reversing symmetries of $\mathbf{P}$, $\mathbf{P}_b$ is also closed. It follows that the background – the set of points $\mathcal{B}(\mathbf{P})$ of $\mathbb{R}^2$ that do not lie in $\mathbf{P}$ – is not empty. Necessarily, the background $\mathcal{B}(\mathbf{P})$ is preserved by all the symmetries of $\mathbf{P}$. The pattern shown in Figure 6 satisfies these conditions (The pattern of Figure 5 does not as there is overlap between $\mathbf{P}_r$ and $\mathbf{P}_b$ along the line b).

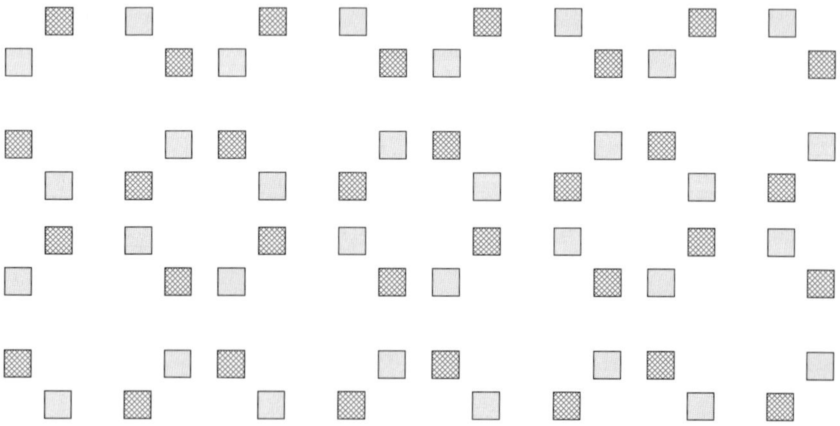

Fig. 6. A 2-colour pattern of type **p4m′m′** (**p4m**/**p4**)

If we denote the wallpaper group of the uncoloured pattern $\mathbf{P}$ by G and the subgroup of colour preserving symmetries of $\mathbf{P}$ by G_p, then it is easy to see that the quotient group $G/G_p \cong \mathbb{Z}_2$ – the cyclic group of order two.

We abstract this example in the following way.

Definition 9 (cf [11]). A two-colour wallpaper pattern consists of a triple $(\mathbf{P}, \mathbf{P}_1, \mathbf{P}_2)$ of wallpaper patterns satisfying

(1) $\mathbf{P} = \mathbf{P}_1 \cup \mathbf{P}_2$.
(2) If g is an element of the wallpaper symmetry group of $\mathbf{P}$, then either $g\mathbf{P}_1 = \mathbf{P}_1$ or $g\mathbf{P}_1 = \mathbf{P}_2$.
(3) There exists at least one symmetry g of $\mathbf{P}$ such that $g\mathbf{P}_1 = \mathbf{P}_2$.

In the sequel we refer to $\mathbf{P}_1, \mathbf{P}_2$ as *subpatterns* of the two-colour pattern $\mathbf{P}$. Note that in our definition we do *not* require $\mathbf{P}_1$ and $\mathbf{P}_2$ to be disjoint. Indeed, we allow $\mathbf{P} = \mathbf{P}_1 = \mathbf{P}_2$.

Without further structure, Definition 9 adds rather little. Let $\mathcal{O}(\mathbf{P}) = \mathbf{P}_1 \cap \mathbf{P}_2$ – the overlap. If the overlap is empty or, allowing for boundaries, is of measure zero, we are in the situation described above. If there is an overlap, then we can assign a new colour to $\mathcal{O}(\mathbf{P})$. In this way, our two-colour pattern uses four colours corresponding to the sets $\mathbf{P}_1$, $\mathbf{P}_2$, $\mathcal{B}(\mathbf{P})$, and $\mathcal{O}(\mathbf{P})$. All symmetries of the pattern $\mathbf{P}$ will preserve the colours on $\mathcal{B}(\mathbf{P})$, and $\mathcal{O}(\mathbf{P})$.

Matters become much more interesting when we add dynamics.

4.2 Two-colour quilts from dynamics

Suppose that we want to construct two-colour wallpaper patterns of type $\mathbf{p}/\mathbf{q}$ (Coxeter's notation). Noting Definition 9, it suffices to construct a subpattern $\mathbf{P}_1$ of type $\mathbf{q}$. Once we know $\mathbf{P}_1$, we can apply a colour reversing symmetry to obtain $\mathbf{P}_2$ and hence $\mathbf{P} = \mathbf{P}_1 \cup \mathbf{P}_2$ – a pattern of type $\mathbf{p}$. In what follows, we work on the associated torus. Everything we say lifts immediately to $\mathbb{R}^2$.

Let J and J_s denote the point group associated to $\mathbf{p}$ and $\mathbf{q}$ respectively. Note that we can and do always regard J_s as an index two subgroup of J.

Let $\mathbf{P}_1$ be a pattern of type $\mathbf{q}$ constructed by dynamics. We can represent $\mathbf{P}_1$ by a pair (X_1, ξ_1), where $X_1 \subset \mathbb{T}^2$ is a J_s-symmetric attractor and ξ_1 is a J_s-invariant density (function) on X_1. It follows by symmetry that if $g \in J \setminus J_s$, then $X_2 = gX_1$ will be a J_s-symmetric attractor and $\xi_2 = \xi_1 \circ g$ is a J_s-invariant density on X_2. Neither J_s nor ξ_2 depends on the particular choice of $g \in J \setminus J_s$.

Set $X = X_1 \cup X_2$. Extend ξ_i to X by setting ξ_i equal to zero on the complement of X_i in X, $i = 1, 2$. Define the vector density $\xi : X \to \mathbb{R}^2$ by $\xi = (\xi_1, \xi_2)$. We also define $\xi^r = (\xi_2, \xi_1)$. Symmetry properties of ξ are given in the following lemma.

Lemma 10 ([11]). *Let $x \in X$, $g \in J$, then*

$$\begin{aligned} \xi(gx) &= \xi(x), \quad (g \in J_s) \\ &= \xi^r(x), \quad (g \notin J_s). \end{aligned}$$

Definition 11 ([11]). A 2-colouring of (X, ξ) consists of a map $C : X \to \mathcal{C}$ that can be factored

$$C = F \circ \xi,$$

where $F : \mathbb{R}^2 \to \mathcal{C}$.

It follows from the definition that the colour assigned to a point $x \in X$ depends only on the densities $\xi_1(x), \xi_2(x)$. Further, it is a consequence of lemma 10 that points related by symmetries in J_s have the same colour. On the other hand, points symmetrically related by symmetries in $J \setminus J_s$ will generally have different colours – unless the map F is symmetric: $F(x, y) = F(y, x)$. In general, symmetries of a dynamically generated two-colour pattern will either preserve colours or symmetrically interchange colours. There is no restriction to *two* colours. In practice, we typically work from a palette of thousands of colours.

There are many different ways of implementing two-colouring algorithms. We refer the reader to [8] for more details and examples. Roughly speaking, our two-colourings depend on symmetry *and* dynamics in the sense that colours measure the probability of lying in one of the subpatterns as well as the statistics of the iteration on the subpattern. For this reason, colour palettes are typically two-dimensional arrays of colour.

4.3 Numerical Algorithms

In the previous sections we have already described all that is required to construct two-colour wallpaper patterns using deterministic algorithms. It turns out that there is an elegant way to construct two-colour wallpaper patterns using non-deterministic algorithms. This method was devised and implemented, in collaboration with Marty Golubitsky, in 1992. An example will suffice to explain the general method. Suppose we want to construct a two-colour wallpaper pattern of type **p4m′m′**. In this case the point group $J = \mathbb{D}_4$ and $J_s = \mathbb{Z}_4$. Let f be a mapping of the torus and $\mathcal{F}$ denote the corresponding iterated function system symmetrized over $\mathbb{D}_4$. Let $\mathcal{F}_s$ be the subset of $\mathcal{F}$ defined by symmetrizing over $\mathbb{Z}_4$. We carry out the usual random iteration but now construct two subsets X_1, X_2 of the torus according to the following rule. Suppose that we have constructed x_m, then $x_{m+1} \in X_1$ if $x_{m+1} = hx_m$, where $h \in \mathcal{F}_s$. Otherwise $x_{m+1} \in X_2$. A particularly nice feature of this scheme is that it allows one to see changes in the structure of the subpatterns X_i without changing the underlying pattern X. The three patterns in Figure 7 were all obtained by symmetrizing the same generating function over different point groups.

Reading from left to right, the subpatterns in Figure 7 are of type **cmm**, **pmm** and **p4**. Ignoring colour, all three patterns are of type **p4m**.

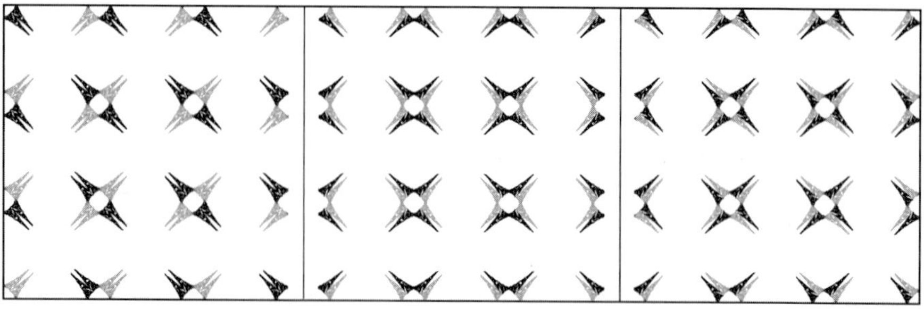

Fig. 7. Two-colour patterns of type **p4′m′m, p4′mm′, and p4m′m′**.

4.4 Examples of two-colour designs

We conclude with some examples of two-colour designs that we have created using the ideas described above. Except for Figure 3, all the patterns we show were computed using deterministic algorithms.

In Figure [MF]1, we show *Armies of the Night*, a two-colour pattern of type **c′m**. The original of this picture is printed on thirty inch square photographic paper and was exhibited in the *Eighth New York Digital Salon.*

Figure [MF]2, shown in the Appendix, is an example of a pattern of type **pgg′**.

The image shown in Figure [MF]3, is an example of a two-colour design of type **pmm′**. This figure possesses a three dimensional quality, characteristic of certain two-colour symmetry types. Our final image, Figure [MF]4, is an example of a two-colour pattern of type **p6′**.

References

1. M. A. Armstrong, *Groups and Symmetry*, (Undergraduate texts in Mathematics, Springer-Verlag, New York, Berlin, Heidelberg, 1988).
2. L. Arnold, *Random Dynamical Systems*, (Springer Monographs in Mathematics, Springer-Verlag, Berlin-Heidelberg, 1998).
3. G. Bain, *Celtic Art: The Methods of Construction*, (Dover Publications, 1973).
4. M. F. Barnsley, *Fractals Everywhere* (Academic Press, San Diego, 1988).
5. P. Chossat, M. Golubitsky, 'Symmetry increasing bifurcations of chaotic attractors', Physica D **32** (1988), 423–436.
6. H. S. M. Coxeter, 'Colored symmetry', In: *M C Escher: Art and Science* (H. S M. Coxeter et al, eds., Elsevier, Amsterdam and New York, 1986), 15–33.
7. K. Falconner, *Fractal Geometry* (John Wiley & Sons, Chichester, 1990).
8. M. J. Field, 'Harmony and Chromatics of Chaos', In: Bridges, Mathematical Connections in Art, Music, and Science, (Conference Proceedings, ed. Reza Sarhangi, Southwestern College, Kansas, 1999), 1–20.
9. M. J. Field, 'Color Symmetries in Chaotic Quilt Patterns', In: Proc. ISAMA 99, (eds. N. Friedman, J. Barrallo, San Sebastian, Spain, 1999), 181–188.
10. M. J. Field, 'The Art and Science of Symmetric Design', In: Bridges, Mathematical Connections in Art, Music, and Science, (Conference Proceedings, ed. Reza Sarhangi, Southwestern College, Kansas, 2000), 53–60.
11. M. J. Field, 'Designer Chaos', J. Computer Aided Design, **33** (5) (2001), 349–365.
12. M. J. Field, 'Mathematics through Art – Art through Mathematics', In: Proc. MOSAIC 2000, (eds. D Salesin and C Séquin, University of Washington, 2000), 137–146.
13. M. J. Field, 'Dynamics, Chaos and Design', In: *The Visual Mind 2* (ed. M. Emmer, MIT Press), to appear.
14. M. J. Field, M. Golubitsky, *Symmetry in Chaos*, (Oxford University Press, New York and London, 1992).
15. M. J. Field, I. Melbourne, M. Nicol, 'Symmetric Attractors for Diffeomorphisms and Flows', Proc. London Math. Soc., (3) **72** (1996), 657–696.

16. B. Grünbaum, G. C. Shephard. *Tilings and Patterns. An Introduction*, (W H Freeman and Company, New York, 1989).
17. S. Jablan, *Theory of Symmetry and Ornament*, (Mathematics Institute, Beograd, 1995).
18. C. S. Kaplan, 'Computer generated Islamic star patterns', In: Bridges, Mathematical Connections in Art, Music, and Science, (Conference Proceedings, ed. Reza Sarhangi, Southwestern College, Kansas, 2000), 105–112.
19. I. Melbourne, M. Dellnitz, M. Golubitsky, 'The structure of symmetric attractors', Arch. Rat. Mech. Anal. **123** (1993), 75–98.
20. D. Washburn, D. Crowe, *Symmetries of Culture*, (University of Washington Press, Seattle, 1988).
21. H. J. Woods, 'The Geometrical basis of pattern design. Part 4: Counterchange symmetry in plane patters', Jn. of the Textile Institute, Trans. **27** (1936), 305–320.

Machines for Building Symmetry

Maria Dedò

Università degli Studi di Milano, Dipartimento di Matematica "F. Enriques", via Saldini 50, 20133 Milano, Italy
<Maria.Dedo@mat.unimi.it>

Abstract. We use the expression "machine for building symmetry" to mean something which, by the aid of mirrors, builds up a virtual image of a plane figure, or of a 3d-object, with a certain pre-determined type of symmetry. The mathematical interactive exhibition *"Simmetria, giochi di specchi"*, recently realized by the Mathematics Department "F. Enriques" of Milan University, is based just on this kind of objects: in this paper we are going to sketch the mathematical background underlying the exhibition and to discuss the different ways to use it with respect to different sorts of public.

1 Introduction

What I am going to describe here is a mathematical interactive exhibition, *"Simmetria, giochi di specchi"*, based on the theme of symmetry, which has been realized in the last few years by the Mathematics Department "F. Enriques" of Milan University.

The main objects of this exhibition are six "machines for building symmetry". Three of them are planar, made up of three mirrors each, all perpendicular to the plane of the table where they are posed, and cutting on this plane – respectively – an equilateral triangle, a right isosceles triangle, a right triangle with angles $\frac{\pi}{6}$ and $\frac{\pi}{3}$ (here we shall call them *mirror boxes*). The other three are 3-dimensional *kaleidoscopes*: each one again is made up of three mirrors, belonging to three planes concurrent in a point O and each one cuts on a sphere with center O a spherical triangle: the first one of these three triangles has angles $\frac{\pi}{2}$, $\frac{\pi}{3}$ and $\frac{\pi}{3}$; the second one has angles $\frac{\pi}{2}$, $\frac{\pi}{3}$ and $\frac{\pi}{4}$, while the third one has angles $\frac{\pi}{2}$, $\frac{\pi}{3}$ and $\frac{\pi}{5}$ (see the first figure [MD]1 in the Appendix)(1).

Each one of the mirror boxes is provided with some tiles, which are the right ones in order to see in the box a given picture (which will be a tessellation of the plane): some of these pictures are shown on the walls of the box, so that the visitor knows in advance in which one of the three boxes he will be able to build that picture; other ones are shown mixed up in a poster, so that the visitor willing to reconstruct that picture has first of all to

1 All the photos have been made by Sabrina Provenzi, and their reproduction here is authorized by the head of the Mathematics Department "F. Enriques" of Milan University.

decide in which one of the three boxes this can be done ([MD]2 and [MD]3). Mathematically, this is a first not trivial problem: this decision corresponds in fact to detecting the symmetry group of the given picture.

The 3-dimensional kaleidoscopes can be handled in an analogous way: some small pieces are provided, and each single piece corresponds to the fundamental domain ([2]) for the action of the symmetry group on some polyhedra. By putting the piece in the corresponding kaleidoscope, the polyhedron is reconstructed.

Both in the 2-dimensional mirror boxes and in the 3-dimensional kaleidoscopes the same mathematical concept is underlined: the classification of something (a planar picture in the first case, a solid object in the second one) with respect to its symmetry group. So a "machine for building symmetry" builds up different things, depending on what is put inside it, but all what is built with the same machine has the same kind of symmetry; while different machines build up different symmetries ([3]).

The actual exhibition contains in fact other objects, which are all functional to this main idea.

The paper is organized as follows: section 2 is more technical and is directed to the reader who likes to have a more detailed idea of the mathematical concepts underlying the objects just described; section 3 contains a discussion about different ways for using these "machines", with different sorts of public. Readers mainly interested to this discussion can skip directly to section 3.

2 The mathematics underlying the "machines for building symmetry"

2.1 Coxeter groups

Coxeter, in a series of papers around 1930, began to study those subgroups of the isometry group in $\mathbf{R}^n$ having the two properties of being discrete and being generated by reflections in hyperplanes. These two properties are exactly what is necessary in order to "see" the group in a system of mirrors. In order to explain what we mean, we shall first illustrate a couple of examples in the plane.

Let G be the group generated by the reflections A and B in two lines a and b, making an angle of $\frac{\pi}{4}$ in point O. G is a finite group, consisting of four rotations (with center O and angles, respectively, $\frac{\pi}{2}, \pi, \frac{3\pi}{2}, 2\pi$) and four reflections, with respect to the two lines a and b and other two lines c and d (see Fig. 1).

[2] D is a fundamental domain for the action of G on P if for any point x in P we can find an element g in G such that $g(x)$ belongs to D and no two points in the interior of D are related by an element in G.

[3] This is in fact just a first rough approximation, which is not quite correct: see the following section for a more precise statement.

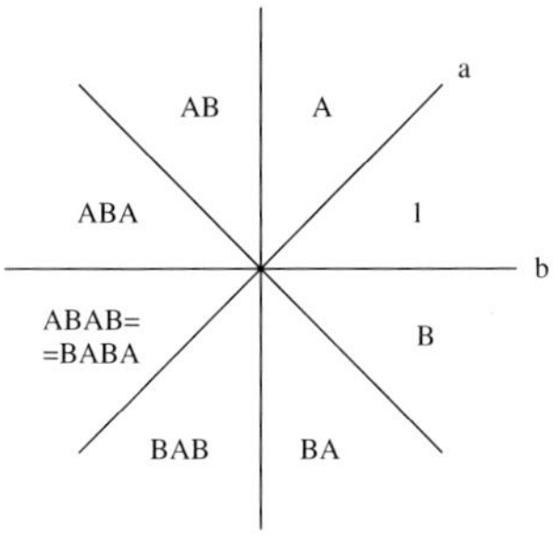

Fig. 1. Chambers and set of hyperplanes for the group $G = \langle A, B : A^2 = B^2 = (AB)^4 = 1\rangle$

The set of these four lines is the *complete set of hyperplanes* related to the group G (that is, the maximal set of hyperplanes such that the reflection with respect to that hyperplane belongs to the group); the connected components of the complement, in the plane, of these four lines are the *chambers* of the group G. All the chambers are equivalent (in the sense that there exists an element of the group sending one into the other), each one of them is an angle and the group is generated by the reflections in the walls of any chamber. There is a bijection between the whole set of the chambers and the elements of the group: fixing one of the chambers R_0, the bijection is obtained by sending any other chamber R into the element g in G such that $g(R) = R_0$.

In order to consider another example, let H be the group generated by the reflections A, B, C in the three sides a, b, c of a right isosceles triangle. H is an infinite group, the complete set of hyperplanes is an infinite grid, and the complement of this grid in the plane is made up of an infinity of isometric triangles, which are the chambers of the group: see Fig. 2.

It is worthwhile observing that, in both cases, the complete set of hyperplanes (and, as a consequence, also the set of the chambers) is just what we see if we start from a set of mirrors (two in the first case, three in the second one) corresponding to the generators of the group.

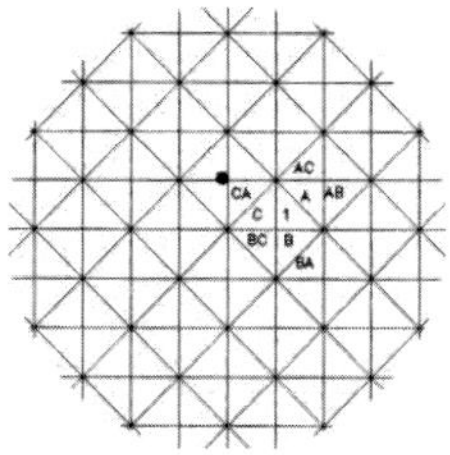

Fig. 2. Chambers and set of hyperplanes for the group $G = \langle A, B, C : A^2 = B^2 = C^2 = (AB)^4 = (AC)^4 = (BC)^2 = 1\rangle$

In fact, when we say that we are "seeing the group", we are not referring only to the fact that the chambers we see are in one-to-one correspondence with the elements of the group, but, also, to the possibility of reading out from what we see the presentation of the group with generators and relations: in the case of Fig. 1, the group G is generated by A and B with relations

$$A^2 = B^2 = (AB)^4 = 1;$$

in the case of Fig. 2, H has three generators A, B, C and relations

$$A^2 = B^2 = C^2 = (AB)^4 = (AC)^4 = (BC)^2 = 1.$$

In [MD]4, there is another example, with three generators A, B, C and relations

$$A^2 = B^2 = C^2 = (AB)^2 = (BC)^3 = (AC)^6 = 1.$$

This situation is quite general. From an algebraic point of view, one can prove that any Coxeter group has a presentation with k generators $g_1, \ldots, g_k$ and relations of the kind $(g_i g_j)^{N_{ij}} = 1$; if $i = j, N_{ii} = 1$, as each g_i is the reflection in a hyperplane; if $i \neq j$, $\frac{\pi}{N_{ij}}$ is the angle between the two hyperplanes associated to the generators g_i and g_j.

From a geometric point of view, one can prove that any (irreducible ([4])) Coxeter group is generated by the reflections on the walls of a chamber which can only be:

- a simplex, if the group is infinite;
- the cone on a simplex, if the group is finite.

There are other bonds on the form of these chambers, which lead to the complete enumeration of Coxeter groups in any dimension; this is reached through an analysis which is quite lengthy in the general case, but, in low dimensions, is essentially a consequence of the observation that, due to the discreteness of the group, the angles between the walls of the chamber must be of the form $\frac{\pi}{n}$, where n is a positive integral number (≥ 2). This leaves very few possibilities, that is:

- for finite groups in two dimensions: two mirrors making an angle of $\frac{\pi}{n}$ (for any n in $\mathbf{N}$); the chamber is the cone on a 1-dimensional simplex;
- for infinite groups in two dimensions: three mirrors forming a triangle, with angles $\frac{\pi}{p}, \frac{\pi}{q}, \frac{\pi}{r}$. As $\frac{1}{p} + \frac{1}{q} + \frac{1}{r} = 1$, the only possibilities are:
  - $p = q = r = 3$: equilateral triangle;
  - $p = 2, q = r = 4$: right isosceles triangle;
  - $p = 2, q = 3, r = 6$: right triangle with angles $\frac{\pi}{3}$ and $\frac{\pi}{6}$ ([5]);

[4] A Coxeter group is irreducible if it is not isomorphic to the direct product of two Coxeter groups acting in lower dimensions.

[5] If we try to construct a plane polygon with all angles of the form $\frac{\pi}{n}$, there is another possibility besides these three triangles, that is a rectangle, with all angles $\frac{\pi}{2}$. This case is less significative among Coxeter groups, because it is a reducible group: in fact, it can be seen as the direct product of two copies of the group generated by the reflections in two parallel lines, each copy acting over $\mathbf{R}$.

– for finite groups in three dimensions (see [MD]5) : three mirrors forming the cone on a triangle; the angles between the mirrors must be of the form $\frac{\pi}{p}, \frac{\pi}{q}, \frac{\pi}{r}$ and, as they correspond to the angles of a spherical triangle, $\frac{1}{p} + \frac{1}{q} + \frac{1}{r} > 1$, which leaves the only possibilities:
  - $p = 2, q = r = 3$: this corresponds to the symmetry group of a tetrahedron;
  - $p = 2, q = 3, r = 4$: this corresponds to the symmetry group of a cube, or of an octahedron;
  - $p = 2, q = 3, r = 5$: this corresponds to the symmetry group of a dodecahedron, or of an icosahedron .

The six cases just obtained in 2. and in 3. are exactly the six "machines for building symmetry" described in the introduction.

In order to complete the list of Coxeter groups acting in dimension less or equal to 3, we have to add also:

– groups in one dimension: there is only one finite group, corresponding to just one mirror; there is also only one infinite group, corresponding to two parallel mirrors.
– infinite groups in three dimensions: these are given by four mirrors forming a tetrahedron with all dihedral angles of the form $\frac{\pi}{n}$; there are three possibilities, whose corresponding chambers are all shown in Fig. 3:

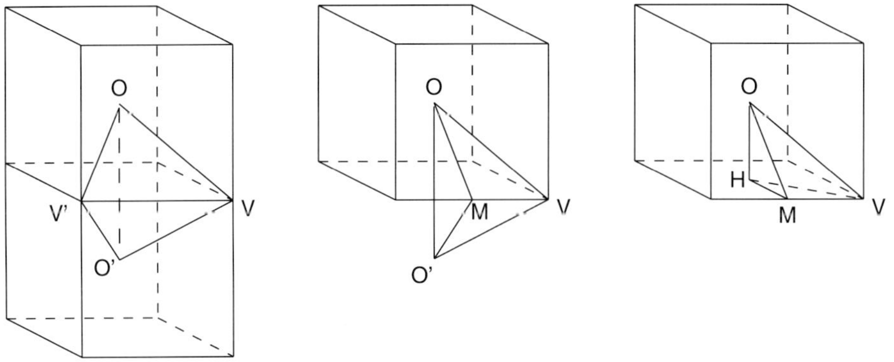

Fig. 3. Fundamental domains for the three possible infinite Coxeter groups in $\mathbf{R}^3$

- the tetrahedron of vertices O, V, O', V', which has two dihedral angles equal to $\frac{\pi}{2}$ and four equal to $\frac{\pi}{3}$;
- the tetrahedron of vertices O, V, O', M, which has three dihedral angles equal to $\frac{\pi}{2}$, two equal to $\frac{\pi}{3}$ and one equal to $\frac{\pi}{4}$;
- the tetrahedron of vertices O, V, H, M, which has three dihedral angles equal to $\frac{\pi}{2}$, one equal to $\frac{\pi}{3}$ and two equal to $\frac{\pi}{4}$.

2.2 Two-dimensional machines and plane crystallographic groups

In the last subsection we found three irreducible cases among Coxeter infinite two-dimensional groups (equilateral triangle, right isosceles triangle, right triangle with angles $\frac{\pi}{3}$ and $\frac{\pi}{6}$) and one reducible case (rectangle). The same four cases appear among crystallographic groups, that is discrete groups of isometries in $\mathbf{R}^n$, whose translation subgroup is generated by n independent translations. For $n = 2$, these are the 17 (up to affine conjugacy) *wallpaper groups*; among them, the ones which are generated by reflections are Coxeter groups (reducible or irreducible), and there are four of them:

p3m1 corresponding to the mirror box with the shape of an equilateral triangle: see Fig. 4;

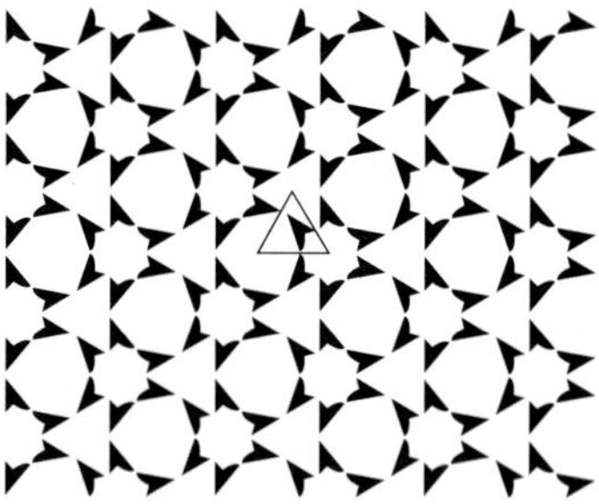

Fig. 4. The group **p3m1**, generated by the reflections in the three sides of an equilateral triangle

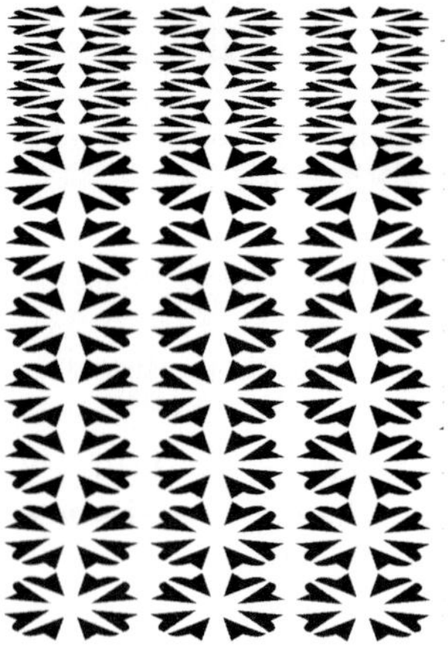

Fig. 5. The group **p4m**, generated by the reflections in the three sides of a right isosceles triangle

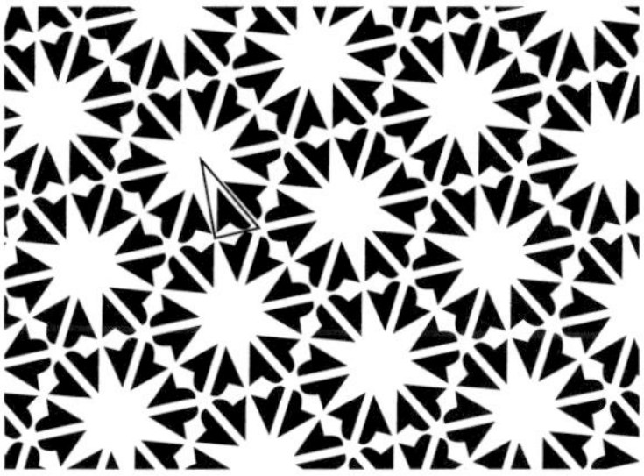

Fig. 6. The group **p6m**, generated by the reflections in the three sides of a right triangle with angles $\frac{\pi}{3}$ and $\frac{\pi}{6}$

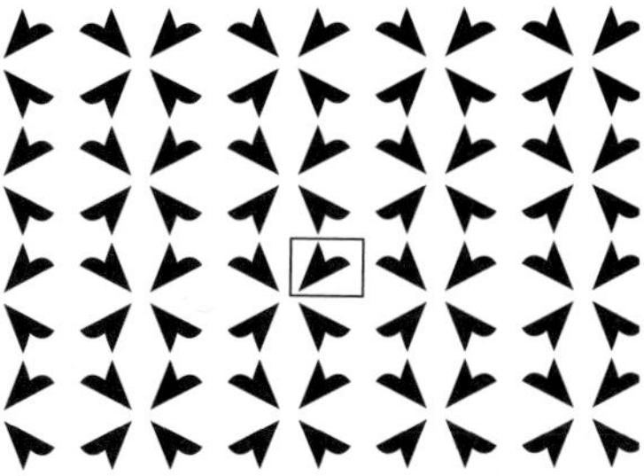

Fig. 7. The group **pmm**, generated by the reflections in the four sides of a rectangle

p4m corresponding to the mirror box with the shape of a right isosceles triangle: see Fig. 5;

p6m corresponding to the mirror box with the shape of a right triangle with angles $\frac{\pi}{3}$ and $\frac{\pi}{6}$): see Fig. 6;

pmm corresponding to the mirror box with the shape of a rectangle: see Fig. 7.

It is also interesting to notice that these four groups are not the only ones – among crystallographic groups – which can be seen in our "machines": in fact, when we put something in a mirror box, the planar picture we get has a symmetry group which *contains* the group G associated to the box (that is, the group generated by the reflections in the walls of the box), but not necessarily coincides with it. If we put in the box something which already has some symmetry, what we get is a group H which properly contains G as a subgroup and we may also "read" the index of G in H ([6]). So, if we want to make a list of which ones of the 17 wallpaper groups may be seen in our machines, we have to add some cases, that is:

[6] The index of G in H is equal to the ratio of the areas of the fundamental domains of G and H: this can easily be read from what we see, keeping in mind that the fundamental domain of G is the box itself.

– **p6m** is not contained as a proper subgroup in any of the other 16 groups; so in the mirror box with the shape of a right triangle with angles $\frac{\pi}{3}$ and $\frac{\pi}{6}$ we can only see something with symmetry group **p6m** ([7]);
– the same happens for the group **p4m** corresponding to the mirror box with the shape of a right isosceles triangle ([8]);
– **p3m1** is contained as a proper subgroup of (minimal) index 3 in the crystallographic group **p31m**; so, in the box with the shape of an equilateral triangle we generally see something with symmetry group **p3m1** (by putting in the box something with no symmetry at all); but we can also see a planar picture with symmetry group **p31m**, by putting in the box something with symmetry group C_3, that is a center of rotation of order 3 in the center of the mirror box: see Fig. 8.

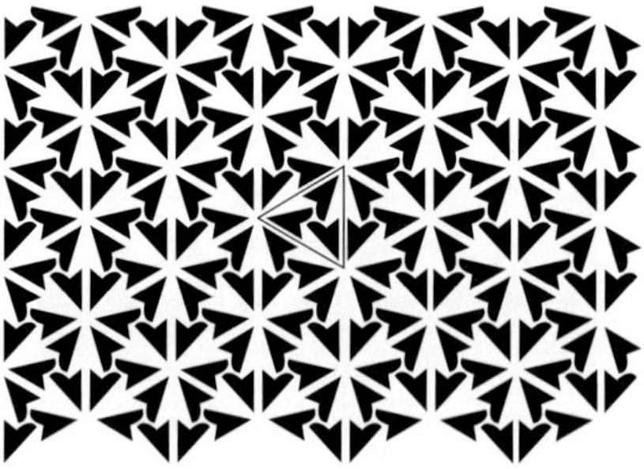

Fig. 8. The group **p3m1** is a subgroup of index 3 in the group **p31m**

p3m1 is also contained as a proper subgroup of (minimal) index 2 in the crystallographic group **p6m**; this is for us less interesting, as it does not give a "new" group to see; however, in order to see the group **p6m** in the equilateral mirror box it is enough to put inside the box something with a bilateral symmetry.
– **pmm** is contained as a proper subgroup
  • of (minimal) index 2 in the group **cmm**: see Fig. 9;
  • of (minimal) index 4 in the group **p4g**: see Fig. 10.

  Besides these two cases, which add two new groups to the list of the ones we can see in mirror boxes, there are other two possibilities, as **pmm** is also contained as a proper subgroup

[7] As **p6m** is a subgroup of itself, what can happen is that by putting something in the box we may get a picture whose symmetry group H is still isomorphic to **p6m**, *but* is different from the group G generated by the reflections in the walls of the box.

[8] As **p4m** is a subgroup of itself, the same phenomenon described in footnote 7 for **p6m** can happen.

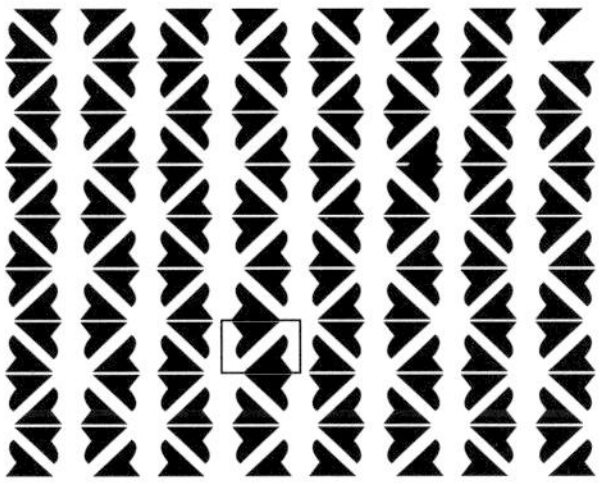

Fig. 9. The group **pmm** is a subgroup of index 2 in the group **cmm**

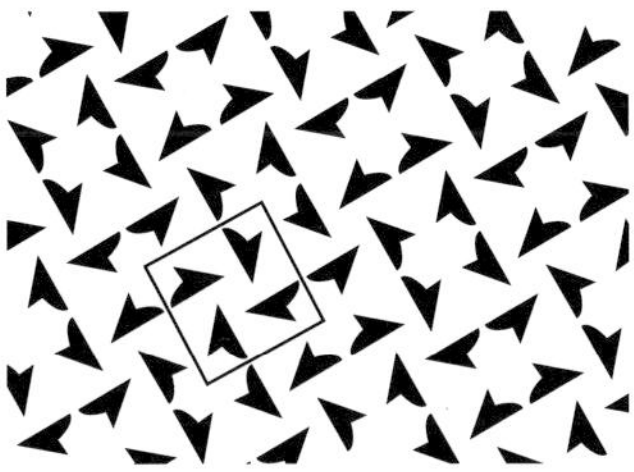

Fig. 10. The group **pmm** is a subgroup of index 4 in the group **p4g**

- of (minimal) index 2 in the group **p4m**;
- of (minimal) index 6 in the group **p6m**.

Thus in a rectangular mirror box we generally see something with symmetry group **pmm** (that is, this happens when we put inside the box something without any symmetry); we manage to get a picture with symmetry group **cmm** when we put inside the box something with a center of symmetry of order 2 in the center of the rectangle.

In order to get a picture with symmetry group **p4g**, we cannot start from any rectangular mirror box, but we need a square one, and in the square box we have to put something with symmetry group C_4, that is with a rotational center of order 4 in the center of the square. If we put in the square box something whose symmetry group is the dihedral group D_4 we get a picture with symmetry group **p4m**, but in fact we can get the same group (as a subgroup of index 2 instead of 8, that is) by putting inside the box a picture with just a symmetry axis along the diagonal of the square. In order to get a picture with symmetry group **p6m**, we cannot start from any rectangle, but we need one such that the ratio between its sides is $\sqrt{3}$, so that it can be divided in six right triangles with angles $\frac{\pi}{3}$ and $\frac{\pi}{6}$, as shown in Fig. 11.

So, the wallpaper groups which can be seen in a mirror box are seven, the four ones which are generated by reflections, and other three, which contain a subgroup generated by reflections.

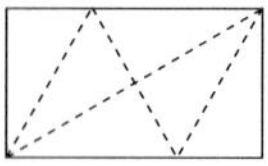

Fig. 11. The group **pmm** is a subgroup of index 6 in the group **p6m**

2.3 Three-dimensional machines and polyhedra

And what about three-dimensional symmetry machines? The ones we described in the introduction realize the only possible irreducible finite subgroups of $Iso(\mathbf{R}^3)$ generated by reflections, which correspond to the symmetry groups of the regular polyhedra. As in the two-dimensional case, the significative mathematical result hidden in these machines is the fact that they are not just an example, but they are in fact the only possible cases.

As in the two-dimensional case, by putting something in the "machine", what one sees is the orbit F of that "something" with respect to the group associated to the kaleidoscope (that is, the group generated by the reflections in the walls of the kaleidoscope); the symmetry group of F contains the group of the kaleidoscope, but is not necessarily equal to it. However, finite groups of isometries in the space are very few, so that for two of the kaleidoscopes (the one associated to the symmetry group of the cube and the one associated to the symmetry group of the dodecahedron) we may be sure that what we see inside always has a symmetry group which coincides with the group of the kaleidoscope: the reason of this is simply the fact that there does not exist any finite group of space isometries containing as a subgroup neither the symmetry group of the cube nor the symmetry group of the dodecahedron.

Instead, in the kaleidoscope associated to the symmetry group G of the tetrahedron we may see either objects having G as symmetry group or objects having the same symmetry group H of a cube (as G is a subgroup of index 2 in H).

Colouring can be used, both in this case and in the planar case, to underline this phenomenon. For example, an octahedron can be reproduced in the kaleidoscope of the cube, but if we colour it with black and white faces and we do want to reproduce that colouring also, then we need the kaleidoscope of the tetrahedron ([9]).

Of course, regular polyhedra are not the only objects one can observe in the three-dimensional kaleidoscopes. To give another example, we can consider the orbit F of a single point x with respect to the group associated to one of the three kaleidoscopes, and the convex envelope of the points in F: in this way, we naturally get a uniform polyhedron (that is, a polyhedron whose symmetry group is transitive on the set of vertices). In particular, we get in this way nearly all the 13 archimedean polyhedra:

[9] In fact, the symmetry group of the octahedron is isomorphic to H, while the coloured symmetry group (that is, the subgroup of those isometries which, besides fixing the object, fix the colouring also) is isomorphic to G.

– in the kaleidoscope of the tetrahedron we get also the archimedean polyhedron (3,6,6) ([10]);
– in the kaleidoscope of the cube we get the archimedean polyhedra (3,4,3,4), (4,6,6), (3,4,4,4), (3,8,8), (4,6,8);
– in the kaleidoscope of the dodecahedron we get the archimedean polyhedra (3,5,3,5), (5,6,6), (3,4,5,4), (3,10,10), (4,6,10).

The only two archimedean polyhedra which can *not* be seen in the kaleidoscopes are (3,3,3,3,4) and (3,3,3,3,5), whose symmetry groups contain only rotations.

The same construction does not only yield the archimedean polyhedra (which, besides being uniform, have regular faces): in general, for any choice of x, we always get a polyhedron whose faces are equiangular; for each one of the described cases, there is just one position for the point x such that the faces of the corresponding polyhedron are also equilateral.

2.4 Elliptic, euclidean (and hyperbolic) geometry

Up to now we spoke about two-dimensional and three-dimensional machines for building symmetry; but there is another way – probably more suitable ([11]) – to look at this situation. It should be noticed in fact that the world of polyhedra can be handled in (at least) two different ways: we can think to a polyhedron as a solid object (homeomorphic to D^3), the analogous in the three-dimensional space of what is a polygon (homeomorphic to D^2) in the plane; alternatively, we can think to a polyhedron as a two-dimensional object (homeomorphic to S^2). From this point of view a polyhedron is not so much the analogous of a plane polygon, but rather the analogous of a plane tessellation. With the first point of view one rather sees the symmetries of the object as isometries in the space; with the second point of view, one thinks more about isometries of the sphere.

This second point of view is probably the best one to underline the similarity between the different situations shown with our six "machines for building symmetry". In any case we handle surfaces, more precisely triangulated surfaces; with the first three machines, the mirror boxes, we are in the world of euclidean geometry, and the restriction on the number of possible machines comes out essentially from the equality

$$\frac{1}{p}+\frac{1}{q}+\frac{1}{r}=1,$$

[10] The notation $(a_1, a_2, \ldots, a_k)$ used for an archimedean polyhedron expresses the fact that each vertex is adjacent to a regular a_1-gone, a regular a_2-gone,..., a regular a_k-gone (in this cyclical order).

[11] The distinction between two- and three-dimensional situation can generate ambiguity in a context where the objects are physically shown (and, therefore, are all necessarily three-dimensional).

implied by the fact that the sum of the angles of a euclidean triangle is equal to π; with the other three machines, the kaleidoscopes, we are in the world of elliptic geometry, and the restriction on the number of possible machines comes out essentially from the inequality

$$\frac{1}{p}+\frac{1}{q}+\frac{1}{r}>1,$$

implied by the fact that the sum of the angles of a spherical triangle is greater than π.

A very natural extension of this would be to have "machines for building symmetry" in hyperbolic geometry; and, in this case, we have a much greater variety, as there exists such a machine for any p, q, r such that

$$\frac{1}{p}+\frac{1}{q}+\frac{1}{r}<1,$$

so there are an infinity of them. Each one corresponds to fixing a hyperbolic triangle whose angles are $\frac{\pi}{p}, \frac{\pi}{q}, \frac{\pi}{r}$ and considering the subgroup G of hyperbolic isometries generated by the reflections in the sides of the triangle ([12]). There is no difficulty in simulating on a computer a virtual hyperbolic machine: it is enough (for example, in the Poincaré model) to substitute reflections with circular inversions. This is of course conceptually identical to a physical realization: however, in a plan for a math exhibition, a real object still makes a great difference, at least in my opinion, with respect to a virtual one (and it does not seem technically easy to construct such a machine).

3 What can be done with the "machines for building symmetry"

3.1 Classification with respect to symmetry, for different visitors

The principal aim of an exhibition based on the objects described in the introduction is to give the visitor an idea of the problem of classifying something with respect to its type of symmetry – an idea which of course will be

[12] In the three cases just described we always get a triangle group $G = G(p, q, r)$, generated by the reflections A, B, C in the three sides of the triangle, and with relations of the following kind: $A^2 = B^2 = C^2 = (AB)^p = (BC)^q = (AC)^r = 1$. However, from an algebraic point of view, the situation is very different in the three cases: in the elliptic case ($\frac{1}{p}+\frac{1}{q}+\frac{1}{r}>1$) the group G is finite; in the euclidean case ($\frac{1}{p}+\frac{1}{q}+\frac{1}{r}=1$) G contains a subgroup, isomorphic to the abelian free group of rank 2, and with finite quotient. Instead, the groups we get in the hyperbolic case ($\frac{1}{p}+\frac{1}{q}+\frac{1}{r}<1$) are much more complicated to deal with: for example, they all contain a subgroup isomorphic to the free non-abelian group of rank 2 (which implies they contain as subgroups free groups of any rank).

at very different levels, depending on the degree of mathematical knowledge of the visitor.

The possibility of giving at least a flavour of this idea, even to someone with no mathematical knowledge at all (like small children) is due to the fact that, given a group generated by reflections, it is possible to express the idea of isomorphic or non-isomorphic groups without any technical algebraic language, but simply by looking at the geometry of the mirrors.

This opens the possibility of making a lot of non trivial considerations, related to the symmetry group of a planar picture or of a solid object, without the necessity of introducing the algebraic language related to groups; for example, planar pictures which can be reconstructed in the same mirror box (or solid objects which can be reconstructed in the same kaleidoscope) have isomorphic symmetry groups, while pictures (or solid objects) reconstructed in different mirror boxes (or kaleidoscopes) have non-isomorphic symmetry groups.

Of course, due to what we observed in the preceding section, the last sentence is in fact not quite correct ([13]). However, we do not think this ambiguity makes a serious problem towards mathematical communication in such a sort of exhibition. In fact, for the minority of the public who can appreciate this difference, the ambiguity is not hidden but, on the contrary, some problems are posed on purpose, in order to provoke questions on the subject, and this gives a new, not-trivial problem to investigate. For the majority of the public, in order to give a flavour of the different types of symmetry, it is enough to make them observe how the pictures coming out from certain mirror boxes are all based on numbers 3 or 6 (and on a triangular grid), while others are based on number 4 (and on a square grid).

Moreover, the same problems can be used for the more and for the less sophisticated public with different purposes. To give just one example, the reconstruction of the same tessellation with different kinds of colouring (which can also lead to different symmetry groups) can be used with the more sophisticated public precisely to provoke the ambiguity we were discussing before, while with the less sophisticated one the same problem can be used for simpler observations.

In fact, one of the reasons why we thought the "machines for building symmetry" were useful for an exhibition is exactly what is exemplified in the previous comment, that is the fact that the same objects can be used to communicate different levels of mathematics to different sorts of public. The fact that they are based on some non-trivial mathematical concepts from one side allows also an interesting communication towards a public with more mathematical background and from another side it is clearly perceivable

[13] For example, we can build in a square mirror box pictures having four possible (non isomorphic) symmetry groups (**pmm**, **cmm**, **p4m**, **p4g**); while a picture with symmetry group **p4m** could be built both in a square mirror box or in a mirror box with the shape of a right isosceles triangle.

also by the public with very little mathematical knowledge: often, it is not necessary to be able to enter deeply inside a problem in order to understand whether the problem has, or has not, such a depth.

3.2 The role of interactivity

Another reason why we think these machines can be a useful example in the direction of finding ways for the popularization of mathematics is the fact that they give occasions of "doing mathematics"; and, in saying this, we think both to the public with less mathematical knowledge and to the public with more technical instruments. Both will have the possibility of putting their hands on the objects and meet a problem they will have to solve: for a seven-year old child the problem can be how to put a given triangle in a mirror box in order to see an hexagon; for a mathematics university student the problem can be that of understanding why a crystallographic group cannot have order 5 rotations, ...: in both cases (as well as with other possible categories of public) the objects can provoke an active reaction by the visitor: which we think is the only effective way to learn some mathematics.

Another aspect, related to this one, is the crucial role played by fancy and creativity, which are human capacities usually (and wrongly) thought to be far away from mathematical capacities. This crucial role comes out from the fact that the main thing to do with our machines is just to look what happens when one puts something inside, and to observe analogies and differences between them. In order to make these observations, any object would do (see [MD]6 and [MD]7): if I take the piece which represents the fundamental domain of the action of the symmetry group of the cube and I put it, in the "right" way, in the kaleidoscope of the cube, I see a cube; but if I put it in a "wrong" way, or I do not put that piece at all, but I prefer to insert a ball (see [MD] 8), or even a dry flower I had in my pocket, that will work equally well; in this case also I can observe that what I see has always the same kind of symmetry of a cube and always a kind of symmetry different from what we see in the other kaleidoscopes.

So, the "wrong" trials to solve the problems proposed are equally well useful to observe what happens and thus familiarize with the concepts involved; this may be very "relaxing", especially for people (unfortunately not so rare) who are paralyzed by a sort of "fear" towards maths.

3.3 The role of mathematical proofs

We spoke of active interaction with the objects, and of fancy, and of playing. But mathematics is also (or mainly?) rigorous proofs. What could the role of proof be in this kind of proposal? In fact, the problem of achieving a rigorous proof is (always, but especially in the context of undergraduate teaching) a matter of subsequent approximations. And the first stages of these approximations are the understanding of what has to be proved, and the consciousness that the given fact is not trivial and has thus to be proved. This seems (and in fact is) an obvious consideration, but it is unfortunately frequently forgotten: pupils are often forced to prove a statement in the moment when they have not yet clear ideas about what it does mean (what it means if it happens to be true, what would happen if it is false, ... etc.); or they may be asked to prove some facts which are eventually not so easy to prove, but whose statement is (or appears) evident.

It is of course one of the main aspects of mathematics the fact that, in a deductive construction, one has to prove everything, also "self-evident" statements, and we perfectly know how some self-evident statements are not at all trivial to prove (Jordan's theorem, just to give an example), and some are even false. But...; but one needs some mathematical maturity to appreciate the need to prove self-evident statements; and it can be useless, or even damaging, to propose their proof in a context where this maturity is lacking.

This brings to the (apparent) paradox that it may be easier to propose the proof of "difficult" statements than that of "easy" ones in secondary school. The role of a mathematical exhibition towards the achieving of proofs could well be (on some categories of public) that of making people conscious, and curious, about some facts to be proved.

Let us exemplify what we are trying to say on two statements: the first one is the fact that a triangle with two equal sides has two equal angles; the second one is the fact that the frieze groups ([14]) are seven. And let us keep in mind two sorts of public, a secondary school student and an adult with no mathematical background after school. It is very unlikely that both the student or the grown-up person manage to become particularly curious about the statement on isosceles triangles; in any case they both believe it is true, and they would use the statement, without realizing it requires a proof, if in a concrete problem they happen to need it. The situation is completely different for the statement about friezes: first of all the statement is strange, and difficult to understand; one does not understand it at once, but has to think about it, to make a list of the possibilities, to reason about the fact that whatever drawing he or she is making, it *has to be* one of those seven. When one grasps what this means, usually this is related to a sense of beauty: the result is beautiful, conceptually beautiful. Moreover, it looks strange; and it is very natural to ask *why* is an eighth case impossible. So, with some time at disposal, it is very easy that people arrive naturally to the consciousness of the need for a proof.

When we have obtained such a consciousness, there are still many intermediate stages which can be significative, before achieving a complete proof: for example, one could begin to observe that, due to the fact that the group is a frieze group, there must be some restrictions on the kind of possible isome-

[14] Frieze groups are the symmetry groups of patterns repeating in just one direction; that is, they are discrete subgroups of the group of plane isometries, whose translation subgroup is isomorphic to **Z**.

tries in the group (rotations may only be of order two; the axis of reflections may only be either parallel or perpendicular to the direction of translations; the axis of glide reflections must be parallel to the direction of translations; ...). This of course is not a proof, but it begins to give some flavour of it, and the statement, which at first sight appeared completely mysterious (*"why just seven?"*) can now be seen as more reasonable (*"I still do not know why they are exactly seven, but I do understand that the situation is not completely free, and there are some limitations to be respected"*).

In fact, these intermediate stages may be exactly what we would like to be grasped about proofs in mathematics undergraduate learning, much more than the particular proof of a particular statement, which is often not relevant in itself (at least at that level).

3.4 The role of beauty

Beauty has – in many different ways – a crucial role in the exhibition just described.

The first aspect regards a problem of motivations. We all know very well that mathematics has on the whole a very bad reputation; it is quite common to meet persons who have a sort of hate and/or fear towards mathematics; there may be also (in the same person) interest or curiosity about maths, besides fear, but it is very likely that fear acts as a sort of block towards curiosity. This block is a concrete problem that any trial of popularizing maths has to consider; one needs a way to overcome it, to be able to *begin* to communicate with the public; and, moreover, this way must be an immediate one, because it has to win an irrational feeling, not a rational one.

The strong impact of beauty is an enormous help in this sense; and we found symmetry a very good subject for popularization of maths also for this reason (besides the ones already discussed).

By saying this, we do not only refer to the trial of involving art, by proposing posters with the reproductions of some masterpieces where symmetry is wonderfully used; but we refer also to the "beauty" of the artificial images that the visitor is invited to reconstruct (and/or to invent) by himself. Of course these two kinds of "beauty" are not comparable, but it is a fact that both have a precious role in disposing people to be willing to interact with maths – a result which is not at all obvious to reach.

A particular role is played by the strong effect of surprise. In our mirror boxes one "can see infinity" and this effect is very strong, very beautiful, and also very unexpected, in all sort of public.

Moreover, this effect of surprise does *not* come from spectacular, enormous, scenic objects, but from very simple ones. This has a double positive effect: the first one is that it magnifies the surprise (if I have to enter into an enormous building, which I see from very far away, with much light and colour, I do expect I shall see something which will surprise me; maybe I have no idea of what I shall see, so I shall still be surprised, but I know in advance that this will happen; instead, if I put my eye on the border of a small object made by 3 mirrors, I do not expect any particular "special effect"); another positive consequence is that the objects are easily reconstructible, so for example teachers realize they can easily build something analogous in their schools.

A last aspect about beauty I would like to remind here is one which has already been mentioned in the preceding section: mathematics is beautiful not only for the beauty of some of its images, but, also, for the conceptual beauty of some of its results. Too often – in my opinion – we do not even try to communicate this kind of beauty: mathematicians seem to lack any confidence about the fact that this could be communicated to someone, unless he or she has done the right number of exams in algebra, geometry, analysis etc. Sometimes this is true, but it is probably much less true than what is generally thought, and it is possible to communicate much more than we think. At least, it is worthwhile trying.

References

1. H.S.M. Coxeter (1961) Introduction to geometry. Wiley
2. H.S.M. Coxeter, J.Moser (1980) Generators and relations for discrete groups. Springer
3. P. Cromwell (1997) Polyhedra. Cambridge University Press
4. M. Dedò (1999) Forme. Decibel-Zanichelli.
5. H. Weyl (1952) Symmetry. Princeton University Press

The Mathematics of Tuning Musical Instruments – a Simple Toolkit for Experiments

Erich Neuwirth

University of Vienna
<erich.neuwirth@univie.ac.at>

Abstract. This paper gives an overview of the (rather simple) mathematics underlying the theory of tuning musical instruments. Besides demonstrating the fundamental problems and discussing the different solutions (only on an introductory level), we also give Mathematica code that makes it possible to listen to the constructed scales and chords. To really get a "feeling" for the contents of this paper it is very important to hear the tones and intervals that are mentioned. The paper also has the purpose of giving the reader a Mathematica toolkit to experiment with different tunings.

More than 250 years ago Johann Sebastian Bach composed "Das wohl temperirte Clavier" (the well tempered piano) to celebrate an achievement combining music, mathematics, and science. Finally, a method of tuning musical instruments had been devised which allowed playing pieces in all 12 major and all 12 minor scales on the same instrument without retuning. Nowadays, we are so used to this fact that we almost lack an understanding for the kind of problems musicians were facing for a few hundred years.

The appendix of this paper contains some Mathematica code. This code allows us to play scales and chords with given frequencies. Using these functions, we will be able to listen to the musical facts we are describing in a mathematical way. (Warning: On slower machines this code may take some time to create the sounds.) The waveform used for this sound is not a sine wave. For musicians, sine waves sound very bad. Therefore, we are using a more complicated waveform, which has been described in [4] and is heavily used in [3]. Our code defines three Mathematica functions, PlayScale, PlayChord, and PlayStereoScale, and we will explain the use of our examples later in the paper. The code will run on any computer with a sound device and a Mathematica version supporting the Play function on this platform. In particular, it will run on PCs with any 32-bit version of Microsoft Windows.

Now let us start historically. The ancient Greeks, and especially the Pythagoreans, noticed that the length of strings (of equal) tension and the musical intervals they produced showed some interesting relationships. Using more modern knowledge from physics we know that the lengths of strings and the frequencies of the tones are inversely proportional. So in the context of this paper we will study the relation between frequencies, frequency ratios, and musical relationships like consonance.

The first fact we note in this respect is that when we compare two tones and the frequency of the second tone is double the frequency of the first tone, we feel that this is "the same tone on a higher level". The interval created this way is called an octave, and it seems to be a universal musical constant in the sense that an octave is perceived as a consonance in every musical culture.

To listen to this phenomenon, we can execute the command PlayScale[{220, 440},1.5]. To listen to these two tones played as a chord, we execute PlayChord [{220,440},1]. We also can hear that this fact only depends on the ratio and not on the absolute frequencies by playing PlayChord [330*{1,2},1.5] or PlayChord[264*{1,2},1.5].

Since we noticed that doubling the frequency produces something musical, we also might be interested in listening to a sequence of tones consisting of the first few integer multiples of a base frequency, for example the sequence 220, 440, 660, 880, 1100, 1320, 1540, 1760.

In our code, we use PlayScale [220*{1,2,3,4,5,6,7,8},1.5].

Listening to this sequence, between the consecutive tones we hear many intervals, which in Western music are considered to be consonant. In musical terms, we hear 7 intervals, and the first five are octave, fifth, fourth, major third, and minor third. The last 2 intervals normally are not used in Western music. Especially the tone with sevenfold base frequency is not used in Western music, but it is used in Jazz.

Considering integer multiples of a base frequency is not just "mathematical aesthetics", valveless fixed length wind instruments like historical horns and trumpets only can produces tones with exactly this property. So asking about the kind of music possible under these restrictions in not just academic, but connected with real wind instruments. The 2 musically most important intervals in our sequence are the major third and the fifth. From our series we see and hear that the fifth corresponds to $3/2$ and the major third corresponds to $5/4$. Playing a base frequency and these two intervals at the same time produces a major triad, probably the most used chord in Western music.

Defining MajorTriad={1, 5/4, 3/2} we can do PlayChord [264*MajorTriad,1.5] and hear that it sounds very consonant.

The sequence of tones having integer multiple frequencies of a base frequency is often called overtone series. Now let us try to construct a major scale by using only intervals we found between neighboring tones in the overtone series we just studied. Using a piano keyboard as our visual aid for constructing a scale we see that we immediately can create the base tone and tones for the third, the fourth and the fifth and, of course, for the octave.

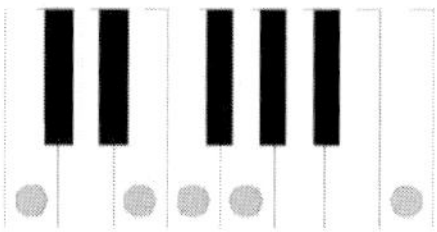

Fig. 1.

Defining PartialScale 1= {1,5/4,4/3,3/2,2} we can listen to PlayScale [264*PartialScale1,1.5].

So we still are missing the second, the sixth, and the seventh. To find the corresponding frequency ratios, we look at the following picture:

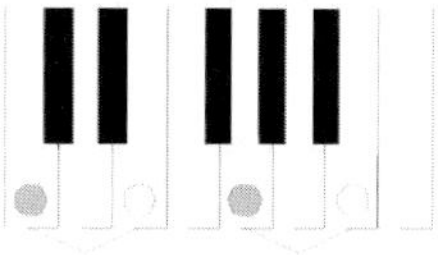

Fig. 2.

We see that the lower interval marked by one dark and one light circle are similar intervals. We know that the lower interval, as a major third, corresponds to a frequency ratio of 5/4. The lower tone of the upper interval has a frequency ratio of 3/2 to the base tone. Therefore, we use a frequency ratio of (3/2).(5/4)=15/8 for the seventh.

We can listen to these intervals: Third = 1,5/4. PlayScale [264*Third,1.5] and PlayScale [264*3/2*Third, 1.5]. We also can listen to these intervals as chords: PlayChord [264*Third, 1.5] and PlayChord [264*3/2*Third, 1.5]. Using this we can define PartialScale2 = {1,5/4,4/3,3/2,15/8,2} and do PlayScale[264*PartialScale2,1.5].

So we have been able to fill one of the holes in our scale. Similarly, we can construct the sixth by noting that the sixth is one third above the fourth:

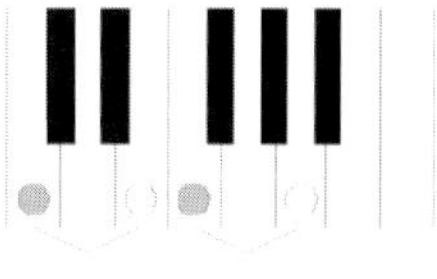

Fig. 3.

So the frequency ratio we need for this tone is (4/3).(5/4)=5/3. We can check the musical quality of these intervals with PlayChord[264*Third,1.5] and PlayChord[264*4/3*Third,1.5].

To extend our scale, we define PartialScale3 = {1,5/4,4/3,3/2,5/3,15/8,2} and do PlayScale [264*PartialScale3,1.5].

Summarizing we see that we have almost all the tones we need for a major scale:

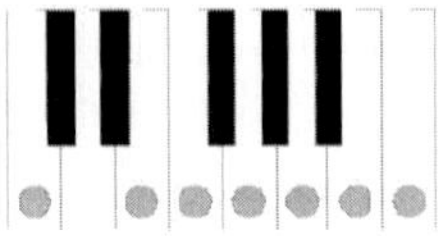

Fig. 4.

The only tone we are missing is the second. We cannot get the second from the overtone series up to the eightfold multiple of the base tone (i.e. within a range of 3 octaves of the base tone). But we can note that by extending the keyboard a little bit and going up 2 fifths:

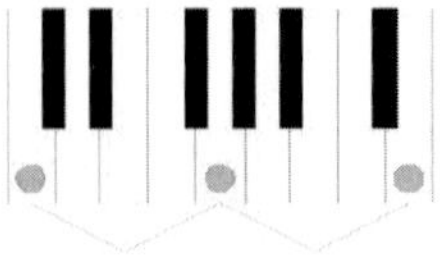

Fig. 5.

we get the tone one octave above the second. Just going down one octave (i.e. multiplying with 1/2) we see that we can construct the second as (3/2).(3/2).(1/2) = 9/8.

So now we have completed our scale, PureMajorScale = {1,9/8,5/4,4/3,3/2,5/3, 15/8,2} and we can listen to the scale with PlayScale[264*PureMajorScale,1.5].

Now we have a musically pleasing scale represented by fractions with rather small numerators and denominators. The tuning building upon this scale is called pure tuning (or just tuning). The major triad over the base tone is the one we already discussed, it has a very simple mathematical description, and it sounds very harmonic. So what we have now seems like a mathematically and musically perfect solution to the problem of tuning instruments. To test the musical qualities of our scale, let us try a few other chords consisting just of thirds and fifths taken from our scale.

PlayChord[264*{3/2,15/8,2*9/8},1.5] is the major triad based on the fifth of our scale, and it sounds musically pleasing also.

This chord, has the same frequency ratios as the major triad on the base tone: $(15/8)/(3/2)=5/4$ and $(9/4)/(3/2) = 3/2$. Now let us look at the triad based on the second of our scale (it is a minor triad).

PlayChord[264*9/8,4/3,5/3,1.5] does not sound musically pleasing.

Let us look at the internal frequency ratios of this chord: $(4/3)/(9/8) = 32/27$, and $(5/3)/(9/8) = 40/27$. These ratios are not related to the intervals we derived from the overtone series. For the "upper" interval, however, we have $(5/3)/(4/3) = 5/4$, and this is a pure major third. We would expect a pure fifth for the ratio between the lowest tone and the highest tone in our triad, so instead of $40/27$ we would need $3/2$. If we try to change the lowest tone of the triad such that we get the pure fifth, we have to take a $(5/3)/(3/2) = 10/9$ for the ratio between the base one and the second.

Listening to a chord containing this tone, PlayChord[264*{10/9,4/3,5/3},1.5] gives us a consonant musical experience again.

We see that for a pure triad upon the fifth we need a second of $9/8$ and for a pure triad on the second, we need a second of $10/9$. So the problem is that when we try to play different chords with the tones taken from one scale, we are getting into musical trouble. The frequency ratio between the two different seconds we need is $(9/8)/(10/9) = 81/80$, and it is called the syntonic comma. It also occurs in a different problem. Musically speaking, when we go up 4 fifths and then go down 2 octaves, we should arrive at the third above the base tone. Up 4 fifths and down 2 octaves corresponds to $(3/2).(3/2).(3/2).(3/2)/4 = 81/64$, one third corresponds to $5/4 = 80/64$, so the ratio occurring here also is the syntonic comma of $81/80=1.0125$. We can say that the syntonic comma is the degree of incompatibility between the pure third and the pure fifth.

Defining SyntonicComma = 1, 81/80 we can listen to PlayScale [264*Syntonic-Comma, 1.5] and PlayChord [264*SyntonicComma ,1.5] to perceive the musical relevance of this difference.

Musically speaking, we would like to have compatible fifths and thirds. Since the fifth is the simplest interval in the overtone series (except the octave, of course), we try to keep the value for the fifth and use a third, which is "compatible" in the sense that one third and 4 fifth essentially produce the same tone. To achieve this, we have to use a frequency ratio of $81/64$ for the third. Since in our construction of the scale we used the third in 3 places, for the third, for the sixth, and for the seventh, we have to change the definition of the corresponding intervals in our scale. The scale built according to these principles is called the Pythagorean scale.

So we define

PythagoreanMajorScale =
{1,9/8,81/64,4/3,3/2,4/3*81/64,3/2*81/64,2}

and we play PlayScale[264*PythagorealMajorScale,1.5]. To hear the difference between the two different thirds (the pure third and the Pythagorean third) we do PlayScale[264*{5/4,81/64},1.5].

We also can listen to the two thirds played on the two stereo channels:

PlayStereoScale [264*{5/4},264*{81/64},1.5]. The audible beats demonstrate that the difference between these two tones really matters musically. Finally, we can listen to the pure scale and the Pythagorean scale played simultaneously on the two stereo channels : PlayStereoScale [264*PureMajorScale, 264*Pythagorean-MajorScale, 1.5].

We already noted that the Pythagorean third does not sound too harmonic when played as a constituent of a major triad.

To demonstrate this, try PlayChord[264*{1,81/64,3/2},1.5].

The pure triad sounds much better.

To demonstrate this, try PlayChord[264*{1,5/4,3/2},1.5].

So, music played in this temperament (temperament in musical context is just another word for tuning) probably should avoid the major triad on the base tone. Generally speaking, also the chords on the fourth and the fifth do not sound too well in Pythagorean tuning. As long as music is played monophonically, this does not really matter that much, but when playing chords, tuning becomes a serious issue. This fact explains why interest in tuning grew in the 16th century: keyboard instruments became popular. Keyboard instruments are used to play chords, and it is rater difficult to change tuning. Therefore, there was the need for a tuning method that would allow playing many different chords reasonably well. As we have seen, Pythagorean tuning and pure tuning both have serious problems with some very basic chords. So people tried to find alternative methods of tuning. One of the basic problems to be solved was incompatibility between the third and the fifth. Pythagorean tuning had enlarged the third to make it compatible with the pure fifth. The alternative is to reduce the fifth to make it compatible with the pure third. This implies that the fifths are not pure any more. To make the fifth compatible with the pure third we note that one third and 2 octaves together produce a frequency ratio of 5. Therefore, to create a fifth with the property that 4 fifths produce the same interval as one third and 2 octaves, the fifth has to have a frequency ratio of the fourth root of 5, $\sqrt[4]{5}$. The frequency ratio for the fourth then is $2/\sqrt[4]{5}$. Using the building principles we applied to create the pure tuning we can create the scale for this new tuning called meantone tuning.

MeantoneMajorScale = {1, 5^(1/2)/2, 5/4, 2/5^(1/4), 5^(1/4), 5^(1/4)* 5^(1/2)/2, 5^(1/4)*5/4,2}. Using this definition, we can do PlayScale [264*MeantoneMajorScale, 1.5].

We also can compare the meantone scale to the pure scale, PlayStereoScale [264*MeantoneMajorScale, 264*PureMajorScale, 1.5], and to the Pythagorean scale PlayStereoScale [264*MeantoneMajorScale, 264*PythagoreanMajorScale, 1.5].

Major triads in this tuning sound acceptable.

To hear this, we do PlayChord[264*{1,5/4,5^(1/4)},1.5].

Comparing this with the major triad in Pythagorean tuning really makes an audible difference.

To hear this, we do PlayChord[264*{1,81/64,3/2},1.5].

There is another problem we have not coped with until now, the circle of fifths. 12 consecutive fifths would bring us back to the original tone, or, in other word, should be the same as 7 consecutive octaves. If this were true for pure tuning, we should have $(3/2)^{12} = 2^7$, which of course is not true. The frequency ratio $(3/2)^{12}/2^7 = 3^{12}/2^{19} = 531441/524288 = 1.01364$ is called the Pythagorean comma and it is the measure of incompatibility between the pure fifth and the octave.

Defining PythagoreanComma = {1, 3^12/2^19}
we can listen to PlayScale[264*PythagoreanComma,1.5]
and PlayChord [264*PythagoreanComma,1.5] to perceive the musical relevance of this difference.

We note that the Pythagorean comma and the syntonic comma have similar orders of magnitude. To make fifths and octaves compatible, we could either make the fifth smaller or make the octave larger. Since the factor 2 for the octave is an almost universal constant, we will change the fifth to be compatible with the octave. To accomplish that, we need a fifth with a frequency ratio of $2^{7/12} = 1.49831$. Since we also want a third compatible with this fifth, we need a third of $(2^{7/12^4}/4 = 2^{1/3} = 1.25992$. Using these values for the fifth and the third, we can construct a new tuning for a scale. For reasons we will mention briefly later this temperament is called equal temperament.

After some easy algebraic transformations this scale can be defined as follows: EqualMajorScale = {1,2^(1/6),2^(1/3),2^(5/12),2^(7/12),2^(3/4), 2^(11/12), 2}.

We can play this scale, PlayScale[264*EqualMajorScale, 1.5]. Like with our previous scale examples, we can also compare it with other scales using stereo sounds, e.g. PlayStereoScale[264*EqualMajorScale,264*PureMajorScale,1.5].

Chords in equal temperament sound reasonably well, we do not have very bad dissonances. On the other hand, we do not have any pure interval except the octave.

This is illustrated by comparing PlayChord[264*{1,2^(1/3),2^(7/12)},1.5] and PlayChord[264*{1,5/4,3/2},1.5]. Our Mathematica toolkit also contains a function Triad allowing to select a triad with a given base note from a give scale. Triad[264*PureMajorScale,2] will construct the frequencies for the triad constructed from the second, the fourth and the sixth of a scale in pure tuning with a frequency of 264 Hz for the base tone of the scale. Using this toolkit, the reader can do extensive comparisons of chord in different tunings and learn how differently notationally identical chords sound in different tunings.

We have seen that the fundamental numbers for tuning are the frequency ratios for the third and the fifth ; all the other frequency ratios are derived from these two. So let us compare these ratios for all the tunings we have studied in a table:

	Pure	Pyth.	Meantone	Equal
third	1.2500	1.2656	1.2500	1.2599
fifth	1.5000	1.5000	1.4953	1.4983

We see that the largest difference occurs for the pure and the Pythagorean third. This at least partially explains why triads in Pythagorean tuning sound very bad. Meantone tuning for the two basic intervals is very similar to pure tuning, therefore the basic major triad sounds rather well in meantone tuning. The third in equal tuning is also quite different from the pure third; therefore the musical characteristics of the major triad in equal temperament are quite different from pure tuning. Since in all our tunings (except pure tuning) the value of the fifth and the third determine each other, the complete scale can be derived from the frequency ration of the fifth alone. In [1], tunings constructed by this principle are called diatonic tunings, and it is shown that possible values for fifths are in the range from $\sqrt[7]{16} = 1.48599$ to $\sqrt[5]{8} = 1.5157$.

Using ScaleFromFifth from our appendix, we can define PythagoreanMajorScale = ScaleFromFifth[3/2], MeantoneMajorScale = ScaleFromFifth[5^(1/4)], and EqualMajorScale = ScaleFromFifth[2^(7/12)]

In the framework of this paper we could only discuss some of the mathematical problems for tuning musical instruments. Especially, we only studied diatonic scales (i.e. scales without using the black keys on keyboards). The problems get much more complicated when tunings are extended to chromatic scales. Detailed discussions of these problems can be found in [1] and [2]. For studying chords and scales along the lines we have described here, [3] gives a very large set of almost 300 sound examples.

References

1. Blackwood E. (1999) The Structure of Recognizable Diatonic Tunings, Princeton University Press
2. Lindley M. Turner-Smith R., Mathematical Models of Musical Scales, Verlag für systematische Musikwissenschaft
3. Neuwirth E. (1997) Musical Temperaments, Springer Verlag
4. Neuwirth E. (2001) Designing a Pleasing Sound Mathematically, in Mathematics Magazine

Appendix: Mathematica code

(also available from `http://sunsite.univie.ac.at/musicfun/tunsound/`)

```
PlayChord[FreqList_, Duration_] :=
    Play[Min[1, 20*t, 20*(Duration - t)]*
    Sum[(Sin[2*Pi*t*FreqList[[i]]]/
    (1.13 - Cos[2*Pi*t*FreqList[[i]]])),
    i, 1, Length[FreqList]],
    t, 0, Duration - 0.001, SampleRate → 22050]

PlayScale[FreqList_, Duration_] :=
    Play[Min[1, Abs[20*(t - Duration*Floor[t/Duration])],
    Abs[20*(t - Duration*(1 + Floor[t/Duration]))]]*
    Sin[2*Pi*t*FreqList[[1 + Floor[t/Duration]]]]/(1.13 -
    Cos[2*Pi*t*FreqList[[1 + Floor[t/Duration]]]]),
    t, 0, Duration*Length[FreqList] - 0.001, SampleRate → 22050]

PlayStereoScale[FreqListLeft_, FreqListRight_, Duration_] :=
    Play[Min[1, Abs[20*(t - Duration*Floor[t/Duration])],
    Abs[20*(t - Duration*(1 + Floor[t/Duration]))]]*
    Sin[2*Pi*t*FreqListLeft[[1 + Floor[t/Duration]]]]/(1.13 -
    Cos[2*Pi*t*FreqListLeft[[1 + Floor[t/Duration]]]]),
    Min[1, Abs[20*(t - Duration*Floor[t/Duration])],
    Abs[20*(t - Duration*(1 + Floor[t/Duration]))]]*
    Sin[2*Pi*t*FreqListRight[[1 + Floor[t/Duration]]]]/(1.13 -
    Cos[2*Pi*t*FreqListRight[[1 + Floor[t/Duration]]]]),
    t, 0, Duration*Length[FreqListLeft] - 0.001, SampleRate → 22050]

Triad[Scale_, BaseTone_] :=
    List[Scale[[BaseTone]],
```

```
    If[BaseTone + 2 > 8, 2*Scale[[BaseTone - 5]],
    Scale[[BaseTone + 2]]],
    If[BaseTone + 4 > 8, 2*Scale[[BaseTone - 3]],
    Scale[[BaseTone + 4]]]]]

ScaleFromFifth[Fifth_] :=
    List[1, Fifth^2/2, Fifth^4/4, 2/Fifth,
    Fifth, Fifth^2/3/2, Fifth^2/5/4, 2]

PureMajorScale = {1, 9/8, 5/4, 4/3, 3/2, 5/3, 15/8, 2}
    PythagoreanMajorScale = {1, 9/8, 81/64, 4/3, 3/2,
    (4/3)*(81/64), (3/2)*(81/64), 2}
    MeantoneMajorScale = {1, 5^(1/2)/2, 5/4, 2/(5^(1/4)), 5^(1/4),
    (5^(1/2)/2)*5^(1/4), (5/4)*5^(1/4), 2}
    EqualMajorScale = {2^({0, 2, 4, 5, 7, 9, 11, 12}/12)
    SyntonicComma = {1, 81/80}
    PythagoreanComma = {1, 3^12/2^19}
```

The Garden of Eden

Charles O. Perry

20 Shorehaven Road, Norwalk Ct. 06855, USA
<www.charlesperry.com>

This is the Garden of Eden. All of the other allegories not withstanding we have been given a whole universe. I am obsessed with the wonderful mysteries of our universe. All my work in art has had some reference to what we were given. My very first water colour was a vain attempt to extol the beauties of the wilds of Montana, where I was born and where I camped with my father. Two years later, when fighting in Korea, I realized that creating things, actually making or inventing things, was always, and always would be, a necessary part of my daily life. I even invented a better telescope for my observation post on the line.

Then on R and R in Kyoto, the beauty of the Japanese Architecture moved me to act. Their reverent use of our natural materials was perfection itself. This forced the issue: I must return to school and study Art and Architecture.

When I arrived at Yale, it was at the time of the Bauhaus, when we were encouraged to experiment with materials to discover "their true nature" which really meant that we were to invent new forms that exploited the properties of these materials.

But let us go back to the Garden of Eden. Today this once simple garden is now a vast array of everything, everything from the sub atomic to the galactic is exploding all around us and we call it the "information explosion." It is more like a peek into Pandora's box. Is this the discovery of the tip of the glacier of eternal secrets?

The news today of scientific discoveries is epic in proportion to just twenty years ago. Just the Hubbell telescope's discoveries and the DNA race give us this peek at our gifts, not to mention the amazing quantum physics which blows everything away with a capricious photon.

I try to grapple with these wonderful happenings all around us and somehow make things that remind us, as in music, of these phenomena. For some reason I insist on inventing or at least being rigidly specific in each piece of sculpture. If I were a painter it would become an other game, but I have to make things with the laws that nature has laid out.

In applying these laws to building things, I soon came across the book "Growth and Form" by D'Arcy Thompson. This was a wonderful book on morphology with enough illustrations to spur my interest and at the same

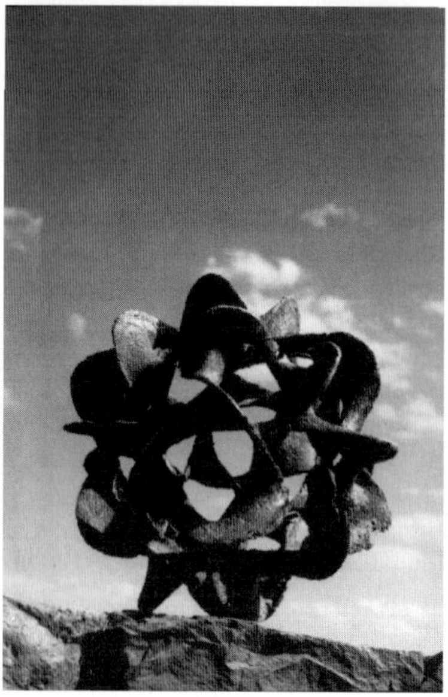

Fig. 1. [CP]1: Star Cinder

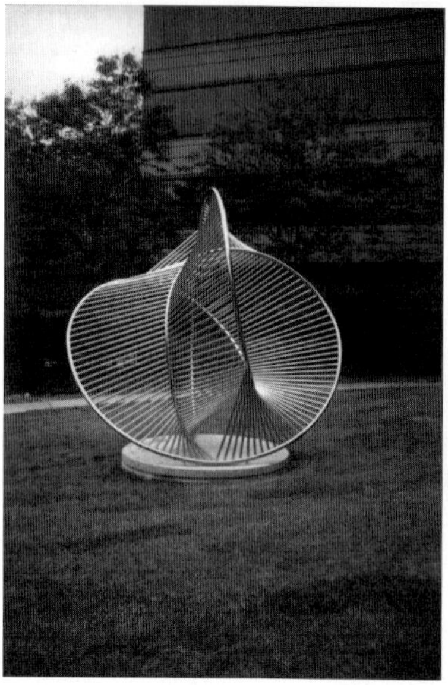

Fig. 2. [CP]2: Ribbed Möbius Mace

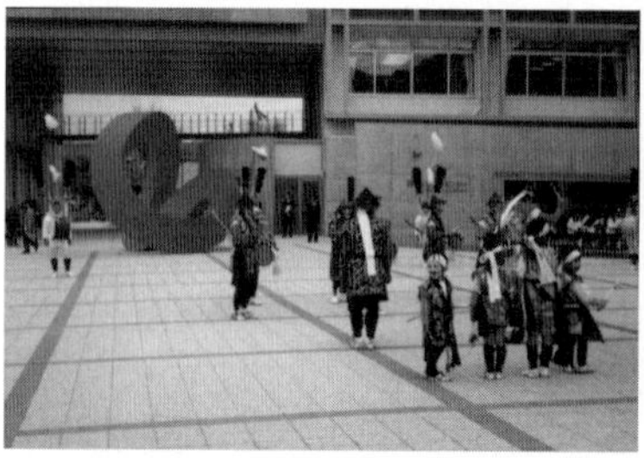

Fig. 3. [CP]3: Regeneration

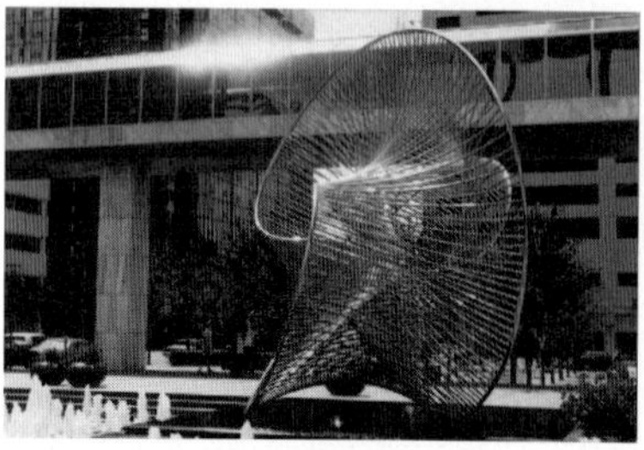

Fig. 4. [CP]4: Solstice

time convince me of the worth of my quest. My ignorance was definitely a big help, because it added to the mystery of what I was trying to do.

Actually that same ignorance has driven me always toward the end of that rainbow. The very misunderstanding of the way tetrahedrons fit together eventually taught me solid geometry. (At least my version of solid geometry). After all I am not a mathematician. I am an artist and the next piece of news in science is always another piece of my puzzle.

What joy it brings to realize how infinitely small we are and how little we know.

Visualization and Dynamical Systems

John Hubbard

Department of Mathematics, Cornell University, Malot Hall, Ithaca
<jhh8@cornell.edu>

1 Introduction

When asked to explain what my mathematics is about, I often answer by showing pictures illustrating the behavior of the dynamical systems I study. Non-mathematicians usually respond with interest, at least polite interest but sometimes much more. They often express amazement that my pictures are in any way related to mathematics. Mathematicians are also often interested, but many (fewer as time goes on) dismiss what I do as some sort of "fad math" devoid of theorems.

Visualization can be essential even if the object is the purest of mathematics. The pictures generated by computers using non-linear dynamical systems may or may not be beautiful; they may or may not be art; there doesn't seem to be a "true" answer to such questions, and each viewer must come to his or her own conclusions. But the pictures most definitely contain information of essential interest to mathematicians.

They help us to formulate conjectures, which in many cases would be inconceivable without the pictures. Experimentation, again using computer graphics, often can guide us to a proof, from which any computer aspect is totally absent. Finally, in the essential task of communicating our discoveries to others, the computer graphics may be absolutely essential.

2 Complex dynamics and visualization

I remember Lars Ahlfors, in 1982, telling me that in his youth his adviser Ernst Lindelöf made him read the memoirs of Fatou and Julia on the iteration of rational functions. These memoirs, he told me, struck him at the time as "the pits of complex analysis." He said that he only understood what they were about when seeing the pictures Mandelbrot and I were showing.

If even Ahlfors, the creator of some of the principal tools in the field, couldn't see what the authors were getting at, what of lesser mortals ? Indeed, those memoirs were practically forgotten for 60 years, waiting for computer graphics to reveal what Fatou and Julia had glimpsed.

3 The dynamics of the pendulum

I will now present a more personal example: the movements of the forced damped pendulum, governed by the "garden-variety" differential equation

$$\theta'' + a\theta' + b\sin\theta = c\cos(\omega t).$$

Since a robot is an assemblage of forced damped oscillators, it is of great interest to understand the dynamical properties of such an object.

Moreover, this differential equation has been studied by generations of students in courses in ordinary differential equations, either as an example in perturbation theory (develop the solutions in power series with respect to b, since the equation is linear when $b = 0$) or an example to test various numerical methods. Apparently, none of these studies led to any precise understanding of the behavior of solutions to the equation.

A bit of computer investigation for the values $a = .1$, $b = c = \omega = 1$ leads to the observation that these exists an attracting oscillation of the system of period $T = 2\pi$, corresponding to the downward equilibrium of the unforced pendulum. But the pendulum can get from an initial state to the attracting oscillation in many ways, and one essential measurement of the difference between two such evolutions is how many times the pendulum "goes over the top" before settling down. Certainly if two evolutions correspond to different numbers of such passages, they must be quite uncorrelated. If we color the plane of initial states (positions and velocities) according to how many times the pendulum goes over the top before settling down, we obtain the following picture.

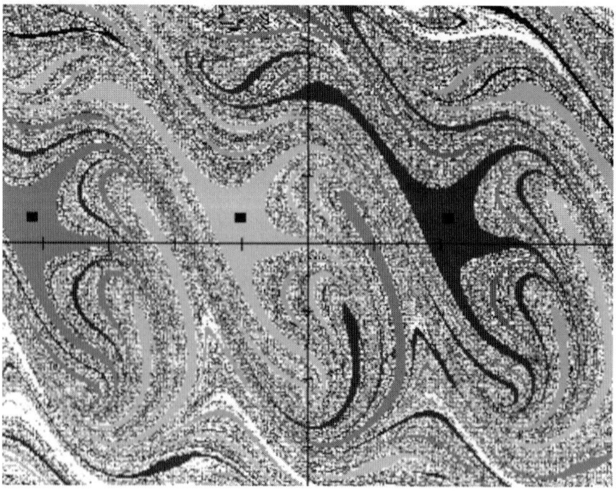

Fig. 1. [JH]1: *The plane of intial states colored according to the number of times the pendulum goes over the top before settling down.*

Contemplating this picture led to the conjecture that the basins form *Lakes of Wada*: every point in the boundary of one is in the boundary of all the others. When originally discovered by Brouwer and Yoneyama [4], this sort of behavior was seen as pathological; I am sure neither thought that such things would show up in mathematical problems of an applied nature.

Drawing the stable and unstable manifolds of the unstable periodic solution corresponding to the upper equilibrium of the pendulum leads to another conjecture: by choosing the initial condition correctly, the pendulum can be induced to go through any sequence of gyrations one wants, for instance turning once counterclockwise, then three times clockwise, then spending time almost vertical, then turning 5 billion times lockwise, and then once counterclockwise, etc.

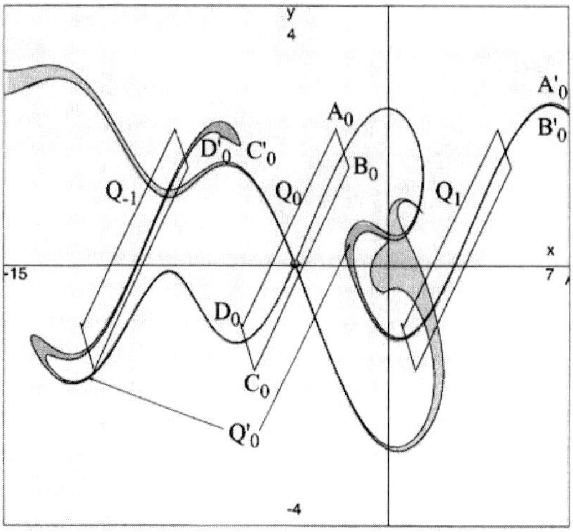

Fig. 2. *A quadrilateral in the plane of initial conditions, with its forward and backward images going through itself three times, forming a Smale Horseshoe. The quadrilateral is "fitted" to the unstable manifold of an unstable equilibrium.*

In both cases, with the help of the computer, the conjectures can be proved, using techniques due to Yorke, Kennedy and Nusse [2] for the first, and to Smale [3] for the second. The details are given in [1].

4 Some conclusions

This story has implications, for mathematics, for science and engineering, and for education. In mathematics, the use of visualization leads to interesting conjectures, which would never have been contemplated without such technology. The entire field of complex dynamics is filled with similar examples:

I am convinced that without computer graphics, the field would simply not exist, and I know that in the parts I have participated in, the motivation provided by computer graphics was essential, even if computers are never mentionned in the proofs.

Further, as Fatou and Julia discovered, even if you can prove theorems, you often can't communicate why they are interesting without the illustrations provided by computers.

The implications for science are probably more important yet. It is clear, from looking at the pictures and from the proofs, that the motion of a pendulum is unstable and chaotic: the challenge is to harness this instability to make robots more efficient. One can imagine two scenarios: a robot moving awkwardly, frequently stopping to reset its position to within prescribed tolerances before going on to the next task, like a beginning skier who stops between turns to regain his balance. But exploiting the instabilities should lead to the robot moving fluidly, expending far less energy, like an experienced skier floating down the slope, constantly out of balance but with effortless ease. At the moment, robots move according to the first scenario; they would be immensely more effective if we could move them to the second.

Where teaching of mathematics is concerned, I have found that computers can be immensely successful in bringing topics to life, and any teacher of math can attest to how essential this is: the mathematical objects students are expected to study often, in fact usually, have no reality in the student's minds, which prevents them from thinking about them effectively. Computers also allow the use of far more interesting examples, much closer to real problems. The pendulum example has been used in several undergraduate classes; this would have been inconceivable without the computer.

References

1. Hubbard J. H., (1999) The Forced Damped Pendulum: Chaos, Complication and Control, Am. Math. Monthly, 106, 8, 741–758
2. Kennedy J. and Yorke J., (1991), Basins of Wada, Physica D 51, 213–255
3. S. Smale, (1965) Diffeomorphisms with many periodic points: in Differential and combinatorial topology, Cairns S. S.Ed., Princeton University Press, Princeton, 63–80
4. Yoneyama K., (1917) Theory of continuous set of points, Tohoku Math. J., 11, 12–43

Solving Polynomials by Iteration

An Aesthetic Approach

Scott Crass

California State University, Long Beach, California
<scrass@csulb.edu>

1 Introduction

One of the classical problems of mathematics is to solve a polynomial equation. One approach to this is to take account of the fact that polynomials have symmetries that can be realized in geometric spaces. The idea is to solve an equation in an elegant way. During the past eight years I have pursued this goal by developing solutions based on iteration. This involves the repeated application of a process that possesses the very symmetries of the polynomial to be solved. An iterative procedure for solving an equation has two aspects:

- **Geometric:** a space where the polynomial's symmetry can be realized
- **Dynamical:** an iterated transformation that respects the symmetry of the equation

This paper discusses these two aspects in the cases of the fifth and sixth degree equations. It concludes with a gallery of graphical images that display both geometric and dynamical properties. Whatever their aesthetic appeal, the pictures play an important mathematical role. They reveal attracting, repelling, and chaotic behavior as well as fractal structures. In so doing, they contain experimental evidence obtainable by no other means.

2 Preliminary Background

2.1 Polynomials

A basic object of mathematical study is the polynomial in one variable: an expression made up of arithmetic combinations of numbers (coefficients) and an unknown quantity (the variable). For instance, the expression

$$x^2 - 3\,x + 2$$

is a polynomial of degree two (the degree is the highest apparent power of the variable x).

Mathematicians have produced a long history of developing methods for solving *polynomial equations*: finding numbers that make the polynomial take on the value zero when they replace the variable. In the example above, the numbers 1 and 2 solve the equation

$$x^2 - 3\,x + 2 = 0.$$

A polynomial has as many solutions – also called *roots* – as its degree, provided that you count them properly.

There is a correspondence between a polynomial and its roots – the roots determine the polynomial. If you know the roots, then, essentially, you know the polynomial. So, we can think of a polynomial in a geometric way. In our example, the two points $(1,2)$ and $(2,1)$ in a 2-dimensional space of ordered pairs of numbers correspond to the *same* polynomial. We can switch the coordinates of either of these points and get a different point but the same polynomial. In this way, every polynomial has symmetry; if you increase the degree, the dimension of the space in which we express the roots also increases as does the amount of symmetry.

2.2 Symmetric and Alternating Groups

Suppose you have five things. The number of different ways that you can arrange them is

$$120 = 5! = 5 \cdot 4 \cdot 3 \cdot 2.$$

If you change one arrangement into another you are performing a *permutation* of the objects. The set of all permutations of five things forms an algebraic structure called a *group* – specifically, the *symmetric group* $\mathcal{S}_5$.

Similarly, if you have n objects, the number of ways you can permute them is

$$n! = n \cdot (n-1) \ldots 3 \cdot 2.$$

These permutations form the symmetric group $\mathcal{S}_n$. A polynomial of degree n typically has $\mathcal{S}_n$ symmetry – the basic idea is that you can permute the roots in $n!$ different ways without changing the polynomial.

The simplest permutation is to exchange two things and leave the other things alone. We can express every permutation as a succession of such *transpositions.* The permutations that break up into an even number of transpositions also form a group – the alternating group $\mathcal{A}_n$. The number of permutations in $\mathcal{A}_n$ is half the number in $\mathcal{S}_n$.

2.3 Group Actions

When you have a set of objects S that you can move around according to the structure of a group, you are using a *group action.* In the case of polynomials of degree five, you can move the points in 5-dimensional space – the set S in this case – that correspond to the roots by permuting their coordinates. For instance, *transform* the point $(1,2,3,4,5)$ into $(2,1,5,3,4)$ by exchanging the first two coordinates and "cycling" the third, fourth, and fifth coordinates. We say that you are "acting on" 5-dimensional space with the symmetric group $\mathcal{S}_5$. If you use only the *even* permutations, you are acting on 5-dimensional space with the alternating group $\mathcal{A}_5$.

The *orbit* under a group action of a single element in S is the set of objects in S to which that element moves when you transform it according to *all* members of the group. For example, under the symmetric group $\mathcal{S}_3$ – the permutations of three things, the orbit of the point $(1,2,3)$ is

$$(1,2,3),\ (3,1,2),\ (2,3,1),\ (2,1,3),\ (3,2,1),\ (1,3,2).$$

2.4 Maps

An operation that takes each point in a space A and "sends it to" another point in A is called a *mapping* (or *map*) from A to A. To illustrate, take a point (x,y) in 2-dimensional space and "send it to" the point each of whose coordinates are the squares of the original:

$$(x,y) \longrightarrow (x^2,y^2).$$

Here, the arrow means "goes to" so that

$$\begin{aligned}(2,3) &\longrightarrow (4,9)\\ (-1,0) &\longrightarrow (1,0)\\ \left(\frac{1}{2},-\frac{5}{4}\right) &\longrightarrow \left(\frac{1}{4},\frac{25}{16}\right).\end{aligned}$$

This sort of map is a *dynamical system*, meaning that you can *iterate* its behavior – apply it repeatedly. For the "squaring map" above, the *trajectory* of $(2,3)$ is

$$(2,3) \longrightarrow (4,9) \longrightarrow (16,81) \longrightarrow \cdots$$

whereas the trajectory of $(\frac{1}{2},\frac{2}{3})$ is

$$\left(\frac{1}{2},\frac{2}{3}\right) \longrightarrow \left(\frac{1}{4},\frac{4}{9}\right) \longrightarrow \left(\frac{1}{16},\frac{16}{81}\right) \longrightarrow \cdots.$$

Some terminology. The *trajectory* of a point x under a map f is the set of points obtained by "applying" f iteratively to x. A point p is *periodic* if its trajectory contains p more than once. A periodic point a is *attracting* when the trajectory of every point near a gets arbitrarily close to a. The *basin of attraction* of a is the set of all points attracted to a. Also, the *attractor* of f is the set of all attracting points.

A fundamental question in *dynamics* is to determine the trajectories of points. In the examples, the trajectory of $(2,3)$ is not attracted to some specific point, but the trajectory of $(\frac{1}{2},\frac{2}{3})$ gets arbitrarily close to $(0,0)$. So, $(0,0)$ is an attracting point.

3 Polynomials, Symmetry, and Dynamics

There are two main ingredients in an iteration-based solution to a polynomial equation:

- a space on which the polynomial's symmetry group acts
- a map on the space that respects the action of the group.

We can harness the dynamics of such a map to break the polynomial's symmetry and thereby, solve the respective equation. The goal is to find maps with especially beautiful geometric and dynamical properties.

For a detailed treatment of the geometry and dynamics involved here as well as how to use both in developing a solution-procedure to the quintic and sextic see [Crass 2001] and [Crass 1999].

4 The Quintic – $\mathcal{S}_5$ Acts in Three Dimensions

If we permute the variables in

$$(x_1, x_2, x_3, x_4, x_5)$$

the equation

$$x_1 + x_2 + x_3 + x_4 + x_5 = 0$$

does not change. This equation defines a 3-dimensional space. (The idea of dimension here is not quite the familiar one. In terms of ordinary dimensions, this is a 6-dimensional space.) Corresponding to the permutations in $\mathcal{S}_5$ is a group $\mathcal{G}_{120}$ of 120 transformations on this space .

4.1 Invariant Polynomials

Similarly, if you permute the coordinates of the point

$$(x_1, x_2, x_3, x_4, x_5),$$

the expression (also called a *polynomial in five variables*)

$$F_2 = x_1^2 + x_2^2 + x_3^2 + x_4^2 + x_5^2$$

does not change. Likewise, the expressions

$$F_3 = x_1^3 + x_2^3 + x_3^3 + x_4^3 + x_5^3$$
$$F_4 = x_1^4 + x_2^4 + x_3^4 + x_4^4 + x_5^4$$
$$F_5 = x_1^5 + x_2^5 + x_3^5 + x_4^5 + x_5^5$$

are $\mathcal{S}_5$-*invariant.* A fundamental fact is that every polynomial that is invariant under the $\mathcal{S}_5$ permutations of variables has a unique expression in terms of these four polynomials. We can use these polynomials to create a *palette* of maps from which we choose some with special qualities.

4.2 A Surface Generated by Lines

The invariant F_2 defines a set that is invariant under $\mathcal{G}_{120}$. This *quadric surface* Q consists of two families of lines. Distinct lines in one family do not intersect while each point on Q belongs to one line in each family.

Furthermore, each *family* of lines – called a *ruling* on Q – has the geometry of the icosahedron. In addition, a transformation in $\mathcal{G}_{120}$ sends lines in one ruling either to lines in the same ruling or to lines in the other ruling. The set of transformations of the former type form a *subgroup* $\mathcal{G}_{60}$ of $\mathcal{G}_{120}$ that amounts to the rotational symmetries of the icosahedron.

4.3 Special Orbits

The smallest orbit under the $\mathcal{G}_{120}$ action consists of five points. Corresponding to each of these "5-points" is a plane. The intersection of each such plane with the quadric Q produces a sphere with the geometry of the octahedron.

Some of the geometry that will have dynamical significance shows up in various collections of lines. First, there is a 10-line orbit that intersects in fours at the 5-points. Figure 1 illustrates this in two ways. The pentagon-pentagram figure displays a 5-fold symmetry while the double pyramid exhibits the 6-fold symmetry of a single "10-line" – represented by the vertical axis.

Within each of the icosahedral rulings on Q there are three special line-orbits. These correspond to the 12 vertices, 20 face-centers, and 30 edge-midpoints of the icosahedron. The 20-line orbits (under $\mathcal{G}_{60}$) in each ruling form ten quadrilaterals at two pairs of 20-points. (See Figure 2 for one such quadrilateral.)

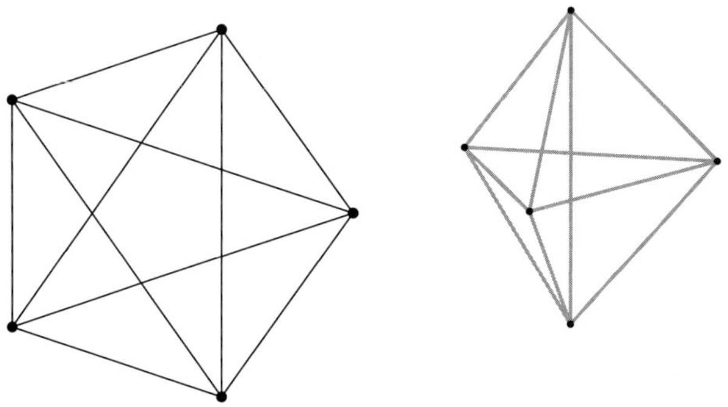

Fig. 1. Configuration of 10-lines and 5-points

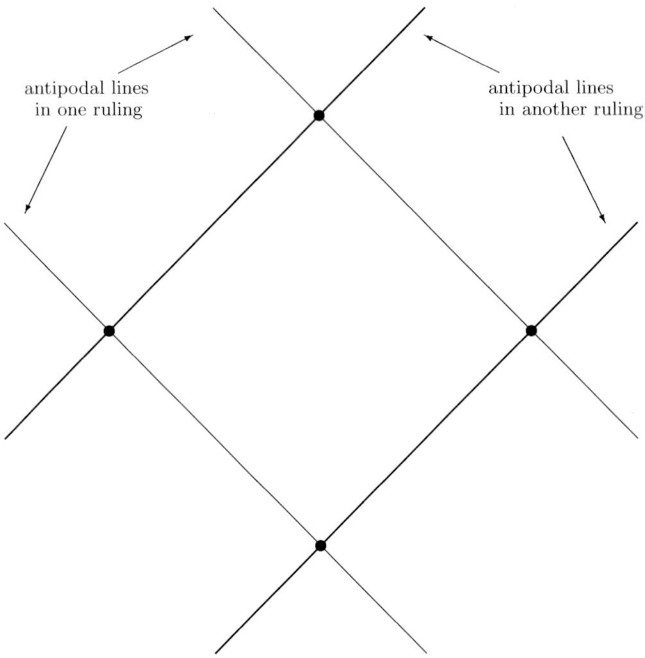

Fig. 2. Configuration of special lines on the quadric surface Q. The pairs of lines correspond to opposite vertices on the dodecahedron formed by the respective rulings.

4.4 Maps with Symmetry

The primary tool to be used in solving a quintic equation is a map that sends points in 3-dimensional space to points in the same space in a way that respects the action of the group of transformations $\mathcal{G}_{120}$. We want to find a $\mathcal{G}_{120}$-*equivariant* map (or simply $\mathcal{G}_{120}$-*equivariant*) with elegant geometry and *reliable dynamics*; this means that its attractor

1. is a *single* orbit under $\mathcal{G}_{120}$
2. has a corresponding basin that "fills up" 3-dimensional space.

By combining the invariants F_2, F_3, F_4, F_5 with four "basic" maps, we can produce all $\mathcal{G}_{120}$-symmetric maps.

4.5 Quadric-preserving Maps

The rich geometry of the quadric Q provides an intriguing setting for dynamical exploration. Are there $\mathcal{S}_5$-symmetric maps that send Q to itself? If so, how do they behave on and off Q? I will describe two species of such maps: one associated with the icosahedron and the other with the octahedron.

Icosahedral Maps Were a $\mathcal{G}_{120}$-equivariant to preserve the rulings on Q, it would move the lines in a ruling as an equivariant under the icosahedral action on the sphere. The lowest degree such a map can have is 11. By computational experimentation we succeed in obtaining maps of this kind.

Restricted to a ruling, the dynamics of each g_{11} is well-understood. Almost every line in a ruling belongs to the basin of one of the ten opposite pairs of the attracting set of 20-lines. (See Figure 2.) Recall that the 20-lines correspond to the face-centers of the dodecahedron. (See Figure 4.5.) Thus, for a random point x on Q, there is a pair of intersections *between* 20-lines in different rulings which the trajectory of x approaches.

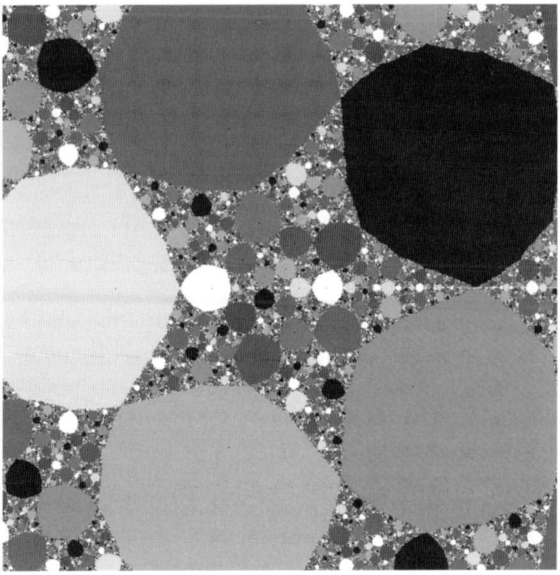

Fig. 3. Dynamics of a special 11-map on a ruling of the quadric surface.

Each of the ten pairs of antipodal dodecahedral vertices – black dots – is a period-2 attractor. (Bear in mind that points in the space of this plot correspond to lines in either ruling on the quadric surface Q.)

An Octahedral Map Since the orbit of the five coordinate planes has fundamental geometric significance, a map that preserves these sets might exhibit interesting dynamics.

The intersection of a 5-plane and the quadric Q is a sphere with the $\mathcal{S}_4$ symmetry of the octahedron (or cube). One of the special equivariants for the octahedral action on the sphere is a degree-5 map that attracts almost every point to the eight face-centers – vertices of the cube. (See [SC]1 in the Appendix.)

The idea is to find a map that sends Q to itself and behaves like the special degree-5 map on each of the octahedral spheres. We also want the octahedral face-centers to be attracting in directions away from Q. Here, we find a degree-11 map h_{11} that attracts points to pairs of opposite face-centers on the octahedron. In the Appendix, Figures [SC]2 through [SC]5, we see pictures that illustrate this map's dynamical behavior. In addition to their aesthetic qualities these images carry valuable mathematical information.

4.6 A Special Map in Degree Six

In the configuration of the 10-line orbit each 5-point lies at the intersection of four lines. (See Section 4.3.) Moreover, these are the only intersections of "10-lines." To take advantage of this structure, we arrange for a map f_6 with "attracting pipes" along the 10-lines. This results in 3-dimensional basins of attraction at the 5-points.

Recalling that a 10-line is a sphere, such a map attracts all points in the northern hemisphere to the 5-point at the north pole and attracts all points in the southern hemisphere to the 5-point at the south pole. The equatorial circle maps to itself chaotically.

We can get a picture of the map's *dynamics* by plotting basins of attraction on certain 2-dimensional "slices" of 3-dimensional space. The graphical evidence supports the claim that the 5-points are the only attractor for this map. (See Figures [SC]6 through [SC]9 in the Appendix.)

5 The Sextic – $\mathcal{A}_6$ Acts in Two Dimensions

5.1 Basics of $\mathcal{A}_6$ and Valentiner's Group

Inside the alternating group $\mathcal{A}_6$ are twelve versions of the alternating group $\mathcal{A}_5$. These twelve *subgroups* decompose into two systems of six.

In the late nineteenth century, Valentiner discovered a group – call it $\mathcal{V}$ – of 360 transformations of 2-dimensional space that has the same structure as the group of permutations $\mathcal{A}_6$. To solve a sixth-degree equation, we seek a 2-dimensional map that is symmetric with respect to $\mathcal{V}$.

5.2 Valentiner Geometry

Icosahedra Each of the $\mathcal{A}_5$ subgroups in $\mathcal{V}$ send a respective sphere to itself. Since the rotational symmetries of the icosahedron correspond exactly to the permutations in $\mathcal{A}_5$, each sphere has the geometry of the icosahedron. The group $\mathcal{V}$ permutes the two sets of six icosahedra in the same way as $\mathcal{A}_6$ permutes six things.

Special Orbits Two spheres *in different* systems intersect in two points. This gives six pairs of opposite icosahedral vertices. These total to a $\mathcal{V}$-orbit consisting of 72 points. Figure 4 illustrates the situation.

As for other special orbits, each of the 45 transpositions in $\mathcal{A}_6$ corresponds to a transformation T in $\mathcal{V}$ that fixes every point on a line associated with T. In addition, T fixes a point that is *not* on its associated line. These give $\mathcal{V}$-orbits of 45 lines and 45 points. An equivariant map typically sends each of these lines and points to itself; however, as we will see (Section 5.4), something quite different can occur. The 72-points have the distinction of being the only special $\mathcal{V}$-orbit that do not belong to the 45-lines.

5.3 Polynomials and Maps with Valentiner Symmetry

Every polynomial that is invariant under the Valentiner group can be expressed as a combination of four *basic* invariants. We can obtain families of $\mathcal{V}$-equivariant maps by combining these invariants with three basic equivariants. Again, the idea is to employ a palette of parameters in designing a geometrically elegant map.

5.4 A Case of Inelegant Dynamics

In degree 16 we find the map of lowest degree with Valentiner symmetry. This map has the property that it squashes a 45-line down to its associated 45-point. Furthermore, it "blows-up" a 45-point to its companion 45-line. This means that the map spreads points near the 45-point over the 45-line. The basin portrait (Figure [SC]9 in the Appendix) fails to reveal the sort of geometric elegance that we seek.

5.5 A Special Icosahedral Map of Degree 19

Associated with the icosahedron is a degree-19 map that takes each of the 20 faces and stretches it around the icosahedron omitting the opposite face.

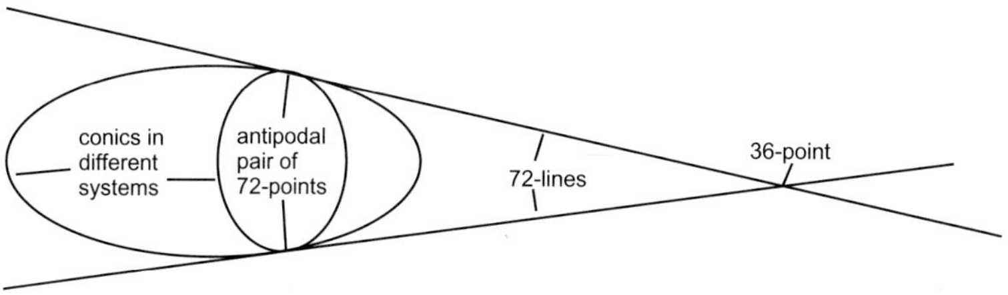

Fig. 4. The triangle of one 36-point and two 72-points.

When iterated, this icosahedral equivariant attracts almost any point of the icosahedron to one of the six pairs of antipodal vertices. (See Figure [SC]10.)

In the higher dimensional case of the Valentiner group, there is a degree-19 map h_{19} that send each of the 12 icosahedral spheres to itself. This means that *on* a sphere h_{19} is the map described above, so that we understand much of its dynamics there. Recall that the vertices of the icosahedral spheres make up the 72 point $\mathcal{V}$-orbit. It happens that away from the sphere, these points are also attracting.

Moreover, h_{19} honors the additional symmetries of certain transformations that exchange the systems of spheres. Therefore, it preserves a plane associated with each of these transformations. The map's dynamical behavior on such a plane appears in Figure [SC]11.

6 Gallery of Basin Portraits

The basin plots that follow are productions of the program *Dynamics* and *Dynamics 2* that ran respectively on a Silicon Graphics Indigo-2 and a Dell Dimension XPS with a Pentium II processor. Its BA and BAS routines produced the images. (See the manuals [Nusse and Yorke 1994] and [Nusse and Yorke 1998].) Each procedure divides the screen into a grid of cells and then colors each cell according to which attracting point its trajectory approaches. If it finds no such attractor after 60 iterations, the cell is black. The BA algorithm finds the attractor whereas BAS requires the user to specify a candidate attracting set of points. Each portrait exhibits the highest resolution available – a 720×720 grid.

References

[Crass 1999] S. Crass. *Solving the sextic by iteration: A study in complex geometry and dynamics.* Experiment. Math. 8 (1999) No. 3, 209–240.

[Crass 2001] S. Crass, 2001. *Solving the quintic by iteration in three dimensions.* Experiment. Math. 10 (2001) No. 1, 1–24.

[Nusse and Yorke 1994] H. Nusse and J. Yorke. *Dynamics: Numerical Explorations* Springer-Verlag, 1994. UNIX implementation by E. Kostelich.

[Nusse and Yorke 1998] H. Nusse and J. Yorke. *Dynamics: Numerical Explorations, 2e* Springer-Verlag, 1998. Computer program *Dynamics 2* by B. Hunt and E. Kostelich.

Mathematical Aspects in the Second Viennese School of Music

Carlota Simões

Department of Mathematics, University of Coimbra, Portugal
<carlota@mat.uc.pt>

Abstract. Mathematics and Music seem nowadays independent areas of knowledge. Nevertheless, strong connections exist between them since ancient times. Twentieth-century music is no exception, since in many aspects it admits an obvious mathematical formalization. In this article some twelve-tone music rules, as created by Schoenberg, are presented and translated into mathematics. The representation obtained is used as a tool in the analysis of some compositions by Schoenberg, Berg, Webern (the Second Viennese School) and also by Milton Babbitt (a contemporary composer born in 1916).

1 The Second Viennese School of Music

> *The 12-tone music was condemned by the Nazis (and forbidden in the occupied Europe) because its author was a Jew; by the Stalinists for having a bourgeois cosmopolitan formalism; by the public for being different from everything else.*
>
> Roland de Candé [2]

Arnold Schoenberg (1874–1951) was born in Vienna, in a Jewish family. He lived several years both in Vienna and in Berlin. In 1933, forced to leave the Academy of Arts of Berlin, he moved to Paris and short after to the United States. He lived in Los Angeles from 1934 until the end of his life.

In 1923 Schoenberg presented the *twelve-tone music* together with a new composition method, also adopted by Alban Berg (1885–1935) and Anton Webern (1883–1945), his students since 1904.

The three composers were so closely associated that the trio Schoenberg-Berg-Webern became known as the *Second Viennese School of Music.*

2 The Twelve-Tone Method

Tonality consists of the relations, melodic and harmonic, between the several notes of a given scale. In tonal music, the most important note is the so called *tonic*: around the tonic gravitates both melody and harmony.

Schoenberg wanted to remove the prevailing role of the tonic, as well as the hierarchy that tonality imposes between the seven notes of the traditional scale. With this idea in mind, in 1923 he established the *Twelve-Tone Method.* This method proposes to give the same merit to each note of the chromatic scale; in a 12-tone composition, all chromatic notes appear exactly the same number of times. The basis for a 12-tone composition is a sequence of the 12 distinct music pitches (without repetitions), appearing in any octave and combined under any rhythm. This sequence is called *12-tone series* or *12-tone row.*

One single series is the basis of each composition; only this basic series or some others related with this one by symmetry can be used in the composition. The composer can use the series in its *original* form, or with its intervals *inverted*, or backwards (*retrograde*), or *transposed* by some half-tones. Imposing as a rule that no series can start before the previous one is finished, at the end of any 12-tone composition all 12 notes have in fact appeared the same amount of times.

The first series of History of Music is the one used in the fifth piece of *Five Piano Pieces, Opus 23*, from Schoenberg, written in 1923: C$^\sharp$, A, B, G, A$^\flat$, G$^\flat$, B$^\flat$, D, E, E$^\flat$, C, F.

3 Mathematical Formalization of Twelve-tone Music

> *People accuse me of being a mathematician, but I am not a mathematician, I am a geometer.*
>
> *Arnold Schoenberg*

In this section we use numbers to represent musical notes. Recall that in a 12-tone row, notes with the same name are considered equivalent, even if they belong to different octaves. We start by identifying consecutive notes with consecutive integers. In this way, if C is represented by the integer 1, then C$^\sharp$ is represented by 2, B is represented by 12 and the following C is again represented by 1. The difference between two integers representing two notes gives the number of half-steps of the interval between the two corresponding notes (ignoring octaves). For instance, C and G can be represented by 1 and 8, respectively, and they define an interval of a perfect fifth, or 7 half-tones (or an interval of $7+12$ half-tones, or $7+12k$ half-tones, for some integer k).

Given an integer p, we define the set $[p] = \{p+k\times 12, \text{ for some integer } k\}$. Notice for instance that $[1] = [49] = [-11]$ and $[5] = [29] = [-7]$. The set $\mathbb{Z}_{12} = \{[0], [1], [2], \ldots, [11]\}$ is called *set of integers modulo 12* and the set $[p]$ is called *equivalence class of p modulo 12.* Addition in $\mathbb{Z}_{12}$ (*addition modulo 12)* is defined as the usual addition in $\mathbb{Z}$, taking further into account that numbers differing by a multiple of 12 are equivalent. For example, we have $(3+8)mod_{12} = 11$, $(5+7)mod_{12} = 0$ and $(7+7)mod_{12} = 2$. Using the equivalence between numbers differing by a multiple of 12, we define *symmetric of an integer modulo 12* obtaining, for instance, $(-5)mod_{12} = 7$, $(0)mod_{12} = 0$, $(-11)mod_{12} = 1$.

We may now define *12-tone series* as a permutation of the integers $0, 1, 2, \ldots, 11$, recalling that each integer has to be seen as an equivalence class modulo 12.

Example 1: In Schoenberg's *Piano Concerto, Opus 42* (1942), the basic row is

$$E^{\flat},\ B^{\flat},\ D,\ F,\ E,\ C,\ F^{\sharp},\ A^{\flat},\ D^{\flat},\ A,\ B,\ G.$$

Considering $E^{\flat} \equiv 0$, $E \equiv 1$, and so forth, this series, which we represent by S, can be re-written as

$$S = (0,\ 7,\ 11,\ 2,\ 1,\ 9,\ 3,\ 5,\ 10,\ 6,\ 8,\ 4). \tag{1}$$

4 Series related to a given series

In order to collect all material to be used in a musical composition, after choosing the basic series (or basic row), it is necessary to list all series which can be obtained by symmetry from the basic one.

Suppose that the *basic row* is defined by

$$P = (a_1, a_2, a_3, a_4, a_5, a_6, a_7, a_8, a_9, a_{10}, a_{11}, a_{12}) \tag{2}$$

The *retrograde* row, which we represent by $R(P)$, is obtained from P playing it backwards

$$R(P) = (a_{12}, a_{11}, a_{10}, a_9, a_8, a_7, a_6, a_5, a_4, a_3, a_2, a_1) \tag{3}$$

The *inverse* row, which we represent by $I(P)$, is obtained from the basic row P maintaining its first note and inverting all intervals between consecutive notes, in such a way that an ascending interval becomes descendent and vice-versa. Denoting the entries of $I(P)$ by a_k^*, $k = 1, \ldots, 12$, they verify the following

$$\begin{aligned} a_1^* &= a_1 \\ a_2^* &= a_1^* - (a_2 - a_1) \\ a_3^* &= a_2^* - (a_3 - a_2) \\ &\ \vdots \\ a_{12}^* &= a_{11}^* - (a_{12} - a_{11}) \end{aligned}$$

If $a_1 = 0$ in the basic row, then the inverse row $I(P)$ takes the form

$$I(P) = ((-a_1)mod_{12}, \ldots, (-a_{12})mod_{12}), \tag{4}$$

that is, the *inverse* series $I(P)$ of P is represented by the substitution of each entry of the series by its symmetric with respect to addition in $\mathbb{Z}_{12}$ (property only valid if we associate to the first note of the basic series the number 0).

The *retrograde inverse* row, which we represent by $RI(P)$, is obtained from P by applying the previous two operations. In case $a_1 = 0$ in P, the row $RI(P)$ is

$$RI(P) = ((-a_{12})mod_{12}, \ldots, (-a_1)mod_{12}). \tag{5}$$

Any *transposition by k half-tones* of the basic row is obtained from P by adding k (modulo 12) to each entry of P. The transposition of P by k half-tones is just

$$P_k = ((a_1 + k)mod_{12}, ..., (a_{12} + k)mod_{12}) \tag{6}$$

In a similar way we obtain the transpositions by k half-tones of the retrograde row, the inverse row and the retrograde inverse row.

5 The Matrix of Series

The rows related with a given basic row P can be organized in one matrix $M(P)$, as follows.

Let P be the series

$$P = (a_1, a_2, a_3, a_4, a_5, a_6, a_7, a_8, a_9, a_{10}, a_{11}, a_{12})$$

with $a_1 = 0$.

Let $M(P)$ be the 12×12 matrix obtained in the following way: the 1st row of $M(P)$ is the basic series P; the 1st column of $M(P)$ is the inverse series $I(P)$; since $a_1 = 0$, the entries of the 1st column are given by (4). The remaining rows are transpositions of P, being that transposition by k half-tones (P_k) if k is the 1st entry of the row. In the same way, the remaining columns are transpositions by k half-tones of the inverse row (I_k), where k is the 1st entry of the column. Reading the 1st row backwards, we obtain the retrograde ($R(P)$) of the basic row. Reading the 1st column up side down, we obtain the retrograde inverse ($RI(P)$). The remaining rows backwards are transpositions of the retrograde row (R_k) and the remaining columns up side down are the transposition of the retrograde inverse (RI_k).

The series related with the basic series are at most 48, being less than 48 if there are some special symmetries in the basic series.

Example 2: In Figure M we have the matrix of series for the Schoenberg's *Piano Concerto, Opus 42* (see Example 1).

	I_0 ↓	I_7 ↓	I_{11} ↓	I_2 ↓	I_1 ↓	I_9 ↓	I_3 ↓	I_5 ↓	I_{10} ↓	I_6 ↓	I_8 ↓	I_4 ↓	
$P_0 \rightarrow$	0	7	11	2	1	9	3	5	10	6	8	4	$\leftarrow R_0$
$P_5 \rightarrow$	5	0	4	7	6	2	8	10	3	11	1	9	$\leftarrow R_5$
$P_1 \rightarrow$	1	8	0	3	2	10	4	6	11	7	9	5	$\leftarrow R_1$
$P_{10} \rightarrow$	10	5	9	0	11	7	1	3	8	4	6	2	$\leftarrow R_{10}$
$P_{11} \rightarrow$	11	6	10	1	0	8	2	4	9	5	7	3	$\leftarrow R_{11}$
$P_3 \rightarrow$	3	10	2	5	4	0	6	8	1	9	11	7	$\leftarrow R_3$
$P_9 \rightarrow$	9	4	8	11	10	6	0	2	7	3	5	1	$\leftarrow R_9$
$P_7 \rightarrow$	7	2	6	9	8	4	10	0	5	1	3	11	$\leftarrow R_7$
$P_2 \rightarrow$	2	9	1	4	3	11	5	7	0	8	10	6	$\leftarrow R_2$
$P_6 \rightarrow$	6	1	5	8	7	3	9	11	4	0	2	10	$\leftarrow R_6$
$P_4 \rightarrow$	4	11	3	6	5	1	7	9	2	10	0	8	$\leftarrow R_4$
$P_8 \rightarrow$	8	3	7	10	9	5	11	1	6	2	4	0	$\leftarrow R_8$
	↑ RI_0	↑ RI_7	↑ RI_{11}	↑ RI_2	↑ RI_1	↑ RI_9	↑ RI_3	↑ RI_5	↑ RI_{10}	↑ RI_6	↑ RI_8	↑ RI_4	

Figure M: The 1st row contains the elements $a_1, \ldots, a_{12}$; the 1st column contains the elements $-a_1, \ldots, -a_{12}$. Since each row i is obtained from the first row by the addition of $-a_i$ (which is the 1st entry of the row i), the element of the matrix in the position (i, j) (row i, column j) is equal to $a_j - a_i$. Obviously, the main diagonal of the matrix $M(S)$ is composed by zeros.

6 Hexachords

If we consider a series just a permutation of the 12 integers $0, 1, 2, \ldots, 11$, then the total number of possible series is $\mathcal{P}(12) = 12! = 479.001.600$. Of course not all permutations can be used as series, since composers are certainly interested in series with special musical properties.

Example 3: Consider the row P

$$P = (0, 2, 7, 5, 10, 9, 3, 4, 11, 1, 8, 6)$$

Transposing P by 6 half-tones, we obtain

$$P_6 = (6, 8, 1, 11, 4, 3, 9, 10, 5, 7, 2, 0).$$

The retrograde of P coincides with the transposition of P by 6 half-tones, that is, $P_6 = R(P)$. Obviously the row P has less than 48 different associated rows.

Proposition: *Let P be a given row. If $R(P) = P_k$, then k=6.*

Proof: Suppose that the retrograde of the series P

$$P = (a_1,\ a_2,\ a_3,\ a_4,\ a_5,\ a_6, ...,\ a_{12})$$

is equal to

$$P_k = (a_1 + k,\ a_2 + k,\ a_3 + k, \ldots,\ a_{12} + k),$$

for some $k \in \{0, 1, 2, \ldots, 11\}$. Then $(a_1 + k = a_{12}) mod_{12}$ and $(a_{12} + k = a_1) mod_{12}$. From $(a_1 = a_1 + 2k) mod_{12}$ we conclude that $k = 6$. ■

Quite often, the choice of a particular series such as the one in Example 3 needs the study of some sub-sets of $\mathbb{Z}_{12}$. We give particular attention to *hexachords*, sub-sets of $\mathbb{Z}_{12}$ containing six elements.

The series P defined in (2) can be divided in two hexachords as follows

$H_1(P)$	$H_2(P)$
$a_1, a_2, a_3, a_4, a_5, a_6$	$a_7, a_8, a_9, a_{10}, a_{11}, a_{12}$

Notice now that in order to have $R(P) \equiv P_6$, neither $H_1(P)$ nor $H_2(P)$ can contain two elements of $\mathbb{Z}_{12}$ with a difference (module 12) equal to 6 and, furthermore, $H_2(P)$ has to be equal to the retrograde of $H_1(P)$, transposed by 6 half-tones.

Proposition: *Divide the series P in two hexachords $P = (H_1(P), H_2(P))$. P verifies the condition $R(P) = P_6$ if and only if*
(a) $H_1(P)$ does not contain any pair of notes differing by 6 half-tones;
(b) $H_2(P) = R(H_1(P) + 6)$.

The amount of series with the property $R(P) = P_6$ is equal to $12 \times 10 \times 8 \times 6 \times 4 \times 2 = 46\,080$.

Definition: Given two series A and B we say that *A and B combine* if both first hexachords together contain all 12 notes, without repetitions.

Let A and B be the series

	1st hexachord	2nd hexachord
A	$H_1(A) : a_1, a_2, a_3, a_4, a_5, a_6$	$H_2(A) : a_7, a_8, a_9, a_{10}, a_{11}, a_{12}$
B	$H_1(B) : b_1, b_2, b_3, b_4, b_5, b_6$	$H_2(B) : b_7, b_8, b_9, b_{10}, b_{11}, b_{12}$

When A and B combine, both sequences $(H_1(A), H_2(B))$ and $(H_1(B), H_2(A))$ also define series.

Obviously, any row combines with its own retrograde. This trivial property has been cleverly used by Webern, as shown in Example 4.

Example 4: Webern's *Piano Variations, Opus 27.*

The basic row used in this piece is

	E	F	C♯	E♭	C	D	G♯	A	B♭	F♯	G	B
P:	0,	1,	9,	11,	8,	10,	4,	5,	6,	2,	3,	7

We separate the row P in two hexachords.

$H_1(P)$	$H_2(P)$
$0, 1, 9, 11, 8, 10$	$4, 5, 6, 2, 3, 7$

The right hand starts playing the 1st hexachord of P followed by the 2nd hexachord of the retrograde $R(P)$, while the left hand plays the 1st hexachord of $R(P)$, followed by the 2nd hexachord of P, leading to the following palindromical structure

right hand:	0 1 9 11 8 10	10 8 11 9 1 0
left hand:	7 3 2 6 5 4	4 5 6 2 3 7

as can easily be seen in Figure 1.

Example 5: Schoenberg's *Piano Piece, Opus 33a.*

The basic row used in this piece is

	B♭	F	C	B	A	F♯	D♭	E♭	G	A♭	D	E
P:	0,	7,	2,	1,	11,	8,	3,	5,	9,	10,	4,	6

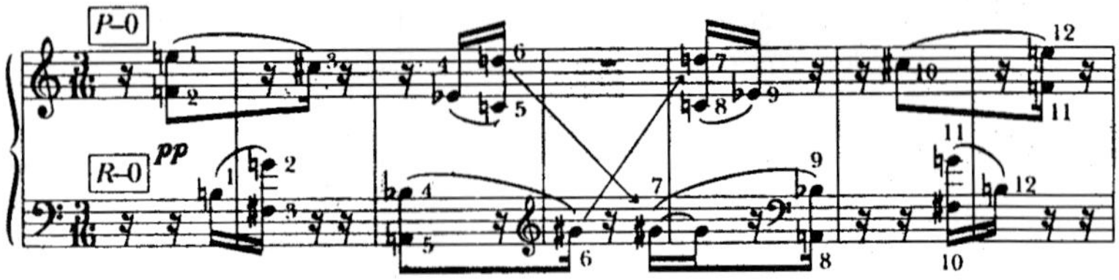

Fig. 1. *Webern's Piano Variations Opus 27*, measures 1–7.

We notice first that P combines with the transposition by 5 half-tones of its inverse $I(P)$. In fact, the first hexachord H_1 of the series P is

$$H_1 = (a_i)_{1 \leq i \leq 6} = (0, 7, 2, 1, 11, 8);$$

the inverse of H_1 is

$$I(H_1) = (-a_i)_{1 \leq i \leq 6} = (0, 5, 10, 11, 1, 4);$$

the transposition by five half-tones of $I(H_1)$ is

$$I_5(H_1) = (-a_i + 5)_{1 \leq i \leq 6} = (5, 10, 3, 4, 6, 9)$$

and since there are no repeated entries in H_1 and $I_5(H_1)$, we conclude that these two hexachords are complementar.

7 Creating 12-tone music

I can tell you, dearest friend, that if became known how much friendship, love and a world of human and spiritual references I have smuggled into these movements, the adherents of programme music (should there be any left) would go mad with joy.

Berg, in a letter to Schoenberg, about the *Chamber Concerto*, 1925.

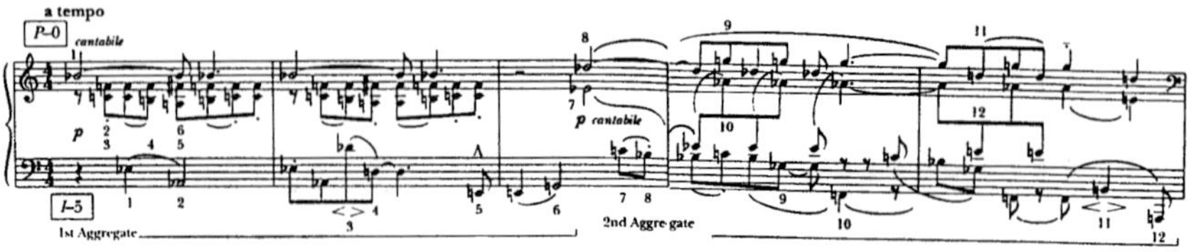

Fig. 2. *Schoenberg's Piano Piece, Op. 33a*, measures 14–18. When both right and left hands have only played the first hexachord of the corresponding series, (P_0 for the right hand and I_5 for the left hand), already both hands together have played all 12 notes of the chromatic scale.

In this section we try to show, by means of some musical examples, that despite their seemingly strict rules, the character of the composer can still be recognized in a 12-tone music composition.

Example 6: Berg's *Violin Concerto*, 1935.

The music of Alban Berg is intensely expressive, almost romantic. In fact, Berg was able to compose 12-tone music without being entirely away from tonality. The work of Berg can be seen as a compromise between traditional principles (tonality) and innovative principles (atonality, 12-tone method), fact that indeed is quite evident in his *Violin Concerto*.

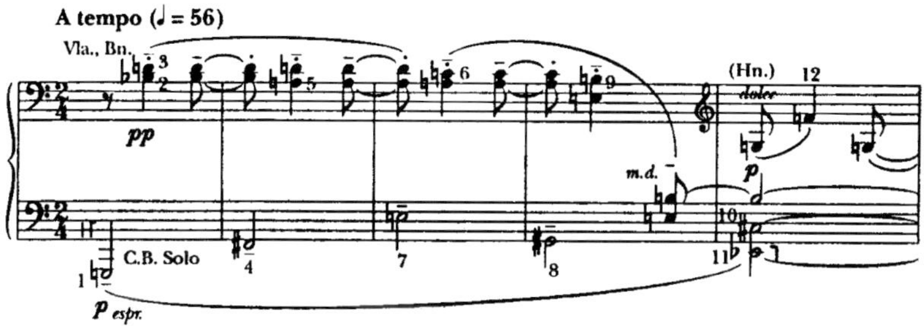

Fig. 3. *Violin Concerto, Berg, 1935*, measures 11–15.

The basic row used in this piece is

	G	$B^\flat$	D	$F^\sharp$	A	C	E	$G^\sharp$	B	$C^\sharp$	$D^\sharp$	F
P:	0,	3,	7,	11,	2,	5,	9,	1,	4,	6,	8,	10

Exceptionally, and to make clear the tonal character of this piece, we study this series using the name of the notes involved.

G minor { G, $B^\flat$, D }
D major { D, $F^\sharp$, A }
A minor { A, C, E }
E major { E, $G^\sharp$, B }
{ B, $C^\sharp$, $D^\sharp$, F }

The basic row crosses several tonalities: the first three notes define the tonality of G minor, moving successively to the tonalities of D major, A minor, E major, finishing with four notes presenting a sequence of three whole tones.

Example 7 Webern's *Concerto for Nine Instruments, Opus 24.*

Anton Webern is, from the Vienna Trio, the one arriving further away from the tonal system. The compositions by Webern are in general brief and dry, while the traditional melodic line is substituted by individual notes, with no melodic connection between them, some times even confined to different instruments, as it is evident in his *Concerto for Nine Instruments, Opus 24.*

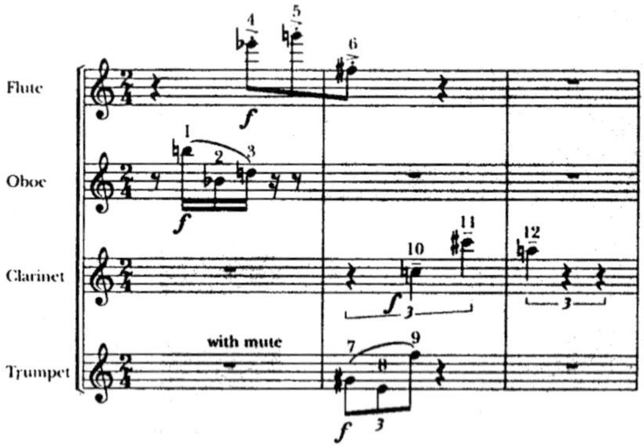

Fig. 4. *Concerto for Nine Instruments, Opus 24,* measures 1–3. The basic row is separated into four small groups of three notes each, which become independent since they are distributed by different instruments and played with different speed.

The basic series for this concert was carefully chosen in order to have an internal structure with interesting symmetries. The basic series used in this piece is

	B $B^\flat$ D;	$E^\flat$ G $F^\sharp$;	$G^\sharp$ E F ;	C $C^\sharp$ A
P:	0, 11, 3;	4, 8, 7;	9, 5, 6;	1, 2, 10

We consider now the first three notes of the series $(0, 11, 3)$, and we deal with this set of notes as a "mini-row" which we represent by T. We have the following

$$\begin{aligned} T &: (0, 11, 3) \\ I(T) &: (0, 1, 9) \\ R(T) &: (3, 11, 0) \\ RI(T) &: (9, 1, 0) \end{aligned}$$

Notice further that

$$\begin{array}{ll} I(T)+1: & (1,2,10) \\ R(T)+6: & (9,5,6) \\ RI(T)+7: & (4,8,7) \end{array}$$

The basic row P can in fact be completely re-written in terms of the mini-row T, as follows

$$P = (T, RI_7(T), R_6(T), I_1(T)).$$

Example 8: Babbitt's *Three Compositions for piano, No.1, 1947.*

After Schoenberg, Berg and Webern, other composers have used the idea of a series in musical elements other than notes. Metric, intensity of sound, rhythm, timbre, among others, have been incorporated into serial structures. In this example we can see how Milton Babbitt (American composer born in 1916) uses the idea of basic series to also establish the duration and intensity of sound.

The basic series used in this piece to describe the notes is

	B♭	E♭	F	D	C	D♭	G	B	F♯	A	A♭	E
P:	0,	5,	7,	4,	2,	3,	9,	1,	8,	11,	10,	6

The transposed of P by 6 half-tones is

$$P_6 = (6, 11, 1, 10, 8, 9, 3, 7, 2, 5, 4, 0).$$

Notice that P combines with its transposition P_6. On the other hand, the retrograde of P is

$$R(P) = (6, 10, 11, 8, 1, 9, 3, 2, 4, 7, 5, 0),$$

the inverse of P is

$$I(P) = (0, 7, 5, 8, 10, 9, 3, 11, 4, 1, 2, 6)$$

and the retrograde inverse of P is

$$RI(P) = (6, 2, 1, 4, 11, 3, 9, 10, 8, 5, 7, 0),$$

leading to

$$RI_1(P) = (7, 3, 2, 5, 0, 4, 10, 11, 9, 6, 8, 1),$$

and we conclude that $R(P)$ combines with $RI_1(P)$.

The structure of the first four measures is as follows

measure	I	II	III	IV
right hand	$H_1(P_6)$	$H_2(P_6)$	$H_1(R)$	$H_2(R)$
left hand	$H_1(P)$	$H_2(P)$	$H_1(RI_1)$	$H_2(RI_1)$

Since P combines with P_6 and $R(P)$ combines with $RI(P)_1$, in each measure the notes of the two hexachords played by both left and right hands are all the 12 notes of the scale.

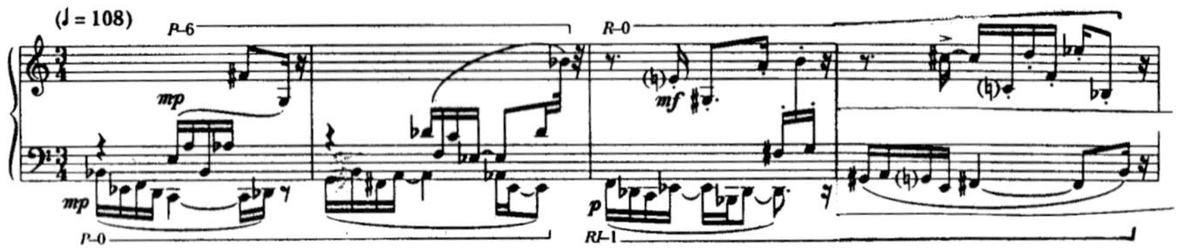

Fig. 5. *Three Compositions for piano, No.1, Babbitt (1947)*, measures 1–4.

For the duration of sounds, a sequence of four numbers is used,

$$D = (5, 1, 4, 2),$$

with operations modulo 6 (in a similar way as $\mathbb{Z}_{12}$, we define the set of *integers modulo 6*, $\mathbb{Z}_6$, with operations defined in such a way that integers differing by a multiple of 6 are equivalent).

Each entry of the row D indicates the quantity of consecutive sixteenths appearing in each group. More exactly, the fact that the initial entry is equal to 5, indicates that the first five notes of the composition must form a group of five consecutive and linked sixteenths. Obviously, the addition of all four entries of D is equal to 12, the number of notes of the series.

Notice further that

$$\begin{array}{ll} D: & 5, 1, 4, 2 \\ I(D): & 1, 5, 2, 4 \\ R(D): & 2, 4, 1, 5 \\ RI(D): & 4, 2, 5, 1 \end{array}$$

Any time the row P (or a transposition of P) is played, the rhythm indicated by D is used; when the inverse of P (or a transposition of $I(P)$) is played, the duration of the notes has to respect $I(D)$; retrogrades are played with rhythm given by $R(D)$, and $RI(D)$ indicates the rhythm of $RI(P)$ (or transpositions of $RI(P)$).

There is also a correspondence between the form of the series and the intensity of the sound; in fact, the row P (or a transposition of P) is always played *mezzo piano (mp)*; $I(P)$ (or a transposition of $I(P)$) is played *forte (f)*; retrogrades are played *mezzo forte (mf)* while $RI(D)$ are played *piano (p)*.

The correspondence between sound, rhythm and intensity is presented in the following table.

notes	rhythm	intensity
P	D	*mezzo piano (mp)*
$R(P)$	$R(D)$	*mezzo forte (mf)*
$I(P)$	$I(D)$	*forte (f)*
$RI(P)$	$RI(D)$	*piano (p)*

8 Final Comments

Something tremendously powerful was lost when composers moved away from tonal harmony and regular pulses... among other things the audience was lost.

John Adams

Many composers from the past, considered revolutionary at their times, just wanted to enlarge the limits of their heritage. Musical taste is in perpetual change. The parallel fourths and fifths, for instance, so pleasant to hear in Gregorian Chant, were completely abandoned by the Eighteenth Century harmony. Many were the composers whom, in all times, surprised the public with new ideas. Schoenberg is just one of them.

References

1. Roland de Candé, (1983) *A Música; Linguagem, Estrutura, Instrumentos*, Edições 70, Lisboa
2. Roland de Candé, (1986) *Convite à Música*, Edições 70, Lisboa
3. Griffiths P., (1986) *Dictionary of 20th-Century Music*, Thames and Hudson, Singapore
4. Holtzman S.R., (1994) *Digital Mantras, The Languages of Abstract and Virtual Worlds*, The MIT Press, Cambridge, Massachussets, England
5. James J., (1993) *The Music of The Spheres. Music, Science and The Natural Order of the Universe*, Copernicus, Springer-Verlag, New York
6. Morgan R. P., (1991) *Twentieth-Century Music*, W.W. Norton & Company, New York
7. De Oliveira João P. P., (1998) *Teoria Analítica da Música do Século XX*, Fundação Calouste Gulbenkian
8. Carlota Simões, (1999) 'A ordem dos números na música do Século XX' *Colóquio Ciências*, Fundação Calouste Gulbenkian, N° 24, 48–59
9. Wilcox H. J., (1987) 'The role of mathematics in atonal music', *The UMAP Jornal*, 8 (1) 83–89

Mathematics and Art: The Film Series

Michele Emmer

Dipartimento di Matematica, Università di Roma "La Sapienza"
<emmer@mat.uniroma1.it>

To Fred Almgren

1 The Mathematics and Art Project

The "Mathematics and Art" project started in 1976. Or better, that year I started thinking of the project. The reasons why I started thinking of it are essentially two, or perhaps three. The first: in 1976, I was at the University of Trento, in the North of Italy. I was working in that area called the *Calculus of Variations*, in particular *Minimal Surfaces* and *Capillarity problems.* I had graduated from the University of Rome in 1970 and started my career at the University of Ferrara, where I was very lucky to start working with Mario Miranda, the favourite graduate student of Ennio De Giorgi; then I met Enrico Giusti and Enrico Bombieri. It was the period in which in the investigations of Partial Differential Equations, of the Calculus of Variations and the Perimeter theory, first introduced by Renato Caccioppoli and then developed by De Giorgi and Miranda, the Italian school of the Scuola Normale Superiore of Pisa was one of the best in the world. And in the year 1976, Enrico Bombieri received the Fields medal. By chance I was in the right place at the right time. All the mathematicians world-wide who were working in these areas of research had to be updated about what was happening in Italy. In July 2000, I participated in the annual congress of the American Mathematical Society in Los Angeles, entitled "Challenges for the 2000". Many of the invited speakers were asked to provide a survey of the researches in the last 50 years. Those who talked of Partial differential Equations, Calculus of Variations and Minimal Surfaces like Karen Ulhenbeck, Haim Brezis and Jean Taylor, all recalled the group of mathematicians from Pisa, born around the famous Ennio De Giorgi, and its great scientific relevance.

Always in the 1976, Jean Taylor proved a famous result that closed a conjecture that was raised experimentally by the Belgian physicist Joseph Plateau over a hundred years before: the types of singularities of the edges that soap films generate when they meet. Plateau had experimentally observed that the angles generated by the soap films are only of two kinds. Jean Taylor, using the Theory of Integral Currents introduced by Federer, and then by Allard and Almgren, was able to prove that the result was true. A few years before, Ennio De Giorgi was able to prove in its generality the existence of the solution of Plateau's problem. To prove that for any chosen boundary it is possible to find a minimal surface that has this boundary. De Giorgi was also able to prove the isoperimetric property of the sphere in every dimension n. In three dimensions, which is the case of soap bubbles, you have a surface with assigned mean curvature that must contain a fixed volume of air. Soap bubbles have always been a great fascination for everybody, from children to scientists. It is sufficient to quote the case of Gilles de Gennes who ended his presentation on "Soft Matter" for the ceremony of the Nobel prize for physics in 1992 with a poem on soap bubbles.

Amusons-nous. Sur la terre et sur l'onde
Malheureux, qui se fait un nom!
Richesse, Honneur, faux éclat de ce monde.
Tout n'est que bulles de savon.

For an exhaustive story of soap bubbles in mathematics, in art, in chemistry, in architecture and in biology, refer to the volume "Bolle di sapone: un viaggio tra matematica, arte e fantasia" (Soap bubbles: a journey into mathematics, art and fantasy).

Let's get back to Jean Taylor and to the year 1976. In 1976, the journal *Scientific American* asked Jean Taylor and Fred Almgren (they got married a few months before) to write a paper on the more recent results on the topic of Minimal Surfaces and Soap Bubbles. A professional photographer was asked to take the pictures for the paper. The same year, Jean Taylor and Fred Almgren were invited to the University of Trento as visiting professors and during the summer, they gave a summer course in Cortona, near Arezzo. I already knew both of them, maybe the one I knew better was Fred, who in his Swedish manner, was always very kind with me. A few years before, in Varenna, on Lake Como (North of Italy), another summer course was held, always on Minimal Surfaces. It was the year of the Olympic Games and the feats of the USA swimmer Mark Spitz. Mario Miranda was carried away with enthusiasm, and proposed a swimming race in the lake. The only participant who accepted was Fred Almgren.

Miranda (at that time I was his assistant) asked me to join the two swimmers. I answered that for ten years I was a professional swimmer so it was not fair on my part to participate to the race. Miranda insisted on asking me to swim, saying that I had been a professional swimmer a number of years before. Of course I won the race leaving the other two swimmers many meters behind. A beautiful satisfaction for a young assistant! When Almgren and Taylor came to Trento in 1976 the issue of the Scientific American had just been published. The pictures of the article and the cover were really beautiful and interesting. I do not remember why, but looking at the pictures I had the idea of making a film on soap film and to show its shapes and geometry in the greatest possible detail, more closely and using the rallenti technique. I spoke of the project with Valeria, who lived in Rome with our two sons, and she was very pleased and attracted by the idea.

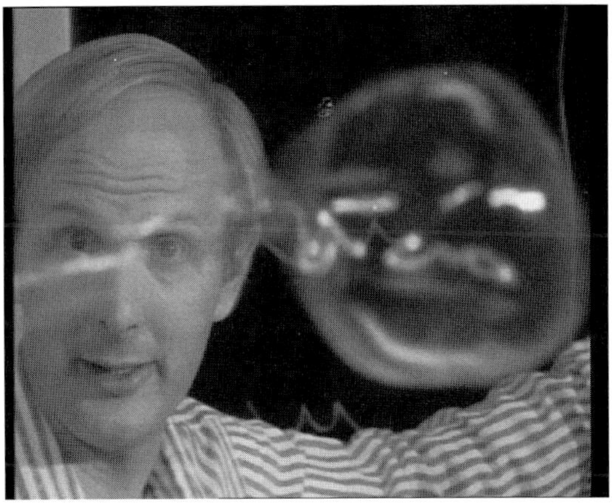

Fig. 1. [ME]1: Fred Almgren in his studio at Princeton University

I must say that for me thinking of making a film was quite natural. My father is a famous Italian film-maker. Mastroianni made his first film with him, "Domenica d'agosto" in 1949. When I was a child and a boy I was always involved in film making, as collaborator, as organiser, even as an actor, in several of my father's movies. Both Almgren and Jean Taylor were very interested in my project. In any case my idea was not to make a "small" scientific film, a sort of scientific commercial just to show some small experiments with soap bubbles and soap films. I have never been able to stand these short films on mathematics (which have fortunately disappeared with the diffusion of computers), made to illustrate theorems or results of plane geometry or similar topics. These films are very boring and not very useful, not even for teaching mathematics at any level. I was attracted by the phenomena of soap films because they were visually interesting and I thought that the technique of filming them would have increased the general interest and fascination about them. I was not at all interested in just filming a lesson by Almgren and Taylor, with them explaining their results, inserting a few images of soap bubbles and soap films here and there. Almgren and Taylor shared my opinion. The project was not making any progress, because the motivation for making a film like this was not clear to me. Which was the purpose, if any; just the fascination of soap films? For which audience. And what did the length of the film have to be?

Now the second reason. I was working at the University of Trento while my family, Valeria and sons, lived in Rome. Every Friday I left Trento to go to Rome (seven hours by train) and then on Monday, I travelled back to Trento. I have been always a lover of art, of any kind, of any culture and period. Of course there are some artists that I prefer. When I was in Trento, I read in a newspaper of an exhibition, in Parma, dedicated to one of the most important artist of this century: Max Bill. I already knew some of the

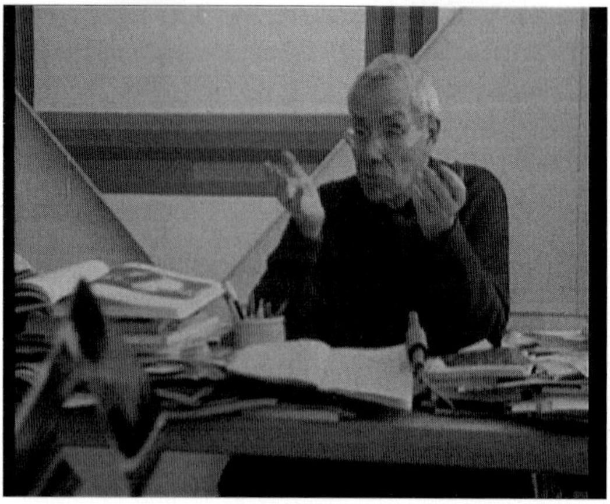

Fig. 2. [ME]2: Max Bill in his Zurich studio

sculptures of the Swiss artist but I had not visited a large exhibition like the one in Parma before. As the town of Parma was more or less on my way from Trento to Rome I decided to stop on my way back to Rome to see the exhibition. The topological sculptures of Bill were a real discovery for me. Years before, I had seen a large exhibition of the works of Henry Moore in Florence and of many other artists, but the ones of Bill almost immediately gave me the impression of *Visual Mathematics.* The *Endless Ribbon*, that Moebius Band, enormous and made of stone, granite, was a real revelation. Its shape, its physical nature, tridimensionally real, making it live in space. A mathematical form, alive. This was the idea that was missing: mathematics, mathematicians in all the historical periods and in all the civilisations have created shapes, forms, relationships. Some of these shapes and relationships are really visual, they can be made visible. The idea for the great success of the use of computer graphics in some sectors of mathematics. In these same years the mathematician Thomas Banchoff was making his first short films in animation of mathematical surfaces but at that time I was not aware of his work.

Coming back from Parma to Rome, I spoke again with Valeria. The project was becoming clearer: to make films, two perhaps, in which to compare the same theme from a mathematical and artistic point of view, asking for the opinion of mathematicians and artists. Not just filming a long discussion between artists and scientists on the theme that is so vague of the connections between art and science, but a real confrontation on the visual ideas of the artists and the mathematicians. To *make visible the invisible* like the artist David Brisson says in the film *Dimensions* made in 1984 with Thomas Banchoff. So the general plan of the project was almost clear: to make two films on the visual relationships of the forms created by artists and mathematicians. The themes of the two films were: soap bubbles, topology in particular the Moebius band. To have more visual ideas and objects to film we finally decided to include the connections between mathematics and architecture, all the other sciences, in particular biology and physics, not excluding literature and even poetry. And, why not, cinema. Just from the beginning of the project there was the idea of focusing on the cultural aspect of mathematics, the influence and the connections of mathematics and culture, of course starting from the point of view that mathematics has always played a relevant role in culture, being an important part of it. All these using the most important visual tool: filming. As these were the general lines of the project, it was quite natural to consider as part of it the organisation of exhibitions (many were made in the next years), congresses and seminars, the publishing of books (with many illustrations!), even theses for students in mathematics, in history of art, in architectures. Today, 25 years later, it is easy to say that the project went far beyond the expectations. Starting from 1997 at the University of Ca' Foscari in Venice, we organised an annual congress on Mathematics and Culture. From an idea that started in Torino after a discussion with Valeria, Odifreddi, a mathematician in Torino, and myself. A first, not so precise idea of such a congress was already included in the project of the seventies. In 1976, the whole project seemed a very absurd one for many reasons:

– to make a film was (and is) very expensive; one thing was very clear to me. I did not intend to make an amateur's film. I wanted to make a real professional movie, of high quality, and all the technicians involved had to be well qualified.

– I had started my professional career at the university and one of the most difficult things to do in an Italian university is to be involved in a field connecting two or more different areas. It can be the very quick end of your work at the university. This is still true today. But I was lucky because I was working on the *Calculus of Variations* and *Minimal Surfaces*, a field of great importance in the seventies. – Trying to obtain the collaboration of Italian mathematicians (for the reasons illustrated in the previous point) was very hard. It was considered not very professional for a mathematician to be involved in such a project. During the last ten years I have been invited to many Italian universities to show and discuss my movies. But when I first showed one of my movies in Rome, in 1981, to a public audience, mathematicians of my department told me that it was not good for the reputation of our department. This is the main reason why almost all my movies have been made abroad, in Europe, in the USA, in Canada, Japan, even in India. And the same is true for the publication of books and proceedings of congresses organised abroad or with the help of non-Italian mathematicians. This is the reason why it has been possible to organise the congress *Mathematics and Culture* in Venice, only in the last five years, not before. And in a few years the congress has become an important traditional meeting for mathematicians and students.

– As it is clear from the previous remarks, to obtain funds and support for the project from the Italian institutions, was a desperate feat (and still is in a sense). Notwithstanding all this, we started the project. The themes of the first two films were clear enough; we were looking for funds, this was the hardest thing to do; then it was necessary to choose the mathematicians and artists who would be involved in the project. Of course, the first thing was to obtain their collaboration. In those years RAI, the Italian public television, had a department specifically dedicated to educational projects called DSE, Dipartimento Scuola Educazione. They accepted the challenge to make a program on mathematics for the first time. Nobody before had proposed something on these topics, except the lessons filmed for strictly educational reasons. But they requested one condition: that we make a series of films. This is a magic word for television: series. If you want to do something for the Italian Television (but I have the impression it is the same everywhere) you have to propose a series. It is not important that you have no ideas about how to do a series, but you have ideas only for one or two episodes. So they asked me to make 8 films. I said no for the reason that I had no ideas to make the other six. Finally we reached an agreement to make 4 films, with an option to make 4 more in the following two years. Each film had to be 27 minutes long. The other topics we were thinking about were: Platonic Solids, Symmetry and Tessellations.

In these same years I had already discovered the works of the Dutch graphic artist Maurits Cornelis Escher. From the first time I saw his engravings my purpose was to make a film only on him. My idea was to use the technique of animation in order to make his works really tridimensional. Something that Escher himself suggested; he personally was involved in a short film with several animations of his works before his death in 1972. I discovered his ideas and the film only a few years later. The financial support of RAI was not sufficient to make my first four movies; it was out of question to make a film on Escher using animations, as this technique is very expensive. Financial support from RAI was enough just to make a film, entirely in a studio, with a person talking all the time; their idea of an educational TV series. My idea was to film all over the world where the artists and mathematicians involved in the film were working. My father, film-maker, was the producer of all of them; so I was able to find more supports. In any case the project of a film on Escher was postponed. I only filmed three works of Escher and inserted them in the film *Moebius Band*.

– It will take me more than ten years to complete the film on Escher in 1990. What we needed for the project was a title; it was quite natural to choose the general title "Mathematics and Art" (even if for RAI the title was changed because the Italian television, like the Italian Universities, cannot consider the idea of treating two topics at the same time; the problem is to have a precise target!).

This added more months to the start, but at the end of 1979 I was able to show a first and preliminary version of *Soap Bubbles* at a Scientific Film festival at the CNRS in Paris. Also the second movie on the Moebius band was almost ready. In the film on soap bubbles I asked the collaboration of the Italian artist Arnaldo Pomodoro, who has always been fascinated by the theme of Spheres, while in the Moebius band's film, apart from Max Bill, I filmed the works of Corrado Cagli, of the French designer Moebius.

I contacted Max Bill writing him a letter. He was very kind; he invited me with my troupe to his house in Zurich and he gave me permission to film everything I was interested in, including his fabulous collection of contemporary art. With one exception: it was strictly forbidden to film a little window in which there was his collections of forms, topological forms, made in paper. Very small objects, his Data Base for future works. He was afraid that someone could see his projects and copy them. We then became friends, we made two exhibitions together, and another film on *Ars Combinatoria.* We both were in the editorial board of the journal Leonardo, at that time published by Pergamon press, then by MIT Press. For my book *The Visual Mind: Art and Mathematics*, MIT Press, 1993 (4th edition), Bill rewrote the title and made same changes to his famous paper originally written in 1949 *A mathematical approach to art.* Two of Bill's works are reproduced on the front and back cover of the book. A new volume The *Visual Mind 2* will be published, always by MIT Press in 2002. The volume will be dedicated to Valeria and Max Bill. Of course it is very hard to describe a film using words, it is almost impossible, even not correct. If it is almost impossible to describe a film using words, it is good, because it means that the film has been made really using a visual technique, mixing, images, sounds, music in an essential and possibly unique way. If a film can be narrated it means that something is not working well from the visual point of view. One thing was really clear to me: in making the films, all words, all explanations had to be reduced to the minimum, or even be absent if possible. Whenever possible, images must speak for themselves. If, for its nature, art does not need explanations, mathematics too has to be presented almost without words. A film is not the best tool to explain and to learn. A film can, in a short amount of time, give ideas, suggestions, stimuli, emotions. A film can generate interest, even enthusiasm. Looking at an interesting, pleasant film can stimulate the audience to learn more, both in the artistic and the mathematics fields. In this sense I consider my films educational, but only with this meaning. This was also the reason why at the beginning the films were refused by RAI.

This, on the contrary, is the secret of their success, as for example for the movie *Soap Bubbles*, even 20 years after the film was made. In fact the most beautiful sequences I have ever made, the soap films dancing to Weber's *RosenKavalier* waltz was included in the VideoMathfestival selection for the World Mathematical Congress in Berlin in 1998, and in the European Congress in Barcelona in 2000.

In the making of the films of the series, in the last 25 years (actually the films are 22), I have had the help of many artists. I now want to give some examples of the kind of collaboration that I had; of course the best thing to understand this, is to look at the films. It is not enough to read what the artists are saying. For the occasion of the congress in Maubeuge I have assembled selected scenes taken from different films, making a selection of the artists, making in a sense a new film, new as each time you take some visual material and you manipulate it, you area creating something new following your sensibility at the time of the new edition. A film that can be called: Results of the last 20 years.

2 Moebius Band

There is no doubt that the clearest approach to the possibility of a mathematical approach to the arts has been formulated by the famous Swiss artist Max Bill. In 1949 he wrote: "By a mathematical approach to art, it is hardly necessary to say I do not mean any fanciful ideas for turning out art by some ingenious system of ready-reckoning with the aid of mathematical formulas.

So far as composition is concerned, every former school of art can be said to have had a more or less mathematical basis. Even in modern art, artists have used methods based on calculation, inasmuch as these elements, alongside those of a more personal and emotional nature, give balance and harmony to any work of art.

These methods had become more and more superficial, for the artist's repertory of methods had remained unchanged, except for the theory of perspective, since the days of ancient Egypt. The innovation occurred at the beginning of the twentieth century: it was probably Kandinsky who gave the immediate impulse towards an entirely fresh conception of art. As early as 1912... Kandinsky in his book *Uber das Geistige in der Kunst* indicated the possibility of a new direction which would lead to the substitution of a mathematical approach for improvisations of the artist's imagination... It is objected that art has nothing to do with mathematics; that mathematics, beside being by its very nature as dry as dust and as unemotional, is a branch of speculative thought and as such in direct antithesis to those emotive values inherent in aesthetics... yet art plainly calls for both feeling and reasoning."

We must not forget that Max Bill was first and foremost a sculptor who believed that geometry, which expresses the relations between positions in the plane and in the space, is the primary method of cognition, and can therefore enable us to apprehend our physical surroundings, so, too, some of its basic elements will furnish us with laws to appraise the interactions of separate objects, or group of objects, one to another. And again, since it is mathematics that lends significance to these relationships, it is only a natural step from having perceived them to desiring to portray them. Visualised presentations of that kind have been known since antiquity, and they undoubtedly provoke an aesthetic reaction in the beholder.

And here is the definition of what must be a mathematical approach to the arts "it must not be supposed that an art based on the principles of mathematics, such as I have just adumbrated, is in any sense the same thing as a plastic or pictorial interpretation of the latter. Indeed, it employs virtually none of the resources implicit in the term *pure mathematics*. The art in question can, perhaps, best be defined as the building up of significant patterns from the ever-changing relations, rhythms and proportions of abstract forms, each one of which, having its own causality, is tantamount to a law unto itself. As such, it presents some analogy to mathematics itself where every fresh advance had its Immaculate Conception in the brain of one or other of the great pioneers." To convince his readers, after having clarified his thoughts, Bill needed to provide some examples which pertained to his point of view as an artist, examples of what he called "the mystery enveloping all mathematical problems, the inexplicability of space that can stagger us by beginning on one side and ending in a completely changed aspect on the other, which somehow manages to remain that self-made side; the remoteness or nearness of infinity which may be found doubling back from the far horizon to present itself to us as immediately at hand; limitations without boundaries; disjunctive and disparate multiplicities constituting coherent and unified entities; identical shapes rendered wholly diverse by the merest inflection; fields of attraction that fluctuate in strength; or, again, the space in all its robust solidity; parallels that intersect; straight lines untroubled by relativity, and ellipses which form straight lines at every point of their curves.

Far from creating a new formalism, what these can yield is something far transcending surface values since they not only embody form as beauty, but also form in which intuitions or ideas or conjectures have taken visible substance.

Thus, the more succinctly a train of thought was expounded, and the more comprehensive the unity of its basic idea, the closer it would approximate to the prerequisites of the mathematical way of thinking. The orbit of human vision has widened and art has annexed fresh territories that were formerly denied to it. In one of these recently conquered domains, the artist is now free to exploit the untapped resources of that vast new field of inspiration. And despite the fact that the basis of this mathematical way of thinking in art is in reason, its dynamic content is able to launch us on astral flights which soar into unknown and still uncharted regions of the imagination."

One of the visual ideas that Max Bill used without realising it was the Moebius ribbon. Max Bill called some of his sculptures *Endless Ribbons* and in fact they are shaped like Moebius ribbons even if at that time Bill has no idea of the Moebius band. The interesting thing to note is that Bill thought he had invented a completely new shape. Even more curious was that he discovered it by twisting a strip of paper, just as the mathematician Moebius had done many years previously. Max Bill's Endless Ribbon was put on display for the first time at the Milan Triennale exhibition in 1936.

As a choice, in all my movies there are questions asked to the artists and mathematicians involved. Actually I prepared several questions, a sort of script to be followed in all the interviews. Then of course all the answers were edited, cut to my choice. The most important thing in a movie is that everything must seem natural, the only possible way of doing it. Of course not, it is the point of view of the filmmaker. So here are the words, better, the words I have used, that Max Bill said in his answers. "The first strip that I made is exactly the same as this one, which is a real Moebius strip, however that one was shorter and had a central support. A few years later I discovered that when making a Moebius strip one could orientate it in different directions. With these different possibilities of positioning one can obtain all sorts of variations of the same shape.

I made the first Moebius strip without knowing what it was. I made it by accident. I wanted to make a decorative object to put above an electric fire. An object that moved, so I tried to make it with paper, like a game for children. I tried to make something that would turn in the air and would give the impression of spiralling, and trying over an over again with the paper I came to a shape like this, a shape with only one surface which has all the characteristics of the Moebius strip. I started by making various works always based on the same idea. They are works where the external edge crosses the surface: for example, I take a surface with six angles, I join these angles and I have a complete circle: that way, all this group of figures takes on new characteristics which no longer have anything in common with the Moebius strip, they are only folded surfaces, cut according to my imagination."

We made another film together, *Ars Combinatoria*. In his words Bill explains what is happening in the part of the painting that does not exists, that he has never painted. Bill: "The idea of this picture is that it is a square balanced on a pivot. It is not a normal square as one can see by doing this like for windows, for doors. Here it is balanced on a pivot. When a figure is balanced on a pivot in space it becomes elongated in all four directions; that is the basic principle behind this square on a pivot. There's the other part, here, there are colours that cross it, there is an area of colours and what is beyond the colours is the same quantity as we have here ... .

The inside there is an octagon, and in this octagon there is this square; this square fixes the whole figure, and these areas cross the figure. And then outside there is an area with colours, and there are these triangles. These triangles are equivalent that has been cut away here on the side. So inside there are rhythms made up of a rotation. It is a rotation from yellow to red, as far as blue and green here on the inside; when one looks at it from the outside one sees the same series, there is yellow, red, blue and green. And on the inside there's this composition using colours, the green in the blue, and here there's yellow in greenand one has a rotation similar to the previous one. So this is the system, this lengthening of strips and one creates a mysterious area because we do not know what is happening here on the outside, one can imagine that there is mixture of colours but we can't be sure because this has been cut. So there is something mysterious about it, because this doesn't exist yet, but there is a tendency that's increasing, that's growing there, somewhere ... .

A few years later I studied the possibility of fixing thought precisely, in a way that wasn't too personal, that in a sense is objective. At that time, I was looking for the basis of the development of figures, and I went back to music knowing that it is based on the laws of mathematics, and that it is possible to find rules behind the organisation of a theme. This was the beginning of the idea of variations. I then tried some types of variations but in a limited system. In other works, I tried to push a theme to its limit. For instance, I did something with squares and, once I had carried out the operation that I had established as the basis of the theme, once it had been repeated eleven times, then it was exhausted, there was no other new variation to be tried."

In the film *Ars Combinatoria* another famous Italian artist, Luigi Veronesi, explains his methods for variations: "How did I came to the subject of variation? I took the advice of Leger, the French artist, when I was studying under him in Paris. This was his advice, Veronesi, don't stop at a single image, but consider the image you are thinking of, that you have in your mind, as a theme on which to work variations, exactly like a musical theme. Working like this is the only way to see an image from every side, from every aspect. It's an advice that I accepted with enthusiasm, so much so, that 50 years later I am still working with idea of variations. I can say that one of my first positive experiences was the series of 14 variations on a pictorial theme that I did in 1936 and that the Italian musician Malipiero later set to music. The research that I have been carrying out for many years is into the relationship between sound and colours, studied on a mathematical basis.

Here are some chromatic variations on a rectangle, considered as a minimal element, amongst the least evocative geometrical figures. For me, the rectangle is one of the minimal elements of my compositions, not only in the resolution of chromatic differences between sound, but also in all my paint-

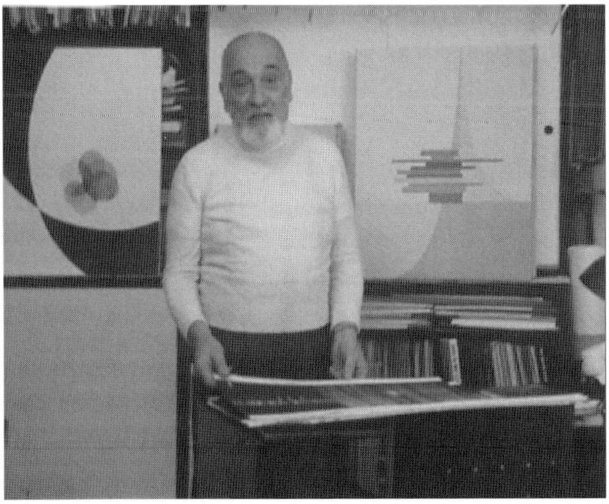

Fig. 3. [ME]3: Luigi Veronesi in his Milan studio

ing. Of course I do not only use the rectangle; I also use curved lines. In fact there's hardly a picture of mine in which there are no circles and other curved lines, parabolas, ellipses, hyperboles and so on. Sometimes I also use triangles and squares but mostly as a counterbalance to the other figures that make up the composition."

3 Labyrinths

In the film on this theme, the Italian painter Fabrizio Clerici realised a very large painting for the film and for the exhibition "The eye of the Horus" that was held in Italy in 1990. I made a film for the Italian public television, RAI, in the series "The great exhibitions of the year" on the exhibition. For the anthology exhibition of Clerici, at the National gallery of Modern Art in Rome I made a short animated movie on Clerici's combinatorial drawings using music with variations by Wolfang Amadeus Mozart. These are the words of Clerici on the theme of the labyrinth: "I think that for an artist who was first educated as an architect, began as an architect and wound up as a painter, for him the labyrinth is a natural step: it's the kind of architecture a painter is continually rebuilding in various ways. I return to it periodically. I think I drew my first labyrinth when I was a child, unthinkingly, without knowing it was a labyrinth; like a geometrical design. But an actual labyrinth, a construction in the form of a puzzle, a kind of trap, that happened much later, around 1945, just after the war. For a long time I was in doubt about what to place in the centre of these corridors, this amphitheatre, and then suddenly the solution came all by itself. Since the edifice was a labyrinth, the personage that must dominate the entire situation had to be the Minotaur; but why the Minotaur all alone in the centre of the arena? Because it is accusing someone and that someone can be no other but the mother which bore this monster.

The theme of the Minotaur accusing the mother was a much wider theme but it didn't stop there. I have played with this theme in many other versions, with or without the Minotaur. Sometimes a labyrinth is made up of letters of the alphabet, or of stones, stones that are created with a tendency towards gently, curving movement, to meandering.I have made several of them and I'll go on making them. We say meander for anything that is a function of the labyrinth so I have decided to create the Meander, right on the river bed itself, a labyrinth as though the river itself were the Meander, that's the word and motif that I've given to this contorted, endless construction."

4 Dimensions

This film was made with Thomas Banchoff and Linda Henderson. The two American artists Harriet and David Brisson were involved in the film. David: "The reason why I am interested in visualisation of the materials is that a lot of it is hard to understand in a verbal sense. But when they were made into visual images then it was possible for me to understand very complicated ideas. And I felt that there were a lot of other people who could profit by making or by having the experience of these ideas expressed to them visually when they were not ready to understand these ideas in purely verbal or mathematical terms. A lot of these mathematical ideas are very interesting and beautiful when converted to 3 dimensional form, such as this projection which is not quite correct, of a 4 dimensional figure, the analogue of a ball, this is a more correct one, which gives you the same form. My personal interest is in visualising ideas so that I can understand them easier because I am a visual person."

5 Final comments

In this paper I was interested in putting together some of the ideas that were the basis for the beginning and the making of the project *Mathematics and Art*: for the exhibitions, for the books, for the congresses and in particular for the films series. It is not so easy to find, in the last twenty-five years, a common and precise line of evolution of the arts that are more closely linked to science, primarily because this common line probably does not exist. In any case the interviews, the filming of the artists and their works, seen some years later, seem to me an interesting and stimulating heritage of how mathematics has had a cultural influence for many artists. This aspect can be of course of general interest for the arts tout-court. In this paper and in the video I have prepared I made a selection of the artists with whom I have been working in the last years. I have not included any mathematicians even if they are the most important motivation for all the project on mathematics and art. This was done first because I have written elsewhere about my collaborations with mathematicians on the project, and also because I was interested in choosing some of the artists with whom I was so lucky to work. With the publication of the volume *The Visual Mind: Art and Mathematics* I have tried, with the help of artists and mathematicians, to indicate the tendencies of evolution of the connections between art and mathematics in the last years. In the year 2002, a new book will be published *The Visual Mind 2: Art and Mathematics*, always by MIT Press. A new occasion to look to the future, to try to understand the new ideas for the new Millennium. In this paper my idea was to look at the recent past and remember some of my fellow-companions; without all of them my ambitious project could not have even been started.

References

BOOKS, VOLUMES AND SPECIAL ISSUES

[1] H.S.M. Coxeter, M. Emmer, R. Penrose, M. Teuber, Eds. *M.C. Escher: Art and Science*, Proceedings of the congress, Amsterdam, North-Holland, 1986

[2] M. Emmer ed., *Visual Mathematics*, special issue "Leonardo", Pergamon Press, Oxford, vol. 25 n. 3/4, 1992
[3] M. Emmer, *La perfezione visibile: matematica e arte*, Edizioni Theoria, Roma, 1991
[4] M. Emmer, *Le bolle di sapone: viaggio tra arte, scienza e fantasia*, La Nuova Italia Editore, Firenze, 1991
[5] M. Emmer, *La Venezia perfetta*, Centro Intern. della Grafica, Venezia, 1993
[6] M. Emmer, *The Visual Mind: Art and Mathematics*, The MIT Press, Cambridge,1993, 4th edition
[7] M. Emmer, V. Marchiafava, eds., *Ricordando Fabrizio Clerici*, Academia di S. Luca, Centro Int. Grafica, Venezia, 1994
[8] M. Emmer, *Matematica e Cultura*, Univ. Ca' Foscari Venezia, Lettera Matematica, Springer, Milano,1998
[9] M. Emmer, *Matematica e Cultura 2*, Springer, 1999
[10] M. Emmer, *Visual Mathematics*, special issue, Int. J. Shape Modeling, vol. 5, n. 1, June 1999
[11] M. Emmer, *Matematica e cultura 2000*, Springer Italia, Milan 2000
[12] M. Emmer, D. Schattschneider, *M.C. Escher: a centennial congress*, to appear Springer

CATALOGUES OF EXHIBITIONS

[1] M. Emmer, C.Van Vlandereen Eds. *M.C. Escher, Catalogue*, Istituto Olandese, Roma, 1985
[2] M. Calvesi, M. Emmer Eds. *I frattali: la geometria dell'irregolare, Catalogue*, Ist. della Enciclopedia Italiana, Roma, 1988
[3] M. Emmer, Ed. *L'occhio di Horus: itinerari nell'immaginario matematico*, Istituto della Enciclopedia Italiana, Roma, 1989
[4] M. Emmer, *L'enigmatico fascino di M.C. Escher, Catalogue*, Futuro-Remoto; Napoli, CUEN, 1989
[5] M. Emmer, *L'Italia di Escher, Catalogue "Escher 1898–1998"*, Ravello, luglio–agosto 1998, Diagonale ed., Roma, pp. 12–13
[6] M. Emmer, *Escher a Roma, ancora, Catalogue "Homage to Escher"*, 24–26 giugno, Museo laboratorio di Arte Contemporanea, Univ. Roma "La Sapienza", Diagonale ed., 1998, pp. 12–13
[7] M. Emmer, Ed., *Homage to Escher: The Leonardo gallery*, mostra virtuale, http://mitpress.mit.edu/e-journals/Leonardo/gallery, aprile 2000

VIDEOS AND FILMS ON ART AND MATHEMATICS

[1] Moebius Band (1979)
[2] Soap Bubbles (1979)
[3] Platonic Solids (1979)
[4] Symmetry and tessellations (1979)
[5] Dimensions (1982)
[6] M.C. Escher: Symmetry and Space (1982)
[7] Spirals(1982)
[8] Helices (1982)
[9] Ars Combinatoria (1984)

[10] M.C. Escher: Geometries and impossible worlds (1984)
[11] Knots (1984)
[12] Geometry (1984)
[13] Flatland (1987)
[14] Labirynths (1987)
[15] Computers (1987)
[16] L'avventura del quadrato (1987)
[17] Figure geometriche (1987)
[18] L'occhio di Horus (1989)
[19] Metamorfosi, di Fabrizio Clerici (1990)
[20] La Venezia perfetta videotape, 20 minuti (1993)
[21] The Fantastic World of M.C. Escher (1994)
[22] F. Armati, M. Emmer, eds. *Ricordando Fabrizio Clerici*, AICS-Accademia di S. Luca, Roma, 1994
[23] *Matematici* in due parti, IDIS Napoli, 1996
[24] *Ennio De Giorgi*, intervista, UMI, 1h 10' (1997)

Guided Tours of Buried Galleries (Inside a Computer)

Jean-François Colonna

CMAP/Ecole Polytechnique, France Telecom R & D, 91128 Palaiseau Cedex
<http://www.lactamme.polytechnique.fr>,
<colonna@cmap.polytechnique.fr>

1 Introduction

The title of this text can seem quite mysterious, but it reveals the double reading it is possible to make of the pictures we will describe. As a matter of fact, most of them are computed in order to be used as scientific or pedagogic tools; but that does not exclude to carry them out under aesthetic constraints while being attentive with proportions, with colors, ... even from time to time by forgetting Science. The word gallery must thus be understood with to quite different meanings which will be defined hereafter.

2 Art Galleries

The art galleries buried inside computers are those which offer pictures for the only pleasure of the eyes, by avoiding the description of the underlying mathematical proceses. But which is then the status of these pictures: are they works of art in the classical sense of this term?

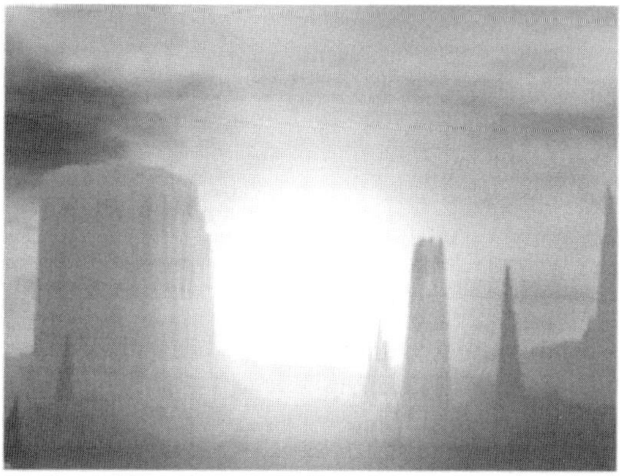

Fig. 1. [JFC]1: *Monument Valley at sunset*
http://www.lactamme.polytechnique.fr/Mosaic/images/PAYS.u9.M.D/display.html

It is advisable to recall initially that they lose here a fundamental property (that gives part of their values): the one of uniqueness; with the pictures known as digital, any copy is as perfect as the original. They lack also a noble medium: here nothing like the Carrara marble ... But beyond these negative characteristics, there is a point which seems very promising: the ermergence of the concept of potential work of art. As a matter of fact, I am myself an "algorithm conceptor" and the result of the creative process is not one or more pictures produced by a certain program, but rather well this last one (the program) which thus must be seen as the potential container of a quasi-infinite (since the infinity does not exist for a computer) number of pictures of the same type. For instance, this is the case of a N-dimensional fractal field generator I have conceived; it is able to produce an incredible variety of natural phenomena: mountains, clouds, ... (see Fig. 1). This concept, which Jorge Luis Borges would certainly have loved, is unfortunately difficult to explain to the general public and thus it is mandatory to produce some "objects" more easily communicable (some pictures) that will be exhibited in actual Art Galleries and more often (unfortunately?) in virtual ones.

3 Mine Galleries

Beside the Virtual Art Galleries buried inside computers, more mysterious Mine Galleries furrow the memories of our digital machines. A mine, in its most common sense, is a place where one digs the ground in the hope to discover wealthes or even treasures; but what is this furrowed ground here? It is made of Mathematics which is a fundamental element structuring our perception of the Reality; these Mathematics have become during the past centuries the language of paramount importance with which Science describes the studied phenomena. Even if this language is certainly not the Ultimate Reality, but rather well a reflection of our own cognitive processes, it enabled us to imagine the Infinity and to see beyond the horizon ...

Virtual Experimentation is a recent scientific approach. It starts with the mathematical model of a certain physical phenomenon, then, in order to solve the equations it contains, some so called numerical methods must be applied. At last, all this is translated into computer programs that, after debugging, are exploited in order to produce results. Most of the time, their analysis is not obvious due to the large quantity of digits they contain; this problem can only be solved using the techniques of picture synthesis. Thus, modifying (and playing) with the model parameters and seeing the model evolution on a screen is a new kind of experimentation.

This approach is very promising as well at the fundamental scientific level as at the industrial one. One must remember the prophetic words of Heinrich Hertz that said during the nineteenth century: "one cannot escape the feeling that these mathematical formulas have their own existence, that they are more erudite than those who discover them and that we can extract more science from them than it was input at their creation". This is also true for the virtual experimentation, as well for the models as for the programs used, but on the very important condition of knowing well the dangers and the limits it contains.

With regard to the mathematical and computational aspects, it is obvious that the real numbers are fundamental. By the way, this omnipresence is quite mysterious: as a matter of fact, for instance when speaking only about distances, what is the usefulness of the infinite precision of the real numbers, whereas the Universe seems to measure only $1.5\ 10^{61}$ (using the Planck length as the measurement unit)? Their use seems to come to the need to have differential equations that are obtained by the means of the manipulation of infinitely small numbers. But (unfortunately) a computer is only a finite machine (its memory, although more and more important, is physically limited) and the manipulated values must be sampled. Thus, for these machines, the infinity and the continuity do not exist, whereas they are essential to us! One of the consequences of this fact is that some fundamental properties are lost: this is the case for the associativity of the addition and of the multiplication of numbers. To forget that can lead, in certain circumstances (case of the problems known as sensitive to initial conditions, see Fig. 2), to some nuisances (euphemism ...).

The picture synthesis, contrary to a naive intuition, is also source of difficulties. As a matter of fact, we are here well far from the applications praising the qualities of a car or of a package of detergent. The so called Scientific Visualization has very often to exhibit abstract objects (mathematical structures for instance) and thus without a priori images. But does that want to say that each "natural" object has its own image? Unfortunately not: this is

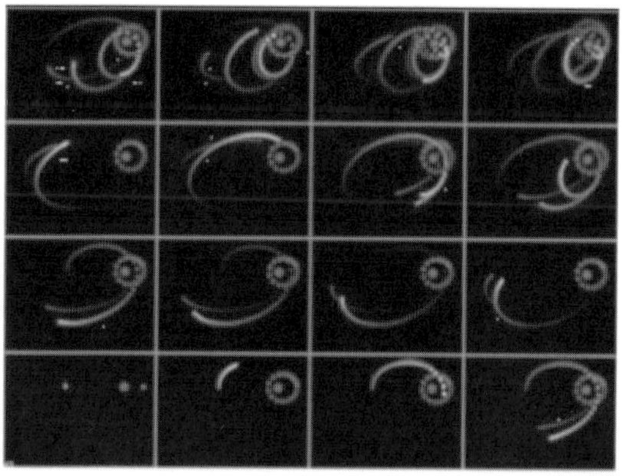

Fig. 2. [JFC]2: *Study of the N-body problem with $N = 4$ (a binary system made of two stars and two planets). The same computation using the same program is made on three compatible computers. Their results are superposed assigning to each computer a fundamental color (Red, Green and Blue)*
http://www.lactamme.polytechnique.fr/Mosaic/images/NCOR.X7.16.D/display.html

obvious in the quantum mechanics realm, but more close to us, what is the color of a pressure field?

This question could seem absurd, but it must be answered in order to display that kind of results! The morality of all that is quite simple: the visualization of scientific "objects" is in most cases arbitrary and thus subjective. Then it very simple to build "orthogonal" (inconsistent) views for a given object (see Fig. 3).

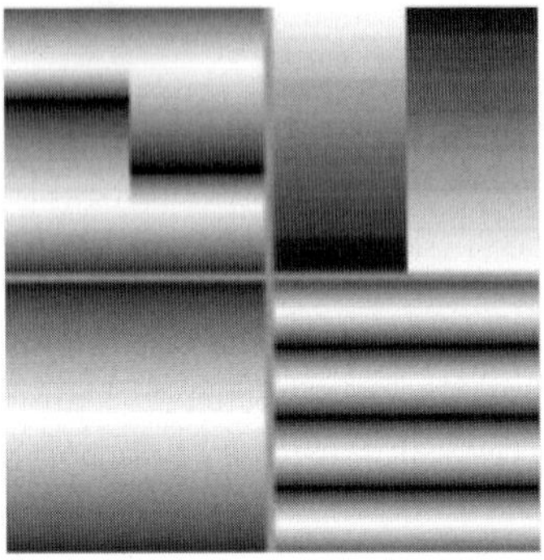

Fig. 3. [JFC]3: *A same bidimensional scalar field is visualized using four different color palettes. The four views thus obtained exhibit incompatible properties when the underlying field is unique!*
`http://www.lactamme.polytechnique.fr/Mosaic/images/PARADOXE.11.D/display.html`

4 Follow the Guide

The techniques of the Virtual Experimentation when using relevant (and refutable) models, robust numerical methods, reliable programming (maybe the most difficult task ...), fast computers and at last the most neutral picture synthesis, make possible the Virtual Space-Time Travel. The exploration of all known scales, from the elementary particles to the whole universe, become possible (see Fig. 4).

5 Conclusion

The Mathematics, benefiting from the fabulous progress of the Computer Science, both play of the spirit and an open window on Reality, are also simultaneously a fabulous object and a tool for creation: object because they propose to the artist new sources of inspiration; but also tool because they give to him, just like to the scientist, new means of expression that can be masterized on condition their limits are well known.

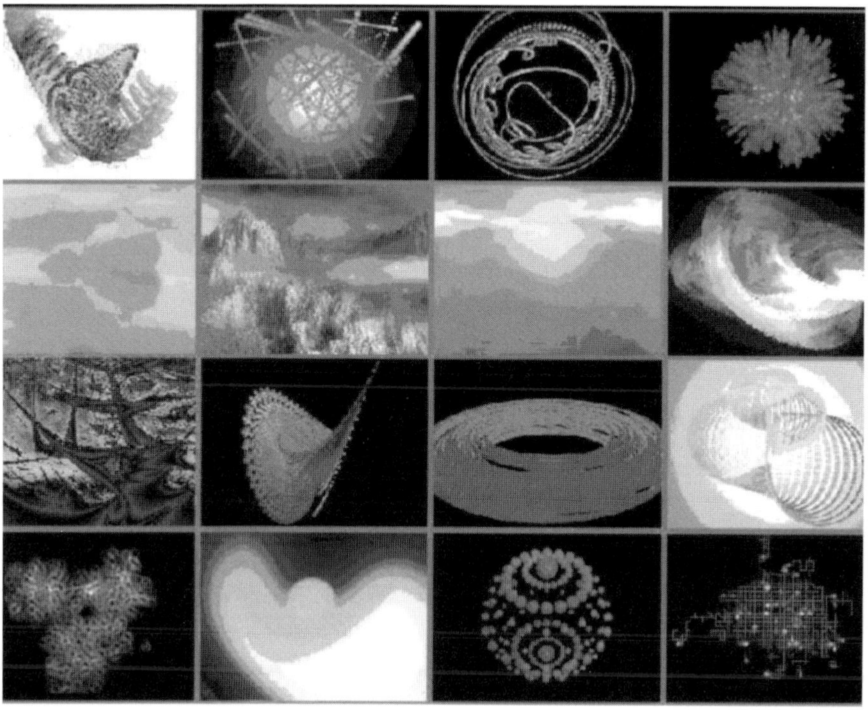

Fig. 4. [JFC]4: *From the quark and gluon structure of the nucleon (bottom left) to the Universe (top right)*
`http://www.lactamme.polytechnique.fr/Mosaic/images/DEMO.31.16.D/display.html`

A Mathematical Interpretation of Expressive Intonation

Yves Hellegouarch

Département de Mathématiques, Université of Caen, Campus 2, Bld du Maréchal Juin, 14000 Caen, France

1 Introduction

In their talks, Erich Neuwirth and Carlotta Simões have given a description of diverse musical temperaments and, respectively, of some uses of equal temperament.

Presently I will describe a mathematical frame in which an old musical practice known as "expressive intonation" by singers and string players can find an explanation which would be impossible in any system of equal temperament. This does not mean that a phenomenon of the sort cannot occur in piano playing, but in this case it belongs to the concept of homonymy: each black key being a sort of "double entendre" as it is in Beethoven's Moonlight Sonata where the black key which is just above C is considered to mean $C^{\#}$ in the first and last movements and to mean D^b in the second.

We will show how players who are not slaves to a fixed pitch, may dispel this ambiguity.

2 "Petite fleur"

I have chosen to let the audience listen to a well-known piece of popular music as an example of this practice: this is "Petite Fleur" played by S. Bechet on an alto saxophone.

The theme of the refrain begins by the top E^b of the minor sixth $\dfrac{E^b}{G}$. According to all treatises this minor sixth is attracted to the perfect fifth $\dfrac{D}{G}$. The musicologist D. Cooke says in [2] that the expectation of the perfect fifth in this context creates a sentiment of anguish, and this can be checked if you pay attention to the meaning of the words underneath the notes of "Petite Fleur".

Each time this refrain reappears, S. Bechet tries to emphasize in his way of playing this sentiment of expectancy. Among other things, the listener notices that his E^b is intentionally flat: too flat according to equal temperament but nevertheless marvellously in tune!

H. Helmholtz ([5], p. 428) was perfectly aware of this shortcoming of equal temperament, when he wrote:

"When the organ took the lead among musical instruments it was not yet tempered. And the pianoforte is doubtless a very useful instrument for making the acquaintance of musical literature, or for domestic amusement, or for accompanying singers. But for artistic purposes its importance is not such as to require its mechanism to be made the basis of the whole music system".

"I think that many of our best musical performances owe their beauty to an unconscious introduction of the natural system, and that we should oftener enjoy their charms if that system were taught pedagogically, and made the foundation of all instruction in music, in place of the tempered intonation which endeavours to prevent the human voice and bowed instruments from developing their full harmoniousness, for the sake of not interfering with the convenience of performers on the pianoforte and the organ".

3 Scale Constructions

The tuning of string instruments is usually based on the pure fifth, which is the interval $\frac{3}{2}$. The reason for this is that it is very easy to produce experimentally, as you only have to check that the harmonic of order 2 of the upper string coincides with the harmonic of order 3 of the lower string.

If we extend this construction mentally in both directions, we find the fractions of a subgroup of $\mathbb{Q}_+^*$ generated by $\frac{3}{2}$, subgroup usually noted $\langle\frac{3}{2}\rangle$ by mathematicians.

If we translate these intervals in the first octave (the interval $[1, 2[$) modulo the powers of 2 (namely the elements of the group $\langle 2\rangle$) we can notice that, although there are near coincidences, all the fractions we get are different.

Expressed in other terms the "equation of the musicians" namely:

$$\left(\frac{3}{2}\right)^x = 2^y$$

with $(x, y) \in \mathbb{Z}^2$, is impossible except for the trivial solution $(x, y) = (0, 0)$.

This was well known to Pythagoras, but at the same time Pythagoras noticed that $(x, y) = (12, 7)$ was a very good approximate solution since $\frac{3^{12}}{2^{19}}$ is very near 1. We will call this quantity "Pythagoras' comma" and denote it by ϖ. As D. Cooke puts it in [2] p. 44:

"We may say that whereas musically we want the equation $\frac{3^{12}}{2^{19}} = 1$, the correct mathematical equation is $\frac{3^{12}}{2^{19}} = \left(\frac{3}{2}\right)^{12} \times \frac{1}{2^7} = 1,013642\ldots$".

Solving "mathematically" this impossible equation can be done in several ways:

1) the official solution, which consists in taking "2" $= 2$ and "$\frac{3}{2}$" $= 2^{\frac{7}{12}} \in \langle r_0\rangle$ with $r_0 = 2^{\frac{1}{12}} \in \mathbb{R}_+^*$.

This can be given a rigourous mathematical meaning if you consider the homorphism h_0 of the free abelian group $\langle 2,3\rangle$ on $\langle r_0\rangle$ given by:

$$h_0(2^x 3^y) := r_0^{12x+19y} \in \langle r_0\rangle$$

The theorem of isomorphism tells us then that:

$$\mathrm{Im}(h_0) = \langle r_0\rangle \cong \langle 2,3\rangle / Ker(h_0)$$

with $\mathrm{Ker}(h_0) = \langle \frac{3^{12}}{2^{19}}\rangle = \langle \varpi\rangle$.

We will say that $\langle r_0\rangle$ is the official tempered scale.

2) the solution of S. Cordier [3] which is similarly constructed when you consider h_c given by:

$$h_c(2^x 3^y) := r_c^{12x+19y} \in \langle r_c\rangle$$

with $r_c := \left(\frac{3}{2}\right)^{1/7} \in \mathbb{R}_+^*$.

As above we deduce that:

$$\mathrm{Im}(h_c) = \langle r_c\rangle \cong \langle 2,3\rangle / \mathrm{Ker}(h_c)$$

with $\mathrm{Ker}(h_c) = \langle \frac{3^{12}}{2^{19}}\rangle = \langle \varpi\rangle$.

We will say that $\langle r_c\rangle$ is S. Cordier's tempered scale.

3) The above isomorphisms suggest to consider an "**abstract scale**" which is:

$$\langle 2,3\rangle / \langle \varpi\rangle$$

when ϖ denotes Pythagoras' comma $\frac{3^{12}}{2^{19}}$. From what we have already said we know that this quotient group is isomorphic to $\mathbb{Z}$ and we can check that it is generated by the class of $\frac{2^8}{3^5}$. In fact it can be verified that

$$h\left(\frac{2^8}{3^5}\right)^{12} = h(2)$$

where h means either h_0 or h_c, and this means that the twelfth power of the class of $\frac{2^8}{3^5}$ is the class of the octave.

These solutions can be summed up in a same diagram:

$$N \hookrightarrow G \underset{\varphi}{\overset{s}{\rightleftarrows}} \mathbb{Z}$$

where G means $\langle r\rangle$ in the case of the tempered scales and $\langle 2,3\rangle$ is the case of the abstract scale, where N means $\langle 1\rangle$ in the case of tempered scales and $\langle\varpi\rangle$ in the case of the abstract scale, where φ is either $r^n \longmapsto n$ or $2^x 3^y \longmapsto 12x + 19y$ and where s is a section chosen such that $\varphi \circ s = \mathrm{id}_{\mathbb{Z}}$. Mathematicians say that G is an extension of N by $\mathbb{Z}$. In both cases the extensions are "trivial" in the sense that we can take for s a homomorphism (the extension is "split").

But, in the second case, musicians do not take for s a homomorphism!

L. Euler helps us to understand the musical choice we have to make by the following remark:

"The sense of hearing is accustomed to identify with a single ratio, all the ratios which are only slightly different from it, so that the difference between them be almost imperceptible".

By "difference" Euler naturally means "interval" or equivalently the quotient of the two ratios.

So, in the case of the abstract scale, we decide to take for $s(\varphi(x))$ the "simplest" ratio ρ in the class of x modulo N.

But what meaning shall we attach to this concept of simplicity?

We will chose ρ as close as possible to 1 for the "harmonic distance" on $\mathbb{Q}_+^*$ (see [4], §1 for the definition of this distance) or, equivalently, we will take $\rho = \dfrac{n}{d}$ (the fraction being reduced in its simplest terms) with $\sup(n, d)$ minimal in the class of x modulo N.

It turns out that this choice of ρ is unambiguous in the sense that there is only **one** element in xN which satifies our condition.

The first twelve values of $s(n)$ are:

n	0	1	2	3	4	5	6	7	8	9	10	11
$s(n)$	1	$\frac{2^8}{3^5}$	$\frac{3^2}{2^3}$	$\frac{2^5}{3^3}$	$\frac{3^4}{2^6}$	$\frac{2^2}{3}$	$\frac{3^6}{2^9}$	$\frac{3}{2}$	$\frac{2^7}{3^4}$	$\frac{3^3}{2^4}$	$\frac{2^4}{3^2}$	$\frac{3^5}{2^7}$
name	C		D		E	F		G		A		B
		D^b		E^b			$F\#$		A^b		B^b	

and we can check that the intervals of the C major scale are exactly the ones given by E. Neuwirth in his empirical construction.

Remarks

1) As we noticed in the introduction there is an ambiguity in the naming of the notes which do not belong to the C major scale. This is the reason why we placed their names on a lower line. We will go back to this later.

2) This construction is the paradigm of an infinite series of Pythagorean chromatic scales which have respectively:

$$1, 2, 5, 12, 41, 53, \text{ etc.}$$

degrees in one octave.

The usual Pythagorean scale is the fourth in this series and the next one (with 41 degrees in an octave) is called Janko's scale. It is given by the diagram:

$$N \hookrightarrow G \underset{\varphi}{\overset{s}{\rightleftarrows}} \mathbb{Z}$$

with $N = \langle \frac{2^{65}}{3^{41}} \rangle$ and $G = \langle 2, 3 \rangle$.

The generator of G/N is the class of $\varpi = \frac{3^{12}}{2^{19}}$.

The "concrete scale" $s \circ \varphi(G)$ begins as follows:

0	1	2	3	4	5	6	7
1	$\frac{3^{12}}{2^{19}}$	$\frac{2^{27}}{3^{17}}$	$\frac{2^8}{3^5}$	$\frac{3^7}{2^{11}}$	$\frac{3^{19}}{3^{30}}$	$\frac{2^{16}}{3^{10}}$	$\frac{3^2}{2^3}$
C			D^b	$C\#$			D

So, in Janko's scale, the Pythagorean diatonic semitone $\frac{D^b}{C}$ is $\frac{3}{7}$ of the whole Pythagorean tone $\frac{D}{C}$ and the chromatic semitone $\frac{C\#}{C}$ is $\frac{4}{7}$ of the whole tone. But we will come back later to the definition of those semitones.

3) The sixth scale is Mercator's scale. It is given by the diagram:

$$N \hookrightarrow G \underset{\varphi}{\overset{s}{\rightleftarrows}} \mathbb{Z}$$

with $N = \langle \frac{2^{65}}{3^{41}} \rangle$ and $G = \langle 2, 3 \rangle$.

The generator of the "abstract" scale G/N is the class of $\frac{2^{65}}{3^{41}}$.

One can check that, in Mercator's scale, the Pythagorean whole tone $\frac{3^2}{2^3}$ is divided in nine parts and that the diatonic semi-tone is $\frac{4}{9}$ of a whole tone while the chromatic semi-tone is $\frac{5}{9}$ of a whole tone.

4) Finally if we consider the ratio of the logarithms of the chromatic semitone and of the whole tone we find: $0,4425\ldots$ and it can be proved that it is the limit of the number of degrees contained in those intervals when the scale tends to infinity in the Pythagoric series referred above.

As a comparison we have:

$$\frac{3}{7} = 0,4285\ldots < \frac{\log C\#}{\log C} = 0,4425\ldots < \frac{4}{9} = 0,4444\ldots$$

So the theories about Holder's comma seem consistent with this construction.

4 Playing "Petite Fleur" in the Pythagorean Scale

The score of "Petite Fleur" shows that the first note is an E^b and that the next chord contains D and $F\#$, so the abstract scale in which it is written is G minor, and this has to be present in the mind of all the listeners from the beginning.

A player like S. Bechet (I mean a good player) will play the E^b in a way which will make it quite different from a $D\#$ (to suppress the homonymy) and the $F\#$ in a way which will make it quite different from a G^b: this can be done in the Pythagorean scale but not in a tempered scale !

And in so doing this player will follow the teaching of masters like P. Casals [1] or the findings of musicologists like D. Cooke [2].

5 Mathematical Interpretation of Expressive Intonation

Sharps and flats are usually introduced by transposing by successive fifths the C major scale. When you want to introduce sharps and flats in the Pythagorean scale you must therefore multiply the ratio by $\frac{3}{2}$ several times (translation).

In fact the practice of transposition of the scale of C major was very popular in the XIXth century, when it was taught systematically by the Tonic Sol-faists (see [5] p. 422-428) as the basis for natural intonation. Let us introduce $F\#, C\#$ and B^b.

Up to now there are no problems of homonymy, but when they do occur we see that:

$$\frac{F\#}{G^b} = \frac{C\#}{D^b} = \frac{G\#}{A^b} = \frac{D\#}{E^b} = \frac{A\#}{B^b} = \frac{E\#}{F^b} = \frac{B\#}{C^b} = \varpi$$

We can observe here the intervention of what the mathematician call the **factor system** attached to the section s_p.

	C	$C\#$	D	E	F	$F\#$	G	A	B^b	B	C
C maj	1		$\frac{3^2}{2^3}$	$\frac{3^4}{2^6}$	$\frac{2^2}{3}$		$\frac{3}{2}$	$\frac{3^3}{2^4}$		$\frac{3^5}{2^7}$	2
G maj	1		$\frac{3^2}{2^3}$	$\frac{3^4}{2^6}$		$\frac{3^6}{2^9}$	$\frac{3}{2}$	$\frac{3^3}{2^4}$		$\frac{3^5}{2^7}$	2
F maj	1		$\frac{3^2}{2^3}$	$\frac{3^4}{2^6}$	$\frac{2^2}{3}$		$\frac{3}{2}$	$\frac{3^3}{2^4}$	$\frac{2^4}{3^2}$		2
D maj		$\frac{3^7}{2^{11}}$	$\frac{3^2}{2^3}$	$\frac{3^4}{2^6}$		$\frac{3^6}{2^9}$	$\frac{3}{2}$	$\frac{3^3}{2^4}$		$\frac{3^5}{2^7}$	

Consider the diagram:

$$N \hookrightarrow G \underset{\varphi}{\overset{s}{\leftrightarrows}} \mathbb{Z}$$

with $\varphi \circ s = id_{\mathbb{Z}}$.

The "system of factors" of s is the application:

$$\mathbb{Z} \times \mathbb{Z} \xrightarrow{\nu} N$$

given by:

$$\nu(m,n) := \frac{s(m)s(n)}{s(m+n)}$$

We can check that it is a $2-$cocycle on $\mathbb{Z}$ with values in N. It is a well-known fact that the homology group $H^2(\mathbb{Z}, N)$ is null (see [7] p. 249). So ν is also a $2-$coboundary (which we already remarked since the extension is trivial).

In the case of equal temperament, $N = \{1\}$ and everything is trivial, so there is **no** expressive intonation in a temperate system.

But in the case of $N = \langle \varpi \rangle$, the choice of the **musical** section s_p gives a non constant **musical** factor system ν_p since:

$$\frac{C\#}{D^b} = \frac{s_p(25)s_p(-24)}{s_p(1)} = \varpi \neq 1.$$

So what musicians do unconsciously is to express their feelings through a factor system!

6 Pure Intonation

The "pure scale" (or Zarlino's scale) can be defined in the same frame:

$$N \hookrightarrow G \underset{\varphi}{\overset{s}{\leftrightarrows}} \mathbb{Z}$$

with $N = \langle \varpi, \delta \rangle$ and $G = \langle 2, 3, 5 \rangle$, where δ means the Didymus (or the syntonic) comma $\frac{3^4}{2^4.5}$. The musical section s is defined in the same way as before and we get a similar system of expressive intonation arising from the factor system of our new s, although there are some minor differences with the Pythagorian expressive intonation. Zarlino's scale seems to be well adapted to the works of certain composers like Mozart (see [4], §6 and [5] p. 327) whose music requires great harmonic purity.

7 Conclusion

When an artist is faced with a particular score it seems that a natural question he should ask himself is to know in which temperament the composer was hearing (in his inner ear) the symbols he was putting on paper.

In the case of the composers of the second Viennese school (Schönberg, Berg, Webern) there is no doubt that they were thinking in the equal temperament.

It may happen that some work of our composer has been recorded with his approval, then the style of this performance can give a clue to our question, as it was the case with "Petite Fleur" and its Pythagorian intonation.

In other cases (a certain quartet by Mozart for example) a study of the score in different temperaments might allow one to make a choice which would give the work a greater coherence (see [4] §6).

But these remarks are just concerned by a single aspect of a much broader problem and the reader is referred to [6] for a deeper study.

References

1. D. Blum, *Casals et l'art de l'interprétation*, Buchet/Chastel, Paris 1980.
2. D. Cooke, *The Language of Music*, Oxford University Press, 1963.
3. S. Cordier, *Piano bien tempéré et justesse orchestrale*, Buchet/Chastel, Paris 1982.
4. Y. Hellegouarch, *Gammes Naturelles*, Gazette des Mathématiciens, juillet 99 n° 81 et octobre 99 n° 82, pp. 25–39 et 13–26, S.M.F.
5. H. Helmholtz, *On the Sensations of Tone*, Dover, New York, 1954.
6. F. Lévy, *Plaidoyer pour une oreille subjective et partisane: une approche "pythagoricienne" de la perception culturelle des intervalles*, in Musique, rationalité, langage, n° 3, L'Harmattan, 1998.
7. D.G. Northcott, *An Introduction to Homological Algebra*, Cambridge U.P., 1960.

Symbolic Sculptures

John Robinson

<http://www.bradshawfoundation.com/jr/>

Curriculum Vitae: Nationality Australian / English, b. 1935. Educated Melbourne and Rugby, Cattle Drover in Northern Australia, Sheep farmer in South Australia, Figurative Sculptor turned Symbolic, lives in Somerset England, Honorary Fellow University of Wales.

Trustee Coordinator of the BradshawFoundation Geneva.

Interests: Pre History Member of Dr Jean Clottes Chauvet Cave team.

Collections: USA and UK, Cambridge, Oxford, Harvard, Aspen Center of Physics, Center for Computational Biology Montana, Isaac Newton Institute, Macquarie University, University of Wisconsin Madison, Australia Sports Centre Canberra, Olympic Museum Lausanne.

Ronnie Brown and John Robinson discussing his Symbolic Sculpture IMMORTALITY in the garden of his Somerset Studio – the sculpture (see [JR]3) has been adopted as their Logo by the School of Mathematics, University of Wales, Bangor:

"I think the one thing that sets us apart from all other forms of life is our Artistic Creativity. The earliest works of art used by our Cro Magnon ancestors to communicate with the Unkown were Symbols, and more often than not these were based on Mathematical patterns. I believe the first paintings and sculptures were DOTs, and the DOT can also be looked upon as the beginning of Geometry." John Robinson

"Our aim is to popularise Mathematics by presenting John Robinson's extraordinary Sculptures and their links with Mathematics and Science. Mathematics is the study of patterns and structures, and the expression and description of these in terms of a language which allows for understanding, deduction and calculation. This is why it yields a necessary language for many aspects of science, technology and human activity, and so is strongly associated with utility and applications. It is also associated with achievement, in unravelling the complexities of the structures it studies. The combination of Mathematics with John's Symbolic Sculptures gives us a sense of excitement and wonder at the beauty and originality of these forms, patterns and structures, and enhances our wish to study them for their own sakes." Ronnie Brown

Acknowledgment: Thanks to John Robinson and Ronald Brown who allowed the reproduction of the page `<http://www.cpm.informatics.bangor.ac.uk/sculmath/main.htm>`. Thanks also to Nick Mee (`VirImages@cs.com`) who made the magnificient pictures, here reproduced by the permission of their authors.

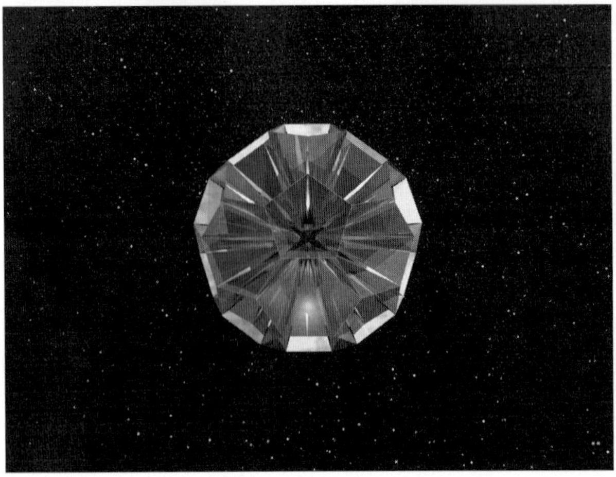

Fig. 1. [JR]1: *Pulse*

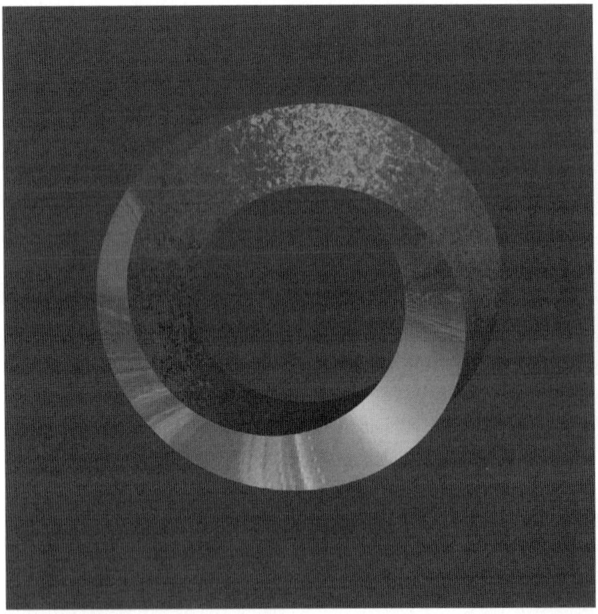

Fig. 2. [JR]2: *Dependent Beings*

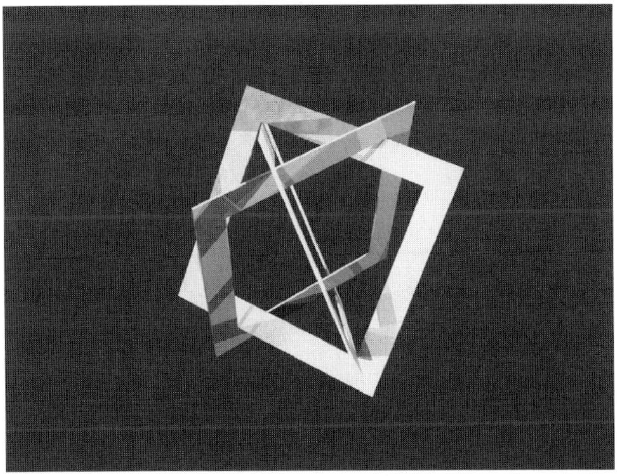

Fig. 3. [JR]3: *Creation*

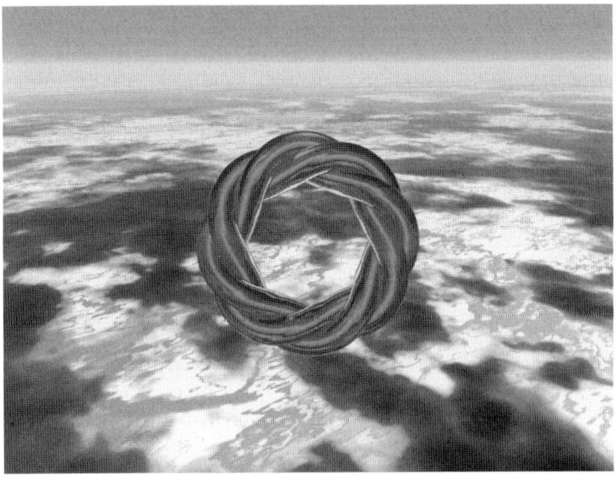

Fig. 4. [JR]4: *Gordian Knot*

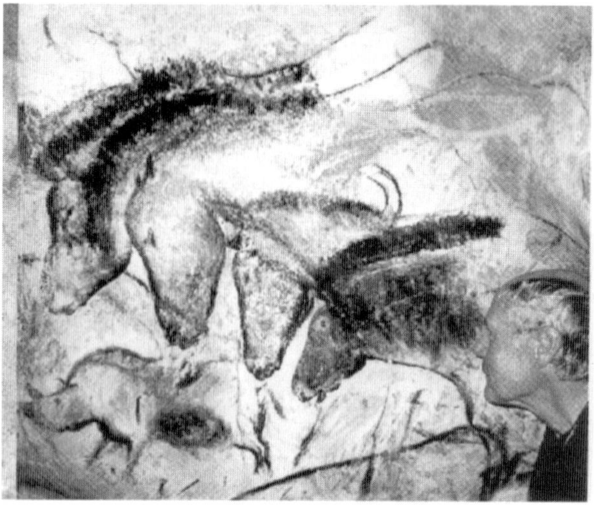

Fig. 5. [JR]5: *Chauvet Cave*

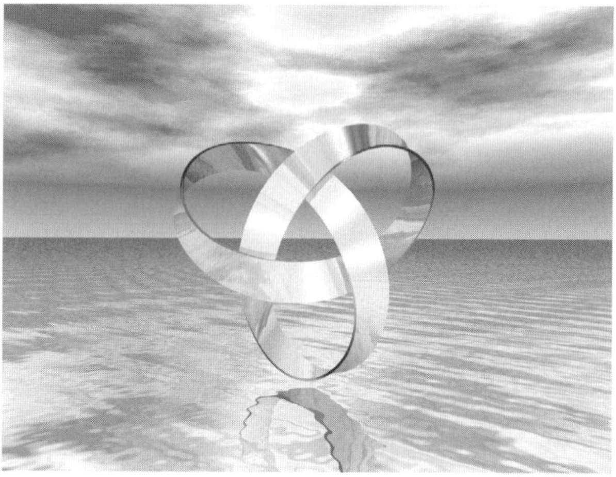

Fig. 6. [JR]6: *Immortality*

FORUM: How Art Can Help the Teaching of Mathematics?

A Short Presentation

Claude-Paul Bruter

Université Paris 12, Créteil, France <bruter@univ-paris12.fr>

The participants in this forum were experienced mathematicians, each one from a different country. Ronnie Brown represented Great Britain, Manuel Chaves Portugal, Michele Emmer Italy, Mike Field who had also worked in Great Britain and Australia, represented the United States, while Konrad Polthier represented Germany.

They told us about some of their experiences attempting to popularize and teach mathematics using various means, in particular through art.

This part of the Colloquium was perhaps a first attempt at establishing common ground on the interplay of Art and Science in education, especially mathematical education.

Henri Poincaré and Hermann Weyl were among the deepest mathematical thinkers of the last two centuries. They were quite convinced that the main goal of education in mathematics was the formation of the mind. I quote from Hermann Weyl famous book *Space, Time, Matter*:

> It seems to me to be one of the chief objects of mathematical instruction to develop the faculty of perceiving this simplicity and harmony.

(p. 23 of the Dover English edition). Quoting Henri Poincaré (*La Valeur de la Science*):

> Le but principal de l'enseignement des mathématiques est de développer certaines facultés de l'esprit.

I will sum up what I believe are the advantages [2] of a well constructed and guided mathematical education. The formation and development of the faculties of analysis and synthesis. The formation and the development of the faculties of observation, reasoning, and intuition. The formation and development of a feeling for intellectual beauty. Classical elementary geometry is perhaps the best tool by which theses aims can be achieved. Indeed, it gives birth to a large diversity of strange and amazing properties, which can be proved in a few organised and well written sentences. In this way, it stimulates the activity of the mind and contributes to the unfolding of all the previous qualities.

---

[2] Cf C.P. Bruter, *Comprendre les Mathématiques*, Odile Jacob, Paris, 1996.

Among the tools which nowadays can be used to improve teaching, the conscious use of Art seems to be new. Here I use the term "Art", in its widest sense to include all the forms it can take. One form is literary art which is definitely missing from the standard teaching books. Formulae and basic language are used: they are insipid and not appealing for a young mind who is impregnated with emotional functions and realism. For such a mind, the abstract discourse does not make sense and can be repelling. We meet here the general tendency of blind modern pedagogy which is to insert the latest discoveries and methods of professionals into introductory courses. We should not forget that children do not have the experienced mind of professionals, and that instruction is an ontogenetic process.

The use of visual art (through fixed objects or animations) is yet in infancy. A valuable, though superficial, use consists in showing beautiful visualisations. They have the ability to give a kind of physical status to abstract objects, and give them some consistency so that the general public can get a better idea of the matter on which mathematicians are working. They do have a power of attraction due to their originality and strangeness. This in turn can stimulate curiosity, due to the strong aesthetic qualities of these visualizations. This power of attraction, inviting the onlooker to look repeatedly at these representations of mathematical objects, induces a familiarity with the objects, and so may help in the understanding of what lies behind them. They can also help others to understand some the aspects of mathematical beauty championed by many professional mathematicians.

A less trivial use of art consists in systematically looking at the mathematics which have inspired, or which may inspire, the realisation of beautiful real objects – some of them being real works of art. Teaching mathematics through art can be useful both in secondary schools and in schools of plastic or musical art. In this regard, although a little has been done, a huge amount of work is before us.

The speakers at this forum have successfully begun to open some doors. However, in order to get a positive result, there are some essential preconditions: an open-minded scientific community, flexible administrative rules, professors dominating all the aspects of their subjects. With these environments, it should be possible to design and create new curricula allowing for the teaching of new, non-traditional traditional, mathematical topics. This could have a major impact on the development of the spirit, and on the acquisition of mathematical knowledge.

We are once more facing the tricky problem of the content of mathematical education: given our aims, what do we have to teach and which programs is it better to set up? We do not forget that additional difficulties arise from the diversity of the audiences we have to sensitize, inform, and teach.

Art can be used at many different levels. In each case with a common goal of fostering intellectual curiosity in a relaxed and stimulating atmosphere, in parallel with the development of an aesthetic appreciation of beauty.

Forum Discussion

Ronnie Brown

School of informatics, Mathematics Division, University of Wales, Bangor
<http://www.bangor.ac.uk/~mas010/>

My work in Public Awareness arouse out of a set of fortunate invitations, decisions and meetings, but motivated by the slogan of 'advanced mathematics from an elementary viewpoint'. I also recall a comment that a major problem in mathematical education is to convey the reality of mathematical objects. This is relevant to the theme of this conference, Mathematics and Art.

Of course not all advanced mathematics can be presented from an elementary viewpoint. Also, in undergraduate teaching, it seems essential that students should have a clear idea of why a topic is studied, so it is desirable to start, if possible, from a clear problem which the student can grasp. So I was led to abandon the teaching of homology, where the motivation is subtle and the examples dealt with at the first level are rather dull.

Instead I developed a course on the theory of knots. The basic problems of how do we know a knot cannot be untied, and indeed what do we mean by this statement, are immediately accessible, and the mathematics which goes some way towards a partial solution is attractive, geometric, and allows for some nice algebra and computations. So when I was asked to give a lecture to the British Association for the Advancement of Science at Sussex in 1983, the subject of knots and the title 'How algebra gets into knots' came to mind. In the teaching process, I had developed several tricks with physical knots and string, so these could be used. The audience, including professional mathematicians, liked the presentation and I was asked to repeat it on many occasions. I am still giving it!

The title became 'How mathematics gets into knots', but even this was not accurate, since the aim was to use the theory of knots to explain some methods of mathematics to the general public. A key point of the lecture was the notion of prime knots and the analogy with prime numbers. After a lecture to children in 1985, a boy, aged maybe 14, asked if there are infinitely many prime knots! I had to say this was a very good question. After another lecture for schools, a teacher told me that nobody in his career had ever before used the word 'analogy' in a mathematical context.

In the course of these lectures, I accumulated material to go in the foyer before the lecture, and we made the rash decision to consider putting this material into the form of a travelling exhibition. The naivety of this decision was of course that we had no design experience, and very little funds. In the end we accumulated £ 4,000 of funding from various sources, and were very fortunate in a series of designers who helped us enormously for little cost.

We also found that the exhibition format is one of the most difficult. Each board has to tell a story in itself, preferably largely through graphics, and each board has to be related to the other boards. It is not enough to say 'This is a nice graphics, let's put it on' – you also have to be clear how the graphics contributes to the story you have to tell. There should be no 'sugaring of the pill' – for this implies that the real mathematics is thought of as a 'pill' to be disguised, rather than a delight to be revealed. Thus the form of the graphics has to contribute to rather than disguise the mathematics.

Further the whole exhibition has to have some clear message or impact – there has to be a decision as to what impression the viewer is supposed to gain from the exhibition as a whole, and this intention has to be implicit rather than explicit. As an example, there is no use in showing weird objects mathematicians study, unless you are trying to show that mathematicians are weird. This may in fact be true, but is not necessarily the impression you wish to convey.

The exhibition was designed to be a travelling exhibition, to be able to sent by carrier, and mounted easily. This militated against hands on material, which has problems of maintenance and security. The exhibition consisted of 16 A2 black and white boards, on bromide paper, mounted on polystyrene with an aluminium surround. A travelling case was designed. It was launched at the Pop Maths Roadshow at Leeds University in August 1989, and then toured the UK with the Roadshow.

The process of design was very instructive to the design team (me, Tim Porter and Nick Gilbert). The necessity of thinking through the basic purpose of what we were hoping to convey has had an impact both on our teaching and our research.

Thus in our teaching we do think more of what should be the impact of the whole course on the student – what is supposed to be the impression of mathematics with which he or she is supposed to leave the course? What sort of qualities are we seeking to assess? There is a danger that courses are designed for an assessment, rather than the assessment is designed to assess the qualities which the course is designed to develop. Employers may want graduates who are good at planning their work, at assessing work done, at formulating problems as well as solving them, and finally at communicating what they have done – as was found by the assessment [6] of graduate mathematicians in employment!

In our research, we are now more inclined to question basic assumptions and to try to conceptualise the reasons for following a particular line of research. This perhaps forces us to think of more fundamental lines of enquiry.

An early aim of the exhibition was to include knots in history, practice and art. This was because of the feeling of presenting a wide context for mathematics – see the arguments in [2]. This is much more exciting to the viewer, who often sees mathematics as a subject isolated from general culture.

The difficulty in designing the mathematics board, and the sheer mass of potential material, made accomplishing this early aim progressively more unreasonable. It so happened that in July, 1985, I passed the Freeland Gallery in Albermarle St., just off Piccadilly in central London. Having time to spare, and attracted by the sculptures of children in the window, I wandered in and was astonished to see sculptures of children happily at ease with strong and beautifully abstract bronze forms, many of a knotted nature. Both sets of sculptures were the work of John Robinson, and it seemed reasonable to me that his abstract work should be drawn to the attention of the mathematical community. In 1988, puzzling over my problems with presenting knots and art, I phoned John and asked him if he would like to contribute an exhibition to the Roadshow. He thought we could something make really good in the time available, and I obtained permission from the organisers for such an exhibition to be included. In April 1989 I met John at his home. The Freeland Gallery had left Albermarle St in 1986, and the sculptures were transferred to his garden, where they looked magical. We chose thirteen substantial sculptures, and drove up to Leeds to decide where they should be placed. On the way up, I agreed to write 'Conversations with John Robinson', giving titles and general mathematical background to the sculptures. When John obtained my text, he added material and 24 colour pictures, and we had a catalogue for the Roadshow! This did sterling service later as a way of introducing John's work to the academic and wider world.

In 1996, we agreed to try to put John's sculptures on the web. This was partly motivated by the greatly increased costs of colour printing, and was stimulated by the work of a student, Cara Quinton, who did an undergraduate project with us on putting mathematics on the web. She carried out the production work on the web site. This work was funded by John's Patrons, Edition Limitée, while her further work of putting the Knot Exhibition on the web was funded by the London Mathematical Society and The Philip Trust.

The web format proved very attractive to us, with its easy capacity for including pictures and animations, and for the use of hyperlinks, both internally and externally. Designing the web sites became an exciting adventure as the material and the complex inter-relationships grew. Thus the web format begins to allow the modelling of the complexity of mathematics.

There are still problems with the web format for wide use in schools because of problems of access to internet linked computers, download time, and cost of telephone calls. Thus the net should be supplemented by CDRoms, and we have produced an RPAMath CDRom [4] based on the web sites. This also allows greatly improved animations, and a more integrated format.

What is the value of linking Art and Mathematics? Generally speaking, people feel that Mathematics is linked with the stifling of the spirit, and Art with freedom of the imagination. For the working research mathematician, it is mathematics which inspires the imagination, as totally new forms and almost unbelievable patterns and structures are revealed. These fruits of the imagination, verified through the tough testing of logic and calculation, have what has been described as an 'unreasonable effectiveness in the physical sciences' (E. Wigner).

However, perhaps mathematicians can learn from the way education in art and design is carried out. A set of objectives for a course read:

'The aims of this course in design are:

1. To encourage independence and creativity.
2. To teach the principles of good design.
3. To give a student a basis of skills to be able to apply the principles of good design in an employment situation.'

I leave the reader to explain why it is, or is not, reasonable to replace the word 'design' in the above by 'mathematics' and whether the resulting aims are desirable, and are realised in current undergraduate courses!

The aims of Art and Mathematics are different. It has been said that Art has a foothold on emotion. This is not the intention of Mathematics, which, in order to find the essence of what is true, usually needs to strip off the inessential for the purpose at hand. Art and Mathematics have a common interest, in form and structure, in geometry and the way parts fit together. Also, it has been said that a major problem in Mathematics Education is to convey the reality of mathematical objects. Vivid realisations of such geometrical forms, with what an artist can give in craftsmanship, proportion, rhythm, and wider significance, can excite, enhance and entrance the imagination.

Mathematicians could also learn from the freedoms found in Art Education, and from the belief in discussion of what is 'good art'. It is often held that discussion of 'what is good mathematics' is inappropriate for an undergraduate education. I totally disagree with this. We have found in students a hunger for discussion on mathematics, and a placing of it in a cultural, technological and social context. It is wrong for students to graduate with a good degree in mathematics without acquiring and being able to conceptualise some professional values such as: the notion of 'good mathematics'; the major achievements of the subject; and the place of mathematics in society. We have attempt to deal with these questions in several ways – I write 'attempted' as we do not claim to have a final solution, and there are surely many modes of doing this.

We have run a course entitled 'Mathematics in Context' – see [2]. After a few years we had a fair objection: 'If Maths in Context is important, why are you running it as an optional third year course?' We have also run a first year course in this area, using [5] as a text. We also ask students to do evaluative projects or assignments of various sizes in other courses, for example 'Write a brief account of the importance of fractals'. In this way students are forced to read books and search for material on the web, and put it together in some way. In a third year course on Groebner bases, students do a small assignment of considering a current paper using Groebner bases, and writing about it. Of course they do not have time to understand the paper properly, but they are asked to make an assessment of the use of these bases, and the practical relevance of the paper.

Two students on the Mathematics in Context course chose to do projects on Mathematics and Art. One wrote a year later that she was still haunted by the project! The other wrote in his project that having chosen this topic, it was about time he visited an art gallery!

It is not at all clear otherwise how we should bring the subjects of mathematics and art together in our curriculum, for those who wish it. Surely many more experiments need to be tried. I am sure that conferences of this kind will encourage such broader debates on mathematics, and on education.

References

1. R. BROWN and T. PORTER, *Making a mathematical exhibition.* In: The popularization of mathematics, edited A. G. Howson and J.-P. Kahane, ICMI Study Series, Cambridge University Press, (1990) 51–64.
2. R. BROWN and T. PORTER, *Mathematics in Context: a new course*, For the Learning of Mathematics, 10 (1990) 10–15.
3. R. BROWN and T. PORTER, *The methodology of mathematics*, Math. Gazette, 79, July (1995) 321–334.
4. R. BROWN and SUMit softwares, *Raising Public Awareness of Mathematics: Knots, Symbolic Sculpture, Mathematics*, CDRom Version 1.1, Center for the Popularization of Mathematics, `<http://www.cpm.informatics.bangor.ac.uk/>`.
5. P. J. DAVIS, R. HERSH, *The mathematical experience* (with an introduction by Gian-Carlo Rota), Harmondsworth: Penguin, 1983.
6. R.R. McLONE, *The training of mathematicians*, Social Science Research Council, 1974.

(The first three articles are available from the articles section of
<http://www.cpm.informatics.bangor.ac.uk/>

Forum Discussion: Presentation of the *Atractor*

Manuel Arala Chaves

Depart. Matemática Pura, Faculdade de Ciências da Universidade do Porto, Porto, Portugal <machaves@fc.up.pt>

The project to which I recently devoted myself in an intensive manner, and that I am going to describe briefly, is a little bit marginal with respect to the theme of the *Forum*: it is not about the teaching of mathematics, at least in the school sense of the word, and art is not present in a systematic way. What it has in common with other initiatives mentioned in this forum, is an attempt to generate interest in mathematics by means which are not necessarily perceived by the public as having a direct relation with mathematics.

Knowing the genesis of the project will help to get a better understanding of the form taken by its realisation. In 1993, under the initiative of a physicist, a group of professors of the Faculty of Sciences of Porto met to try to lay down the foundations of a future museum of sciences. The intention was that the museum would be characterized by a high degree of interactivity. Although at the time, there was general scepticism about the possibility of creating "interactive" realisations that would be interesting at a mathematical level, I joined this group at the outset since I was convinced that something could be done, and that it was desirable that mathematics should not be absent from such a museum.

My presence in that group has been very important (for me!), not so much for the work which has been done – in the first temporary exhibitions which happened in 1994 mathematics was only represented in a rather modest manner – but because since this date, though preparing other things, I always kept in the background the desire to explore this path. I took notes, I got some ideas, and I participated in several annual organization which coordinates the European scientific museums) and I visited a several scientific interactive museums. The first impression I kept from these visits was the fact that in general mathematics was not much represented in these museums. And gradually, my conviction became firmer that matters could be different. In my Faculty, for several reasons, the project of the museum of sciences has not advanced as rapidly as was expected at the beginning. On the other hand, I became convinced that there should be enough material to realize a Centre entirely devoted to Mathematics. Meanwhile, a building in Ovar (between Porto and Aveiro) has been put at our service, and, in 1998, I sent a general circular invitation. The reception was excellent and several university institutions (Aveiro, Coimbra, Lisbon, Porto), the *Sociedade Portuguesa de*

Matemática, the *Associação de Professores de Matemática*, and also the city of Ovar joined into an Association (formed in April 1999) for the creation in Ovar of an Interactive Centre devoted to mathematics: the *Atractor* (see *http://www.fc.up.pt/atractor*). A small group of colleagues got interested in the project in an active way. Also, the Ministery of Science and Technology decided to support it financially, and to integrate the future Centre into its network of *Centros de Ciência Viva*. On its side, the Ministery of Education came to its help by arranging for two teachers to work on the project at full time. During the years 1999 and 2000, several interactive exhibitions were organised in different cities. A big impulse was given to the realisation of the project in the form of an invitation from the Ministery of Science and Technology (Division *Ciência Viva*). Confirmed in March 2000, this invitation called us to organise, in November 2000, in the *Pavilhão do Conhecimento* (the Knowledge Pavillion) in the city of Lisbon, an Exhibition integrated into the setting of the commemorations of the World Mathematical Year. It was a great challenge to be taken by the *Atractor*, because of the media importance of the place, of the fact that it was the first exhibition in the *Pavilhão* entirely conceived in Portugal, and because of the importance of the financial means put at the disposal of the *Atractor*.

The challenge has been accepted, in spite of the short allowance of time for the preparation of this exhibition. Its preparation, which held us during all these months, had a very good secondary effect on the *Atractor*: a certain number of colleagues, who previously had not contributed to the project, joined us, bringing ideas and work. The support of the *Atractor* has been considerably enlarged. The Lisbon's exhibition will remain open for several months. All its content will be later integrated into the permanent exhibition of the future Ovar Centre, which will open its gates after the indispensable work of renovation and of adaptation needed by the building.

The philosophy of the *Atractor* project is first not to choose the topics according to their presence or absence from the school programs: the aim is not to create a kind of mathematical laboratory to support the courses given in the school institutions. That settled, an important part of the visitors of the exhibitions will be young pupils and their teachers. One has been able to notice that among the latter, some strive to relate certain "modules" of the exhibition to the content of the programs of their pupils, thinking about new visits they will make later. These teachers often used to judge the interest of the "module" according to its proximity to the programs. This first reaction, somewhat natural, was followed in most cases by a new mental receptivity on the philosophy of the exhibition when the purposes of the *Atractor* were better understood.

An other important point is the definition of the ordinary target visitor: which *level of knowledge* do we have to assume and which *language* to use? In general, the question of the level is not so important as one might think: we have remarked that a great number of modules could be in fact valued at very different levels of understanding, and, at each level, bring something new and attracting. The question raised by the language to be used is of a different nature, and is not easy to answer. We have a plan but it will have of course to be tested by experience: our intention is to restrain and to simplify to the utmost the printed instructions that are presented with the modules, and to insert, into a network of computers, all the necessary complement of information.

It is easy to include, inside the numerized support, texts corresponding to levels of different preparation. In this way, one allows anyone to choose easily which materials fit them the best. It also allows, in a way combined with the exhibited objects, the possibility of introducing *applets* or other interactive programs which reinforce the participating attitude that we want to encourage among the visitors.

We attribute a great importance to the integration of (interactive) objects with the presentations (equally interactive whenever it is possible) one can get on the network. When it is completely worked out, it will allow visitors: first to prepare in advance their visit which they will be able to anticipate somewhat by consulting the *site* of the *Atractor*; second, to go back in more detail to the elements that they were able to appreciate during their visit, also to lighten the points that have remained intricate.

From the point of view of this integration, we are in the presence of three types of situations:

- In the first one, the "physical objects" are present, but their interactive numerized *translation* is not yet done. Examples of such objects: a non rigid polyhedron, showing that the rigidity of the faces does not imply that of the polyhedron; or still, a great balloon and a "ring", on which one can draw triangles, measure their angles, in each case compute their sum, compare it with the one of triangles located in the plane, evaluate how the difference evolves with the area of the triangles (see [MC]1 and [MC]2 in the Appendix).
- In the second one, physical objects and interactive numerized objects are present. For instance one has polyhedra, including the dual ones that one can create and observe through triedric kaleidoscopes where sometimes they give birth to very beautiful images (see [MC]3 and [MC]4 in the Appendix).

They can be interactively handled and animated with especially adapted *applets*: see [MC]5, [MC]6, and

http://www.fc.up.pt/atractor/mat/duais.html.

- In the last type of situation, the creation of virtual objects and their manipulations goes before the creation of physical objects; these are yet in a phase of conception, or of construction of prototypes. Example of object being in a conceptual phase: the fact that, *in order to illuminate all the points of the boundary of a convex plane figure, only three directions*

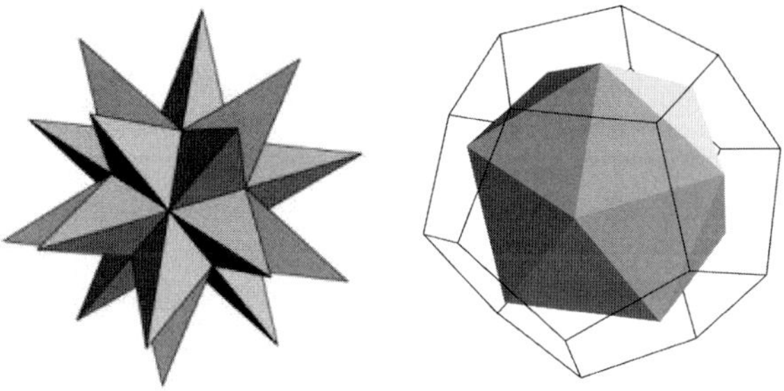

Fig. 1.

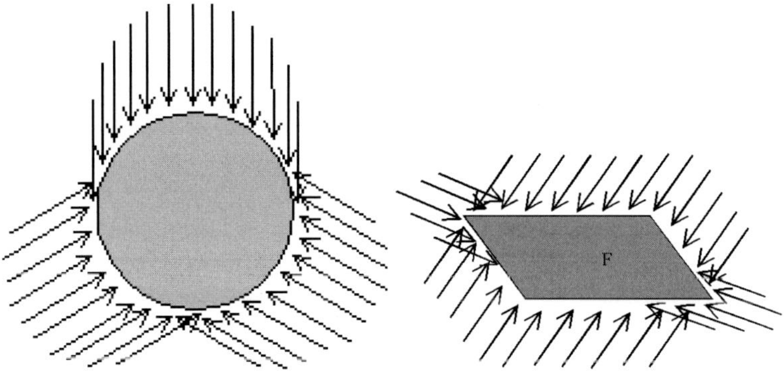

Fig. 2.

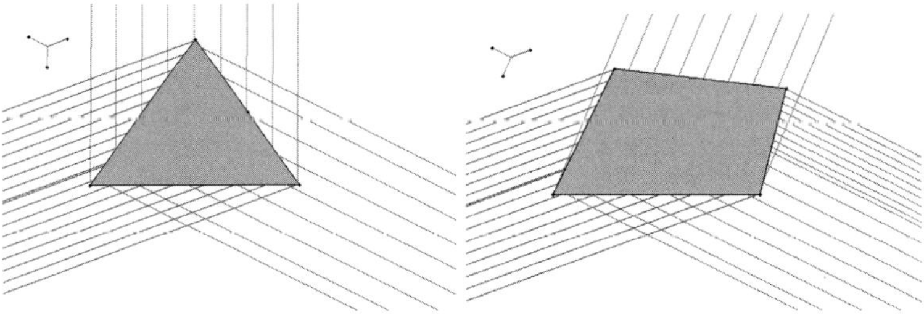

Fig. 3.

of luminous rays are sufficient except for the parallelogram where four are needed, is at first surprising.
However, through the manipulation of the interactive *applet*

http://www.fc.up.pt/atractor/mat/GeomConv/ilumina.htm

it will be within the scope of non specialists to be convinced that the assertion is well-grounded, even if it will be sometimes difficult to write a proof in acceptable terms from the mathematical point of view. To find an interesting manner to present the elements of the proof in terms of real objects which can be handled is a challenge to be taken up.

- An example where the prototype is already constructed but not yet exhibited concerns the concept of *orientation of a surface* (and the one of *non-orientable surface* as well), and the distinction between this concept and the one of *one-sided surface* (in space):

see

http://www.fc.up.pt/atractor/mat/orient.html

Same or opposite orientations?

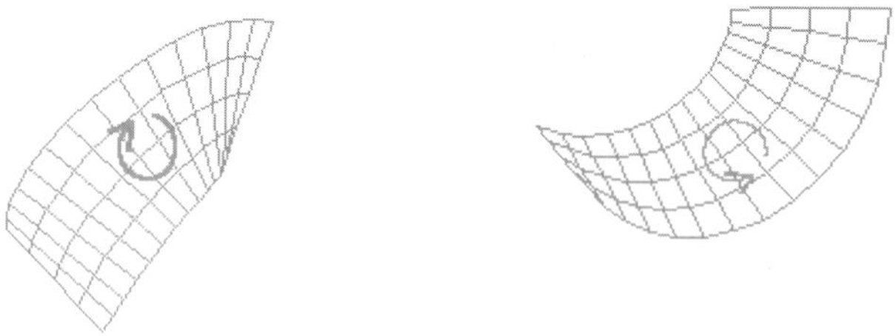

Fig. 4.

and the tableau in:

`http://www.fc.up.pt/atractor/mat/Moebius/Construcao-Moebius.html#Quadro-icons`

These examples might lead one to believe that applications of mathematics are not represented in the *Atractor*. That does not agree at all with the reality: among some ten other modules, which already have been developed, some are concerning applications of mathematics. And, as one has to expect, one will also find these frequently very beautiful objects *attractors*, which will do justice to the name of the Institution – *Atractor*. A model of an orbit of the *Lorenz attractor* has been built by a process of stereolithography, from files prepared with some *packages of Mathematica* (see the figures [MC]7 and [MC]8).

And each visitor, by the personal creation of a point, will be able to contribute to the construction of the *Sierpinski attractor*, a large module that will be present at the Lisbon exhibition.

Non-orientable Möbius Band? Why? And if the middle circle is taken out?

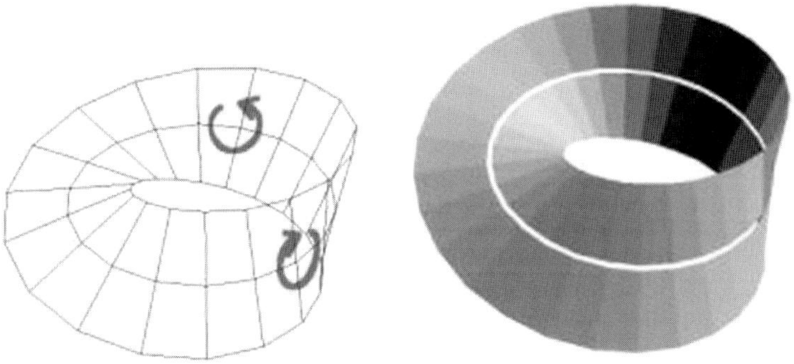

Fig. 5. Möbius band (*left*). Möbius double band (*right*).

The *Atractor* intends first to *attract* a large public to mathematics: the success of this *attraction* will be a second manner to justice to its name. But the Atractor also wants to contribute by giving to people in general, and to pupils in particular, a more positive image than the one which is usually propagated; also a more enlarged image than the one found in many places ("mathematics are used to make calculations"). Finally, it wants to allow interested people to learn something, and, if possible, to show also that in mathematics, as in other sciences, the progress is constant: whereas the public knows about the discoveries in other sciences, it often shares the idea that in mathematics *everything has been done before*.

The *Atractor* is almost a new-born: thus it is yet too early to see how it will succeed in reaching its objectives that I recognize to be very ambitious.

Forum Discussion

Michele Emmer

Dipartimento di Matematica, Universita la Spienza, Roma
`<emmer@mat.uniroma1.it>`

When, in 1977, I began my "Art and Mathematics" project (and it is unimportant which word comes first), I had something that was very clear in mind, I was not at all interested in making "educational" films like the ones being produced at that time, which tried to explain what a result, a theorem, in mathematics was. Films that were very boring, and essentially very short, and in my opinion completely useless. I was not convinced, and I am still not convinced today, that a film, a video, a software can act as a substitute in teaching, and in particular in the case of mathematics, to a direct contact with the teacher. One of the things that the media will never be able to do, is to react to the faces of the persons that are listening to what you are saying. Anyone who has ever taught knows that the faces of the students are a very important thing. A film therefore cannot be used for this, it cannot be used to replace studies that must be individual, with books, exercises software or CD-roms, that can always and only be used as a material support to the "physical" contact with the person who is teaching, with the one who suggests, informs, explains, clarifies. Besides, there is also another very important aspect, a famous Italian mathematician, who died a few years ago, said a number of times that teaching is the best way to learn deeply. However I was convinced that a film, a video, could be very useful to strike one's fantasy, to stimulate imagination, in other words to make one feel the need to understand, to study further. And not only a film, but also an exhibition, a book, a show. From the very beginning of the project I thought of making films, exhibitions, books, meetings, all these things together, and after a number of years I can say that the project has worked. It has worked because one of the great fortunes of art and mathematics is that of being universal. If one speaks of mathematicians or artists from any part of the world, one can be understood if he succeeds in finding the right interpretation key. Why am I speaking of art? Because from the very beginning, as the first idea was that of realising videos, "visual" images had to be used. And therefore why not also ask the collaboration of artists, besides mathematicians? I am also convinced that it is not possible to speak of "Visual Mathematics", in all the sectors of mathematics, just as it would be absurd and ridiculous to make art seem to be always tied to scientific or mathematical ideas. Just like it would be absurd, today, to privilege the computer, the Internet, and make art seem to be "new" because it utilises new technologies. In mathematics and in science, perhaps, we can speak of progress, in art it is totally absurd. Technology serves art in the same way as it serves mathematics, but neither art nor mathematics are pure technology, pure method, pure calculation. Creativity, invention, are essential both in art and in mathematics. Which surely does not mean that the task of "showing" the ties between art and mathematics is a desperate one. So the idea from which I started was to realise videos (and then exhibitions, books, meetings), in which the "visual ideas" that artists and mathematicians (and not only them, architects, biologists, musicians) used, would be highlighted. Ideas that can be compared, "described" in some way, always trying to use mostly a visual language. And so, beside a number of mathematicians, from D. Coxeter to Roger Penrose, from T. Banchoff to De Giorgi, I asked the collaboration of many artists, from Max Bill to Luigi Veronesi, to De Rivera, to Bruno Munari and many others.

2. In order to privilege the images, the spoken part, the explanations were cut down to minimum, and these are not even "real" explanations. A typical example is the film "Soap Bubbles" in which Jean Taylor and Fred Almgren speak of the theory of minimal surfaces, in particular, of the result on the singularities of minimal surfaces, but the video can be seen by children, even if they are very small, and they will interpret it differently from, for example, the students of the University of Princeton, where the same film was shown. Therefore art, artists, can be very useful to mathematics, as, in a certain sense also mathematics, the images of mathematicians can be very useful to art, to artists. Always bearing in mind that each one is doing his own job, that even if creativity seems to have some features that are common in all man's activities, however each discipline, both artistic and scientific, has its own language, its own technical means of expression. Otherwise, one risks making people believe that anyone can be an artist, anyone can be a mathematician. It is not sufficient to know how to "play" with a computer in order to become an artist, it is not sufficient to invent a new technique. As in mathematics, there certainly must be intuition, creative capacity, but there also must be the capacity to prove, to explain what one is doing, right to the end. Given the above, I believe that artists can be very useful in teaching mathematics, and also in making people understand how mathematics is a discipline with its history, with its evolution, with its mistakes and its successes, in conclusion, that mathematics have given and continue to give a great contribution to Culture. Without expecting, as faculty members always do, to teach the school-teachers, from kindergarten to the high schools, what they must teach and how. Those who do not have an experience in teaching in a school, and I do not have it, should be very cautious in suggesting solutions that seem to be the panacea for all the troubles of teaching. Undoubtedly most faculty members always look from above, downwards upon the teachers, and this attitude is not helpful. Also from this point of view, artists can teach mathematicians a lot of things.

Forum Discussion

Michael Field

Department of Mathematics, University of Houston, Houston, TX 77204-3308, USA
<mf@uh.edu>

1 Background

This informal essay is based on an introductory statement I gave in the forum 'How art can contribute to the teaching of mathematics?' It is not intended to be a closely reasoned statement of my 'philosophical position' on the teaching of mathematics (I have no such position).

I would like to start by describing some of my background and experience related to the topic of the forum and explain how a mathematician came to be using ideas based on art (and computers) in his teaching.

1.1 *Prism*

About eleven years ago, I started to develop a computer program called *prism* (PRogram for the Interactive Study of Maps). While the program arose out of research into symmetric dynamics [1,5], the main use of *prism* was for the design, creation and colouring of symmetric planar patterns. Examples of patterns designed using *prism* are shown in the book *Symmetry in Chaos* [4] written jointly with Marty Golubitsky in 1992.

Over the years, *prism* has continued to develop and now includes a multiplicity of algorithms for all seventeen of the wallpaper (planar crystallographic) patterns as well as the forty-six 2-colour wallpaper patterns [2]. Characteristically, the designs created using *prism* exhibit a richly complex and (of course!) symmetric structure. Colouring – closely related to the underlying dynamics – is not routine and requires considerable input from the designer. Some examples of recent designs created using *prism* may be found in in my article in the Maubeuge Proceedings [3][1].

1.2 Symmetry in Design: An Interdisciplinary Course for Art Students

About four years ago, Angela Patton (Department of Art, UH) suggested that I develop an interdisciplinary course on patterns, designs and symmetry based on *prism*. Although I had largely developed *prism* for personal use, I

[1] See also my web site `http://nothung.math.uh.edu/~mike/`.

was intrigued as to how art and design students would respond to an 'art' program from a mathematician. There was also the challenge of communicating some of the underlying mathematical ideas about geometry and symmetry to 'math-unfriendly' students. In the event, I developed a new course 'Patterns, Designs and Symmetry' for Junior and Senior level students at UH. The course has been given three times and is now part of the regular course schedule at UH. The year 2000 course has a mixture of art, graphics, photography, architecture, and mathematics students.

Implementation and Results One of the aims in giving the 'Patterns, Designs and Symmetry' class was to expose students to the mathematical concept of symmetry and thereby enhance their visual perception and design skills. This had to be done in a context where it was not acceptable to develop the mathematics in a formal didactic way. Indeed, many of the students were not only alarmed by the prospect of being taught by a mathematician but some were also nervous about the prospect of learning about and using computers. Equations and formulae were largely banned from the class and characters like 'X' and 'Z' were introduced for their symmetry properties rather than as symbols to be used in algebraic manipulation. Geometry was, however, strongly emphasized. By the end of the course, students were expected to be able to find the glide reflection symmetries of a planar pattern, distinguish, for example, between wallpaper patterns of type **p3m1** and **p31m**, and identify the symmetries of a polyhedron (including the regular stellated polyhedra). The approach that I followed was a mix of theory and practice. Practice was vital. Students designed symmetric patterns, created informative posters and made solids. For example, some of the students made symmetrically coloured models of the regular stellated dodecahedra. As a text we used the book by Washburn and Crowe [7] supplemented by my own notes and websites such as the geometry junkyard (URL: `www.ics.uci.edu/~eppstein/junkyard/`). About 40% of the course was spent in a computer lab. At the outset, I attempted to sensitize the students to symmetry. In Houston, this was easily accomplished by initiating an active and ongoing study of the symmetry of (car) wheels. Using wheels as a model, we discussed the different types of (bounded) symmetry in the plane. Symmetry was introduced as an operational property: an object is symmetric if you can pick it up, move it around and put it back in the same space as it was originally – but with a different orientation. Along the way, it was possible to introduce some ideas from algebra – to help with the composition of symmetries – and to point out how order of composition matters. Concurrent with the investigation of bounded symmetry, students were learning how to log on to a computer and use software packages. Their first project using *prism* was to design, and later colour, a bounded symmetric image in the plane. The successful completion of this design was an essential, highly motivational, part of the learning process – a movement from theory (of symmetry) to producing a design. And the designs were often very attractive.

(See the URL: `http://nothung.math.uh.edu/~patterns/indexart.html` for some examples from the 1999 class.)

1.3 Mathematics Through Art?

In 1999, I organized a seminar 'Symmetry, patterns and designs' for the inaugural year of the *Houston Teacher's Institute* (HTI). The HTI is part of a national demonstration project led by the Yale-New Haven Teachers Institute and supported by the DeWitt Wallace-Reader's Digest Foundation. (For further information, see the URL: `www.cis.yale.edu/ynhti`.) Roughly speaking, the idea of the HTI is that a group of about twelve teachers (the 'Fellows') participate in a collegial seminar led by a faculty member at UH. Each Fellow producing a curriculum unit related to the topic of the seminar (see the HTI website, URL: `www.uh.edu/hti`, for more details).

Most of the Fellows in my seminar were mathematics teachers in local middle or high schools. I used *prism* as a motivational tool tool to introduce ideas about symmetry and chaos. Part of the seminar involved each Fellow producing their own graphic. Some of these graphics were used as the basis for a quilt pattern which was used as the inside cover for the published set [6] of curriculum units for the entire 1999 Houston Teacher's Institute.

Implementation When planning the seminar, I had to keep in mind that content had to be interesting and stimulating for the teachers. Practically speaking, that meant that what we discussed had to have the potential for *practical application* in a high or middle school classroom. In particular, it was essential that the material could be used to facilitate the understanding of geometry (or algebra – if that were the teacher's specialty). The situation for teaching general mathematics courses at college level is broadly similar. It is not simply a question of making mathematics appealing or user-friendly. The message has to have content – and that content must lead to an enhanced understanding of the world that the student perceives.

In the event, none of the teachers had any background in symmetry. This was an advantage as it meant there were no preconceptions to overcome. Just as for the art class, the topic of the symmetry of wheels was interesting to the teachers – but now because they knew that this was one of the topics teenagers could get excited about (we did not discuss the possible relation between symmetry, beauty and sexual attractiveness – though that came up in some of the curriculum units). Another area of great potential was the use of symmetry in textiles – again dress is important to the target group. Architecture and art provided other possibilities. One of the most interesting suggestions made was to combine art and mathematics classes as a way of allowing the students to use ideas about symmetry in their art – this for a group of students who were almost totally alienated from mathematics and who, on good days, spent their time in mathematics classes drawing (or doodling). Overall, I felt the class was quite successful and the teachers produced units that far exceeded my initial expectations (the units may be seen on the HTI website).

2 Conclusions and Comments

2.1 Alienation from Mathematics

Since I started teaching in the United States, I have often encountered a strong alienation from mathematics among university students. One explanation for this alienation is the custom in the United States of requiring students to complete 'core' courses in 'college algebra' if they have not already done these classes in high school. Many of these so-called 'core' courses are unmotivated terminating courses that are intensely disliked by the students (correctly disliked in my opinion). In my view the forced learning of college algebra at university is an optimal strategy for the development of hostility towards mathematics in the general population.

I believe another reason for the negative attitude towards mathematics lies in the attitude of elementary school teachers towards mathematics. Indeed, if elementary school teachers are hostile towards mathematics,that attitude will surely propagate to their students. I have certainly found strong negative attitudes amongst (prospective) elementary school teachers. For example, a few years ago I took a summer class on geometry at UW-Madison for a group of prospective elementary school teachers. When I met the class for the first time, I found they were uniformly hostile towards mathematics and resentful that they had to take the class. They also told me – correctly, I believe – that they felt the class (syllabus) was irrelevant to their classroom needs. In the event, I developed a syllabus that focussed on developing classroom materials that the teachers could use. I felt that the students completed the class with a strong positive attitude to at least part of mathematics.

2.2 The Right Solution Is that there Is No 'Right' Solution

In my opinion, there is no 'right' solution for the teaching of mathematics at school or college level. For example, it is well known that while some students respond well to visual materials, other do not and instead feel more comfortable with an abstract/algebraic approach. I believe the enthusiasm and involvement of the teacher is the *key* factor in the success of a new approach to the teaching of mathematics. Pilot/experimental programs are frequently successful just because they have the active involvement and support of the teachers who developed the program. I have been fortunate at UH in that I have been allowed and encouraged to develop experimental programs. If instead, I had *directed* others to develop and teach programs in 'art and mathematics', these programs would probably have failed because of the lack of commitment and involvement on the part of the teachers carrying out my instructions. In short, I am sceptical of a 'top-down' approach to the teaching of mathematics that is based on edicts from on high (whether administration or senior academics) as to what the teachers ought to be teaching. Rather, I believe that initiatives should come from the teachers and should be supported (facilitated) by the administration. Even if projects eventually fail, the effort and enthusiasm for mathematics shown by the teachers will propagate to the students.

A particular obsession amongst mathematicians is that of 'error'. By this I do not mean numerical error, but rather the fear that mathematical concepts will not be accurately presented to the students. When this view dominates, it is difficult to allow any teacher initiatives – such as the preparation of curriculum units I described above. In extreme form, the result is that every detail of the syllabus has to be approved by government appointed committees of 'experts' – lest the children or students be exposed to an erroneous idea or concept. The reality is that people – especially children – are incredibly robust to error and are quite capable of handling conflicting or incomplete conceptual information. Better that the children learn to enjoy and appreciate geometry, algebra and statistics than that they be alienated from mathematics for ever by the joys of rote long division, factorization or the nuances of the law of the contrapositive.

In brief, we need to trust our teachers and faculty and support and facilitate them in their efforts to develop new ways to teach mathematics.

References

1. P. Chossat and M. Golubitsky, 'Symmetry increasing bifurcations of chaotic attractors', *Physica D* **32** (1988), 423–436.
2. M. J. Field, 'Designer Chaos', *J. Computer Aided Design*, **33**(5) (2001), 349–365.
3. M. J. Field, 'The Design of 2-Colour Wallpaper Patterns Using Methods Based on Chaotic Dynamics and Symmetry', these proceedings.
4. M. J. Field, M. Golubitsky, *Symmetry in Chaos*, (Oxford University Press, New York and London, 1992).
5. M. J. Field, M. Golubitsky, 'Symmetric Chaos: How and Why', *Notices of the Amer. Math. Soc.* **42**(2) (1995), 240–244.
6. *Curriculum Units by Fellows of the Houston Teachers Institute, 1999* (published by the Houston Teachers Institute, Houston, Texas, 1999).
7. D. Washburn, D. Crowe, *Symmetries of Culture*, (University of Washington Press, Seattle, 1988).

Getting Out of the Box and Into the Sphere

Dick Termes

1920, Christensen Drive, Spearfish, SD 57783, USA
`www.termespheres.com`
`<termes@blackhills.com>`

1 What Is a Termesphere

When you look at Termesphere paintings you get a sense of geometric order. Why would a painting of Notre Dame in Paris on a sphere give you the feeling of order? What are these orders?

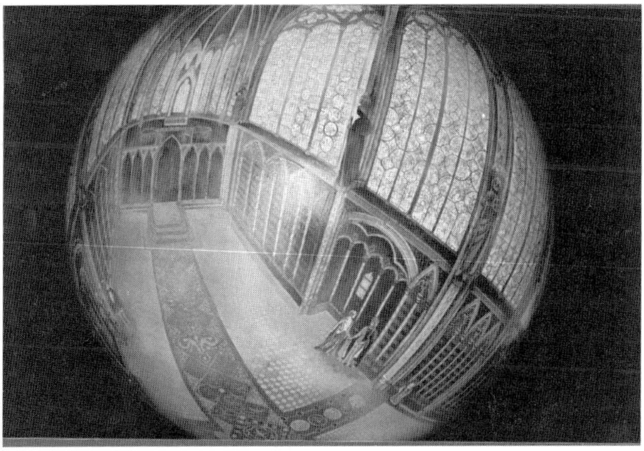

Fig. 1. [DT]1: SAINTE CHAPELLE IN PARIS is a 24″ diameter sphere. You are above the floor some thirty feet so you can look straight across at the stain glass windows. This shows how detailed a basic cubical room can get.

What you are seeing when you look at a Termesphere painting is an inside-out view of a total physical world around you on the outside surface of the hanging and rotating sphere. If you were on the inside of the sphere this painted image around you would seem normal but I make you read it from the outside. From any point you look at the spherical painting, the image reads correctly. Termespheres capture the up, down and all around visual world from one revolving point in space. Most of the time these spheres are painted on the outside so it takes a six point perspective system to keep all of this environment around you organized.

I have been painting spherical paintings since 1968-9 when I received my Masters Degree in Art at the University of Wyoming. I continued to do my thesis on the Termesphere at Otis Art Institutes in Los Angeles for the next two years. I received my MFA in design there.

2 How this Concept Was Found from Inside a Box

The idea of capturing total visual space around you on the sphere or six point perspective came to me in a very odd way. I had been studying and teaching perspective for a number of years. It advanced up to four point perspective but I wasn't satisfied. The main development happened when I focused my attention on perspective while working on my Masters Degree in art at the University of Wyoming. After having come up with some drawings and designs on paper which showed different ways to think of six point perspective, I attempted to put it to use. I decided I would build a small cubical object I could crawl into. This cube would have an exaggerated perspective built into it like my drawings showed. My thought was if you could pull the corners of the cube in toward the center of the cube it would give you the illusion it was a much larger box. I started to imagine what this cube would look like. A fellow student told me it would look like a cube painted on a sphere. The cube would become the sphere if you pulled its corner in far enough. I went back to my studio and tried to construct a cube on a kid's ball. I realized in order to draw the cube accurately on the sphere all the lines of the cube must aim to the center of the adjacent faces of the cube. All the lines became greater circles. If each of the lines of this cube were drawn all the way to the point you would construct a rhombic dodecahedron (twelve diamonds). The points in the center of the cube would create the vertices of the octahedron.

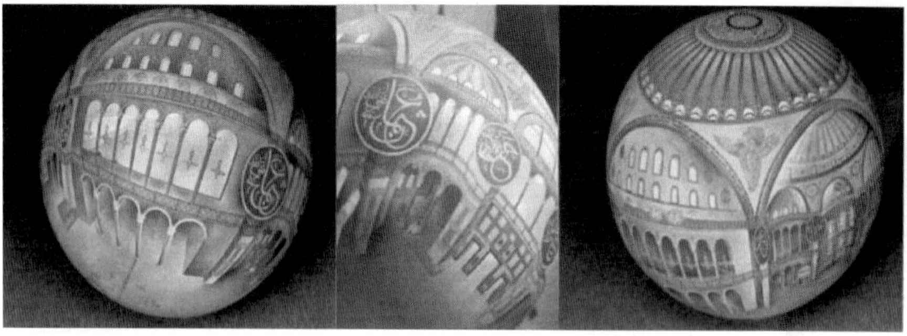

Fig. 2. [DT]2: HAGIA SOPHIA is a 24″ diameter sphere. This painting allows you to float about 40 feet above the floor to view the inside of this wonderful Christian and Moslem structure. This building is based upon the cube, a cube with a dome on the top.

When I looked at this cubical structure on the sphere as a drawing, I had the feeling I could be inside this cube. What if I thought of this cube as a room and I were inside of it? If this cube were thought of as a drawing of a room, the points in the center of each face would be thought of as vanishing points in perspective. The rule in Renaissance one and two point perspective says that all parallel lines converge at the same vanishing point. These parallel line on my cube were projecting in two directions to the center of the adjacent faces. Counting the points of projection it came to six, the same number as the faces of the cube. The lines which came from the four corner of the center square projected out by me. This thinking was inside out but the bottom face of the cube was the floor, the top face of the cube was the ceiling and the four other faces were the four walls. If you put tables and chairs on the floor, hang light fixtures from the ceiling tile and cut windows and doors into the walls, you have a room rather than just a cube and it is the inside of a room on the outside of the spherecube.

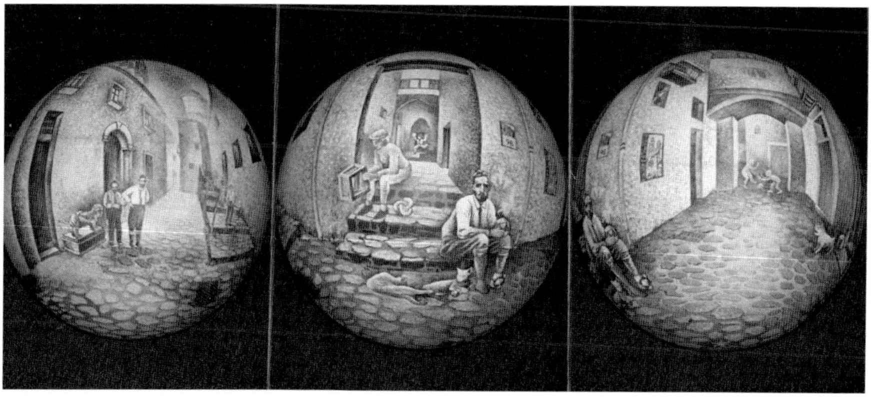

Fig. 3. [DT]3: INSPIRATIONS FOR ESCHER is a 16″ diameter sphere which hangs and rotates from a ceiling motor. This shows M. C. Escher in his favorite small town of Ravillo, Italy. Around him are found ideas which later become his great works of art.

3 The Geometry

What are some of the unique geometries going on in this room on a sphere? If the furniture and other objects in the room are parallel to the room, every line which creates this room or furniture for the room would bisects the sphere if it were continued. Every cubical object created for this room projects to all six equal distant vanishing points. All parallel lines go to two vanishing points. If a railroad track vanishes to a point in front of you, it will also vanish to one behind you. All you have to do is turn around and look.

4 Inside or Outside the Spherecube

The thing that helps to make this all more sensible is to think of yourself inside the sphere looking out. If your eye is in the direct center of the sphere and you are looking at this cube room painted on the sphere it would look very normal. If you were inside and in the center of a real room and your head were inside a transparent sphere and you copied what you saw outside the sphere onto the inside surface of the sphere, it would look exactly the same as the cube painted on the outside of the sphere.

5 In the sphere

Another experiment is to imagine having a transparent sphere on your head and place six equal distant large dots on it like the vertices of the octahedron. With the sphere on your head, arrange the dots so that one dot is above your head and another dot is below your head. The other four dots are parallel to the horizon around you. Now, walk into a cubical building. With one eye in the center of the sphere rotate the sphere until one of the points on the sphere overlaps with one of the vanishing points of the building. Keep the four dots on the sphere parallel to the ground. With your eye in the center of the transparent sphere, notice where the other five vanishing points of the building are located. All of the building's vanishing points will line up with

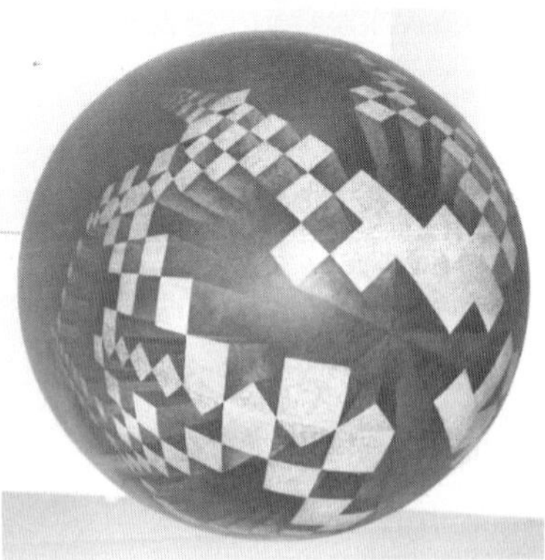

Fig. 4. [DT]4: FISSION OF THE CHECK is a 24″ diameter spherical painting. Cubes grow from a checker board pattern. Another set of cubes also grow from the diagonal of that first check system. This sphere could be said to hold ten points of perspective.

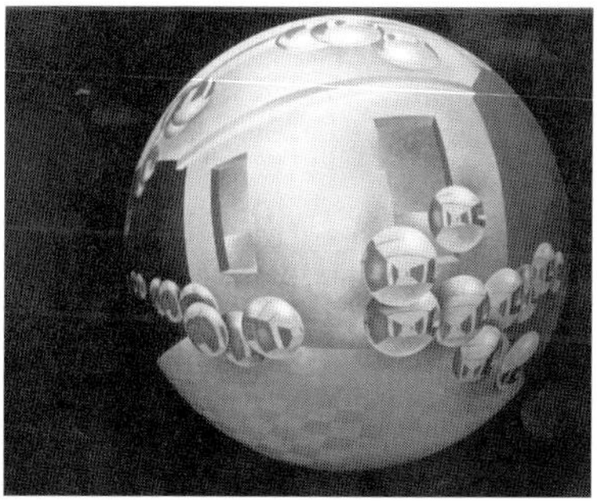

Fig. 5. [DT]6: CONCAVE BUBBLES is a 24″ diameter sphere. There are 100 reflective balls that float through this room each reflecting the room from the location they are within the room. If they are floating high in the room they reflect mostly the floor. If they are floating low in the room they reflect mostly the ceiling. Each bubble is slightly different.

the dots you put on the sphere. This is the way it would work. This will work this way no matter where you locate yourself within the building. Can you see why people see mathematics when looking at the Termesphere?

The following Termespheres illustrate my point. Some are real places, some are imaginary environments and some are geometric worlds. Think of yourself inside the sphere.

Constructing Wire Models

François Apéry

Département de Mathématiques, Université de Haute Alsace, 68093 Mulhouse
<f.apery@uha.fr>

1 Introduction

There is a long tradition of mathematical models building up. Without going back to Archimedes, one can mention the famous Martin Schilling Catalogue which registers and describes about four hundred models in 1911, a few of which still surviving behind the glass cases of some mathematics institutes. One can get a glimpse of them in the beautiful book edited under the direction of Gerd Fischer ([6]).

Most of the models are surfaces for obvious reasons. The ruled surfaces are realized on a framework crossed through by sewing-threads which materialize the generating lines. By using a variable framework, one can also generate a homotopy of ruled surfaces, for instance the deformation of a cone of revolution into a one-sheeted hyperboloid. One of the fascinations of such models comes from the surface not being realized physically, as for clay or wood models, or even grid models, but appears only through its apparent contour, which seems to levitate in space, somewhat similar to a caustic generated by light rays reflected in a mirror.

That is the attribute we wish to extend to surfaces generated by other family of curves, and first of all by conics, by using the mechanical properties of the wire, in particular, those of piano wire. Such a steel is commonly used in orthodontics under the name of rush (jonc in French). As an effect of its elasticity, such a wire doesn't keep the trace of a deformation, provided that it is not too drastic. A wire of a given length submitted to mechanical constraints takes an equilibrium shape which realizes a particular curve.

Imagine for instance that the ends are joined together smoothly at a given point (at which the angle of the two branches is forced to be flat), then it means four constraints, and the equilibrium state is a circle. Now, if in addition we impose the wire to go through a second point in the plane of the circle, which means a fifth constraint, then the equilibrium state of the wire is a convex plane curve not so different from an ellipse. From this came the idea of building a surface generated by ellipses by means of a framework on which wires are subject to at least five constraints.

The surface represented this way gives the impression of existing only virtually through its apparent contours. In addition the one-parameter family of conics, just as the frame of lines on a ruled surface, conveys a first suggestion on the method to parametrize the surface. As a matter of fact, the task is to

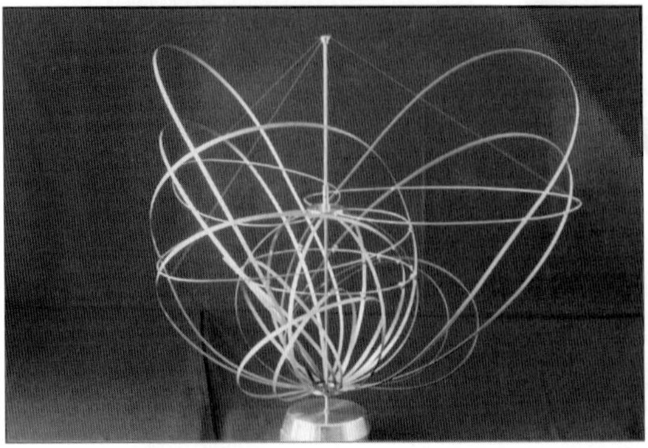

Fig. 1. [FA]1: Boy surface in steel strips realized by F. Apéry

represent a parametrized surface rather than a geometric surface viewed as a subset of the space $\mathbb{R}^3$. So that, if the model has a variable framework one will visualize a homotopy rather than a deformation of topological objects.

The first model I constructed according to this principle is a surface of order six generated by ellipses passing through a fixed point, namely the Boy surface (see figure (1) the and photograph [FA]1 in the appendix) [1]. This model is exhibited at the mathematics department of Cagliari university in Italy. Here I would like to show two other models. The first one, the closed halfway model of the sphere eversion, is a real algebraic surface of order eight generated by a family of ellipses passing through a fixed point. The second one is a variable model of the so-called **gastrulation**, a quasigeneric homotopy between the standard embedding of the sphere and the antipodal one.

Prior to describe these models, it seemed to me useful to show the mathematical context in which one can study the equilibrium state of a springy inextensible wire. The question belongs to calculus of variations and is a problem sufficiently elementary: that theory can be found in [5] for instance.

The elasticity of a wire can be measured by a modulus directly proportional to the speed at which the wire resumes its initial shape after relaxing the constraints. Since Daniel Bernoulli, one defines the potential energy of a length of springy wire as the product of the elasticity modulus by its total squared curvature (or the difference between the curvature under constraint and the relaxed curvature, in case of the relaxed wire is not straight as it happens when it is wound on a reel). The more the wire is twisted or bent, the greater the curvature, and the more the potential energy increases.

The universal Principle of Least Action requires the wire to minimize its potential energy. In order to keep the paper as self-contained as possible, the following section is devoted to the setting of this principle in equation according to Euler.

2 The variational problem

2.1 The functional space

We represent a springy inextensible wire by a path in $\mathbb{R}^3$. We consider the real vector space E_2 of the class C^2 paths traced in the euclidean space $\mathbb{R}^3$, with origin at O, equipped with the norm

$$\|x\| = \sup_I \|x(t)\| + \sup_I \|x'(t)\| + \sup_I \|x''(t)\| .$$

It is a Banach space on which the functional of length

$$\lambda(x) = \int_0^1 \|x'(t)\| \, dt,$$

is continuous, as well as the functional

$$\mu(x) = \int_0^1 x''(t)^2 \, dt,$$

where the exponent 2 denotes a scalar square. Let F be the closed subset of E_2 of the unit length paths parametrized by arc length. According to Ascoli theorem, F is compact with respect to the uniform convergence topology, but not to that of E_2. We shall discuss the extremals for μ in restriction to F.

2.2 The potential energy of a springy inextensible wire

The curvature of a path x in E_2 is given by the formula

$$\kappa = \|x' \wedge x''\| \cdot \|x'\|^{-3},$$

so that, on F, its square yields

$$\kappa^2 = x''^2.$$

Then the functional μ, in restriction to F as well, writes down

$$\mu(x) = \int_0^1 \kappa(s)^2 \, ds.$$

That is the form chosen by Euler to define the **potential energy** of a springy inextensible wire figured by x.

2.3 The energy of the derived path

It is convenient to look at the derivative x' as a path traced in $\mathbb{R}^3$, that is, as an element of the Banach space E_1 of the class C^1 paths (whose origin is no longer assumed to be O) equipped with the norm

$$\|x\| = \sup_I \|x(t)\| + \sup_I \|x'(t)\| .$$

The condition to belong to F amounts to lay out the derived path (which is no longer assumed to be of unit length) on the unit sphere of $\mathbb{R}^3$. The functional to be minimized on the sphere writes down

$$\nu(x) = \int_0^1 x'(s)^2\, ds.$$

That is, by definition, the **energy** of the path on the unit sphere. The functional μ reads as the energy of the derived path. Therefore, an extremal of μ on F is nothing more that an extremal of ν on the unit sphere (no longer assumed to be of unit length). In other words it is a geodesic of the sphere, that is to say an arc of great circle, traversed at unit speed. In particular x' stays in a fixed plane passing through the origin. There exists a constant vector u such that $u \cdot x' = 0$. Then the scalar product $u \cdot x$ is constant, something which proves that x is traced on a plane orthogonal to u and passing through the origin because of the initial condition $x_0 = O$. By integration, one finds that x is an arc of circle possibly degenerated into a line segment. What are the extremals of μ in the space F of paths y subject to satisfy some extra conditions, we shall see in the next sections.

2.4 Computing the derivative of ν

It could be useful to see that ν is of class C^1 and on the same hand to obtain its derivative. Let us define $E \times I \xrightarrow{\psi} \mathbb{R}$ by $\psi(x,t) = \varphi(x'(t))$, where $\varphi(v) = v^2$, so that

$$\nu(x) = \int_0^1 \psi(x,t)\, dt.$$

In order to prove the derivability of ν on E_1, we shall use the theorem of differentiation inside the sign of integration. The map ψ decomposes in the form $\psi = \varphi \circ \Phi$ where the map $E \times I \xrightarrow{\Phi} \mathbb{R}^3$ is given by $\Phi(x,t) = x'(t)$. The map Φ is linear and continuous with respect to x, so that it is differentiable with respect to x and its partial derivative is continuous and writes

$$\partial_x \Phi(x,t) \cdot y = \Phi(y,t) = y'(t) .$$

On the other hand, φ is of class C^1 and its derivative is

$$D\varphi(v) \cdot w = 2v \cdot w,$$

so that ψ is differentiable with respect to x and its partial derivative is continuous and writes, by the chain rule,

$$\partial_x \psi(x,t) \cdot y = (D\varphi(\Phi(x,t)) \circ \partial_x \Phi(x,t)) \cdot y = 2x'(t) \cdot y'(t).$$

Eventually ν is of class C^1 and its derivative can be written as

$$D\nu(x) \cdot y = 2 \int_0^1 x'(t) \cdot y'(t)\, dt.$$

2.5 Constraint of the half tangents at the extremities

Freezing the half tangents at the extremities of an extremal of μ is equivalent to fix the extremities of the corresponding extremal of ν on the unit sphere. We take $y_0 = (1,0)$ as an origin and $y_1 = (\exp i\theta_1, 0)$ as an end, where $0 \leqslant \theta_1 < 2\pi$ and $\mathbb{R}^3$ is identified to $\mathbb{C} \times \mathbb{R}$. A peculiar extremal of μ is given by the horizontal arc of circle centered at i/θ_1 starting at O, ending at $i(1 - \exp i\theta_1)/\theta_1$ and counterclockwise oriented. There is as well the horizontal arc of circle centered at $i/(\theta_1 - 2\pi)$ starting at O, clockwise oriented, and ending at $i(1 - \exp i\theta_1)/(\theta_1 - 2\pi)$. As a matter of fact, there are infinitely many solutions given by the arcs of circles running possibly several laps, centered at $i/(\theta_1 + 2k\pi)$ starting at O, ending at $i(1 - \exp i\theta_1)/(\theta_1 + 2k\pi)$ and oriented according to the sign of k.

2.6 Isoperimetric constraint

We impose conditions at the extremities and the half tangents at the extremities as well. It amounts to look for the extremals of ν on the space of paths y starting at $y_0 = (1,0)$, ending at $y_1 = (\exp i\theta_1, 0)$, traced on the unit sphere and subject to satisfy the condition

$$\xi(y) = x_1, \tag{1}$$

where

$$\xi(y) = \int_0^1 y(s)\, ds.$$

Such a constraint is said to be **isoperimetric**. Since we impose extra conditions, there are fewer paths satisfying them and the extremals are no longer to be geodesics on the sphere. However if a geodesic on the sphere satisfy these extra conditions, then of course it is one of the requested geodesic. The maps

$$y \longmapsto y(0), \quad y \longmapsto y(1) \quad \text{and} \quad y \longmapsto \int_0^1 y(t)\,dt$$

are linear and continuous on E_1, so that the equations $y(0) = y_0$, $y(1) = y_1$ and (1), define a closed affine subspace A_1 of E_1. Therefore we are looking for the extremals of ν on the closed subset $G = S \cap A_1$, where $S \subseteq E_1$ denotes the set of paths traced on the unit sphere. If G is not empty, one can choose an origin $\overline{y}$ on G in order to make A_1 a Banach space, namely the vector subspace $E_{1,0}$ of E_1 defined by the equations

$$y(0) = 0, \quad y(1) = 0 \quad \text{and} \quad \int_0^1 y(t)\,dt = 0. \tag{2}$$

The functional ν in restriction to A_1 is of class C^1 since it is so on E_1 (see section 2.4).

2.7 Euler-Lagrange equation

The isoperimetric constraint (1) introduces the **Lagrange multiplier** 2Λ (factor 2 shall allow a latter simplification) in the Euler-Lagrange equation. Let us write

$$\nu(y) = \int_0^1 N(s, y, y')\,ds \quad \text{and} \quad \xi(y) = \int_0^1 M(s, y, y')\,ds,$$

with

$$N(s, u, v) = v^2 \quad \text{and} \quad M(s, u, v) = u.$$

The variational derivative of $N + 2\Lambda M$ is equal to (assuming that y is of class C^3)

$$\begin{aligned} [N + 2\Lambda M]_y &= \frac{\partial}{\partial u}(N + 2\Lambda M)(s, y, y') - \frac{d}{ds}\frac{\partial}{\partial v}(N + 2\Lambda M)(s, y, y') \\ &= -2y'' + 2\Lambda. \end{aligned}$$

The fundamental lemma in the calculus of variations says that an extremal of ν on G is given by a path $y \in G$ such that

$$[N + 2\Lambda M]_y \cdot z = 0 \quad \forall z \in E_{1,0} \cap T_y, \tag{3}$$

where T_y denotes the vector space of the paths z in $\mathbb{R}^3$ such that, for every s, the vector $z(s)$ be tangent to the unit sphere at $y(s)$, that is

$$y \cdot z = 0.$$

Therefore we have

$$y \cdot z = 0 \Rightarrow (-y'' + \Lambda) \cdot z = 0,$$

something which gives the **Euler-Lagrange equation**

$$y \wedge y'' = y \wedge \Lambda,$$

or equivalently

$$x' \wedge x''' = x' \wedge \Lambda. \tag{4}$$

Considering the initial condition $x_{s_0}, x'_{s_0}, x''_{s_0}$ at the time s_0 not assumed to vanish, the latter equation integrates in the form

$$x' \wedge x'' = (x - x_{s_0}) \wedge \Lambda + x'_{s_0} \wedge x''_{s_0}.$$

We deduce that the scalar triple product

$$[x', x'', \Lambda] = \left[x'_{s_0}, x''_{s_0}, \Lambda\right]$$

is constant, and therefore, thanks to (4), we have

$$[x', x'', x'''] = \left[x'_{s_0}, x''_{s_0}, x'''_{s_0}\right].$$

This shows that the path is planar as soon as the torsion vanishes at one point. The path x is planar only if the vectors $x_1 - x_0$, x'_0, x'_1 are dependent, this condition being necessary but not sufficient.

2.8 The example of the circular helix

Let us consider the whorl of circular helix

$$x = \frac{1}{2\sqrt{2}\pi} \left(e^{2i\pi s}, 2\pi s\right).$$

We do have $x'^2 = 1$. Moreover we have $x'_0 = x'_1$, thus

$$[x_1 - x_0, x'_0, x'_1] = 0.$$

On the other hand

$$[x', x'', x'''] = 2\pi^3\sqrt{2} \begin{vmatrix} -\sin 2\pi s & -\cos 2\pi s & \sin 2\pi s \\ \cos 2\pi s & -\sin 2\pi s & -\cos 2\pi s \\ 1 & 0 & 0 \end{vmatrix} = 2\pi^3\sqrt{2}.$$

We also have

$$\mu(x) = \int_0^1 x''(s)^2 \, ds = 2\pi^2.$$

The whorl x is really an extremal of the studied isoperimetric problem. Actually we have

$$x' \wedge x''' = x' \wedge \Lambda,$$

with $\Lambda = (0, 0, 2\pi^2\sqrt{2})$. The corresponding path y on the unit sphere is a circle of radius $1/\sqrt{2}$, so that it is not an arc of great circle. Nevertheless x does not afford a minimum to μ, for there are plane curves on which the value of μ is lower. In particular, there is one of the Euler elastic curves (see figure (2), and figure (4) from which one can get the requested elastic plane curve) satisfying the initial conditions.

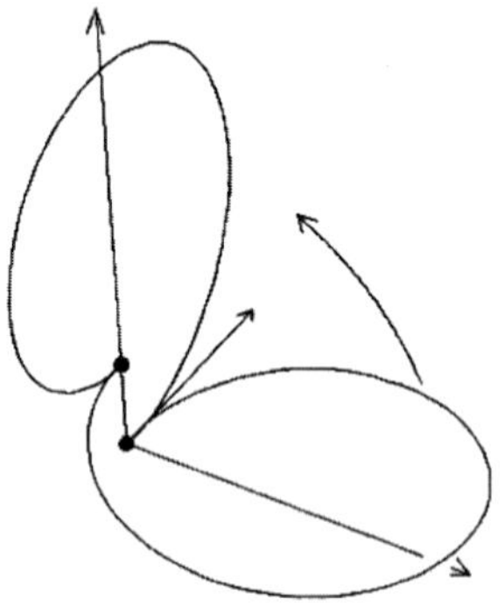

Fig. 2. The horizontal whorl of circular helix and the vertical Euler elastic curve satisfying the same initial conditions. The arrow indicates the decreasing potential energy.

3 The case of plane curves

3.1 Reduction of the Euler-Lagrange equation

Even if it does mean rotating the frame of the plane, one can assume that $\Lambda = \lambda^2 \in \mathbb{R}_+^* \subseteq \mathbb{C}$, the plane being identified to $\mathbb{C}$. We shall recover all the solutions by applying a rotation around the origin. For instance, if we wish to extend a solution on both sides of the origin, it suffices to match it with those the half tangent of it is given by $-x_0'$. We have not considered the case $\Lambda = 0$ already discussed since it corresponds to the absence of isoperimetric constraint. As an intermediate parameter one uses the angle α between the derivative x' and the horizontal axis, so that

$$x' = e^{i\alpha(s)}. \tag{5}$$

Let us set $\alpha(0) = \alpha_0$. The Euler-Lagrange equation (4) takes the form of the pendulum equation

$$\alpha'' + \lambda^2 \sin\alpha = 0,$$

which is known to be integrable by elliptic integrals. There are the trivial solutions $\alpha \equiv 0$ (π) giving rise to line segments. After multiplying by α', a first integration gives

$$\frac{\alpha'^2}{2} - \lambda^2 \cos\alpha = \lambda^2 (2\mu - 1),$$

or equivalently

$$\alpha'^2 = 4\lambda^2 \left(\mu - \sin^2 \frac{\alpha}{2}\right),$$

with $\mu > 0$, the case $\mu = 0$ giving rise to line segments only. It is convenient to set $\beta = \frac{\alpha}{2}$, so that

$$x'(s) = e^{2i\beta}$$

and the former equation now reads

$$\beta'^2 = \lambda^2 \left(\mu - \sin^2 \beta\right). \tag{6}$$

3.2 Case $\mu = 1$:

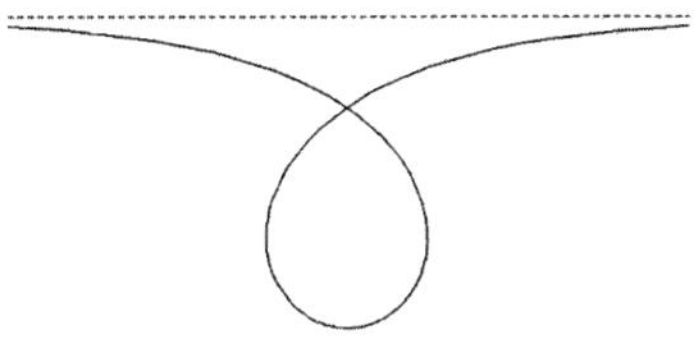

Fig. 3. Elastic curve, $\mu = 1$

The differential equation reads

$$\beta' = \lambda \cos\beta,$$

and the general solution is given by

$$\operatorname{tg}\left(\frac{\beta}{2} + \frac{\pi}{4}\right) = \nu e^{\lambda s}.$$

We make the following change of parameter

$$x(s) = X(t) \quad \text{with} \quad t = \nu e^{\lambda s}.$$

We get

$$x'(s) = \lambda t X'(t),$$

so that the relation (5) now reads

$$\lambda X'(t) = \frac{1}{t} e^{i\alpha(s)} = \frac{-t^4 + 6t^2 - 1}{t\left(1 + t^2\right)^2} + 4i \frac{t^2 - 1}{\left(1 + t^2\right)^2}.$$

Considering the initial condition $X(\nu) = 0$, we obtain

$$\lambda X(t) = -\operatorname{Log} \frac{t}{\nu} - \frac{4(1 + it)}{1 + t^2} + \frac{4(1 + i\nu)}{1 + \nu^2}.$$

Referring to the parameter s, one has (Fig. 3)

$$\begin{aligned} x(s) &= -s - \frac{4\left(1 + i\nu e^{\lambda s}\right)}{\lambda\left(1 + \nu^2 e^{2\lambda s}\right)} + \frac{4(1 + i\nu)}{\lambda(1 + \nu^2)} \qquad (7) \\ &= -s - \frac{4\left(1 + i\nu e^{\lambda s}\right)}{\lambda\left(1 + \nu^2 e^{2\lambda s}\right)} + \frac{2}{\lambda}\left(1 + i \exp i \frac{\alpha_0}{2}\right), \end{aligned}$$

with the initial conditions $x_0 = 0$ and

$$\begin{cases} \operatorname{tg}\left(\dfrac{\alpha_0 + \pi}{4}\right) = \nu \\ \operatorname{tg}\left(\dfrac{\alpha_1 + \pi}{4}\right) = \nu e^{\lambda} \end{cases}.$$

3.3 Case $\mu > 1$:

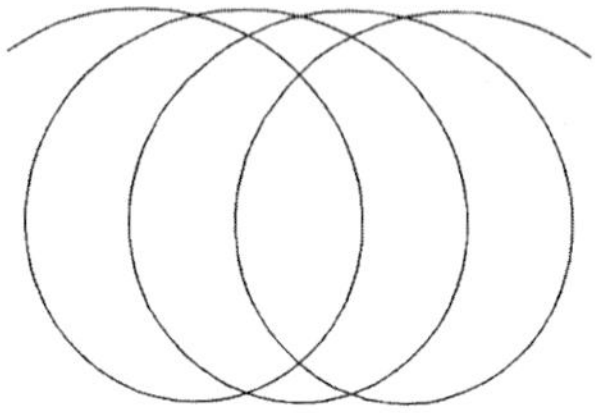

Fig. 4. Elastic curve, $\mu = 2$

Let us set $\mu = 1/k^2$. The equation (6) reads

$$\beta' = \frac{\lambda}{k} \sqrt{1 - k^2 \sin^2 \beta}.$$

The arc length is given in function of β by the Legendre-Jacobi's elliptic integral of the first kind with modulus k

$$s(\beta) = \frac{k}{\lambda} F(\beta) - s_0,$$

where

$$F(\beta) = \int_0^\beta \frac{d\gamma}{\sqrt{1 - k^2 \sin^2 \gamma}} \quad \text{and} \quad s_0 = \frac{k}{\lambda} F(\beta_0),$$

setting $\beta_0 = \alpha_0/2$. Indeed, the origin of the arc length is taken at the time 0 in such a way that $s(\beta_0) = 0$, a value which differs a priori from s_0. This function like the latter depends on the modulus k although, in order to lighten the notation, we have slipped it. It is continuous and strictly increasing. The inverse function is the amplitude

$$\beta = \operatorname{am} F(\beta)$$

It yields to the expression of β in terms of the Jacobi's elliptic functions $\operatorname{cn} = \cos \operatorname{am}$ and $\operatorname{sn} = \sin \operatorname{am}$:

$$\begin{cases} \cos(\beta) = \operatorname{cn} \frac{\lambda}{k}(s + s_0) \\ \sin(\beta) = \operatorname{sn} \frac{\lambda}{k}(s + s_0) \end{cases}.$$

Then we deduce

$$x'(s) = \operatorname{cn}^2 \frac{\lambda}{k}(s + s_0) - \operatorname{sn}^2 \frac{\lambda}{k}(s + s_0) + 2i \operatorname{sn} \frac{\lambda}{k}(s + s_0) \operatorname{cn} \frac{\lambda}{k}(s + s_0).$$

Using the Jacobi's function dn (the derivative of the amplitude) which satisfies

$$\operatorname{dn}^2 = 1 - k^2 \operatorname{sn}^2,$$

one can integrate the imaginary part of the previous expression:

$$\begin{aligned} \operatorname{Im} x(s) &= \frac{2}{\lambda k}\left(\operatorname{dn}\left(\frac{\lambda}{k} s_0\right) - \operatorname{dn}\left(\frac{\lambda}{k}(s + s_0)\right)\right) \\ &= \frac{2}{\lambda k}\left(\sqrt{1 - k^2 \sin^2 \beta_0} - \operatorname{dn}\left(\frac{\lambda}{k}(s + s_0)\right)\right). \end{aligned}$$

For the real part, we consider the Legendre-Jacobi's elliptic integral of the second kind with modulus k

$$\mathcal{E}(u) = \int_0^u \operatorname{dn}^2 t dt,$$

so that

$$\operatorname{Re} x(s) = s\left(1 - \frac{2}{k^2}\right) + \frac{2}{\lambda k}\left(\mathcal{E}\left(\frac{\lambda}{k}(s + s_0)\right) - \mathcal{E}\left(\frac{\lambda}{k} s_0\right)\right).$$

Eventually

$$\begin{aligned} x(s) &= s\left(1 - \frac{2}{k^2}\right) \qquad (8) \\ &+ \frac{2}{\lambda k}\left(\mathcal{E}\left(\frac{\lambda}{k}(s + s_0)\right) - \mathcal{E}\left(\frac{\lambda}{k} s_0\right) - i \operatorname{dn}\left(\frac{\lambda}{k}(s + s_0)\right) + i \operatorname{dn}\left(\frac{\lambda}{k} s_0\right)\right). \end{aligned}$$

3.4 Case $0 < \mu < 1$:

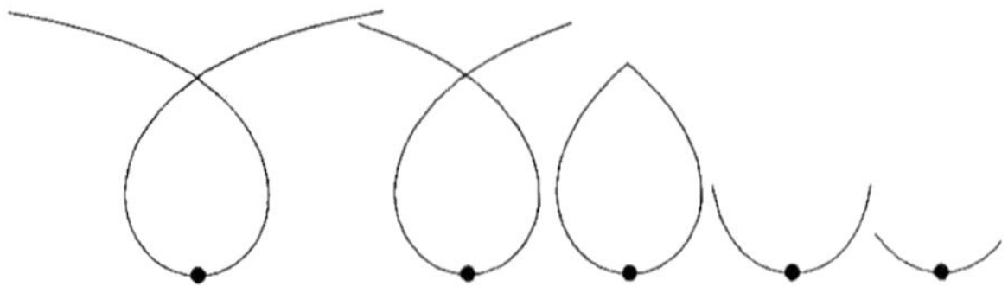

Fig. 5. Elastic curves for $\mu = 0.994$; 0.98; 0.908; 0.5; 0.25

Set $\mu = \sin^2 \beta_1$ with $0 < \beta_1 < \frac{\pi}{2}$. The equation (6) implies $\beta \in [-\beta_1, \beta_1]$ modulo π, in particular, we can assume that $\beta_0 \in [-\beta_1, \beta_1]$. We have

$$\beta' = \frac{\lambda}{k}\sqrt{1 - k^2 \sin^2 \beta},$$

with

$$k = \sin^{-1} \beta_1.$$

Just as above, the arc length is given in function of β by the Legendre-Jacobi's elliptic integral of the first kind

$$s(\beta) = \frac{k}{\lambda} F(\beta, k) - s_0,$$

but this time, the modulus k is greater than 1, and $\beta \in [-\beta_1, \beta_1]$. In addition, the modulus k appears in the notation of F, for we shall need to modify it. We still have

$$\begin{cases} s_0 = \frac{k}{\lambda} F(\beta_0, k) \\ \operatorname{dn}\left(\frac{\lambda}{k} s_0\right) = \sqrt{1 - k^2 \sin^2 \beta_0} \end{cases}$$

The integral F converges when β takes the value $-\beta_1$ or β_1. The function F is odd, continuous and strictly increasing on $[-\beta_1, \beta_1]$. Its inverse function is the amplitude

$$\operatorname{am} : [-F(\beta_1, k), F(\beta_1, k)] \to [-\beta_1, \beta_1].$$

The function $x(s)$ is still given by the formula (8), but s must stay in the interval

$$\left[-\frac{k}{\lambda} F(\beta_1, k) - s_0, \frac{k}{\lambda} F(\beta_1, k) - s_0\right].$$

Notice that

$$\operatorname{dn}\left(\frac{\lambda}{k}(s(\beta_1) + s_0)\right) = \operatorname{dn} F(\beta_1, k) = \sqrt{1 - k^2 \sin^2 \beta_1} = 0,$$

so that

$$\operatorname{Im} x\left(s\left(\beta_1\right)\right)=\frac{2}{\lambda k}\sqrt{1-k^2\sin^2\beta_0}. \tag{9}$$

Case $\mu=\frac{1}{2}$:

Observe that in this case, which is nothing more than a peculiar occurence of the former case, the linear part disappears in the formula (8):

$$x(s)=\frac{\sqrt{2}}{\lambda}\left(\mathcal{E}\left(\frac{\lambda}{\sqrt{2}}\left(s+s_0\right)\right)-\mathcal{E}\left(\frac{\lambda}{\sqrt{2}}s_0\right)-i\,\mathrm{dn}\left(\frac{\lambda}{\sqrt{2}}\left(s+s_0\right)\right)+i\sqrt{\cos\alpha_0}\right).$$

The equation (6) takes the following form

$$\alpha'=\lambda\sqrt{2}\sqrt{\cos\alpha} \quad \text{with} \quad \alpha\in\left[-\frac{\pi}{2},\frac{\pi}{2}\right],$$

which recalls the one of the Bernoulli's lemniscate

$$\alpha'=3\sqrt{\cos\frac{2}{3}\left(\alpha-\frac{\pi}{2}\right)}.$$

3.5 Summing up

The differential equation (6) that determines the extremals of the isoperimetric problem yields, when Λ is fixed, to a local condition. Thus it is possible to obtain the extremals by gluing together some of the above solutions involving the same value of Λ or λ. The elastic curves that are solutions have to be of unit length. On the other hand, we had assumed that $\Lambda\in\mathbb{R}_+^*$, up to rotation of the frame in the plane. Therefore we have the possibility to rotate the former curves in order to get additional extremals. By construction, they have O as starting point. The half tangent at the origin x_0' gives α_0. The two last initial conditions, x_1 and x_1' (actually three conditions, two for x_1 and one for x_1' namely α_1), determine λ and μ and the rotation to be applied around the origin.

For instance, in the case $\mu=1$, we can determine the elastic loop of unit length (Fig. 6). The condition $x_1=0$ carried into (7) yields to the values $\lambda=\lambda_0$ and $\nu=e^{-\lambda_0/2}$, where $\lambda_0=3,83\ldots$ is the unique positive solution of the equation

$$\lambda=4\,\mathrm{th}\,\frac{\lambda}{2}.$$

The angle at the apex of the loop is

$$\theta=\pi-8\,\mathrm{Arctg}\,e^{-\lambda_0/2}\simeq 113^\circ.$$

Thus, if the springy inextensible wire is forced to take the shape of a loop with an angle about 113° at the apex, it will adopt the shape of the figure (6). In return, if the forced angle at the apex differs from 113°, then we will find a loop corresponding to a value of μ different from 1. If the forced angle is equal to 180° we get a convex curve which is close enough to an ellipse to make wire models of surfaces generated by conics definitely realistic.

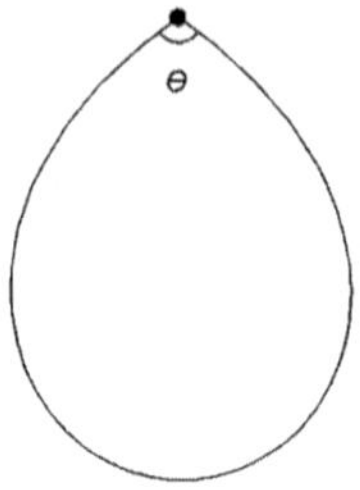

Fig. 6. Elastic loop

4 Wire models

4.1 The halfway model

Equations. The surface in question occurs at the halfway step of a sphere eversion that is symmetric. We call such a surface a halfway model. The geometric construction of such an eversion and therefore of its halway step is due to Bernard Morin ([8] and [7]). A notion of generic transition between two immersions can be defined in such a way that makes it possible to encode sphere eversions by specifying the sequence of its generic transitions (see [2]). Thus one can distinguish several kinds of halfway models, according to the number of generic transitions occuring simultaneously at the time 0, that is at the halfway step of the eversion. The so-called open halfway model gathers two generic transitions together, namely a quadruple point and a transition called D_1 which performs a surgery on the curve of double points.

In order to construct algebraically the surface corresponding to the halfway model, it is worth considering the case where the number of generic transitions is as big as possible. Indeed, such a transition creates an algebraic degeneracy, and the more of these there are, the easier it is to derive the equation of the surface. Hence the notion of **closed halfway model** which grasps five generic transitions: one quadruple point and five D_1. Thus we can define the closed halfway model to be a real algebraic surface of order eight, that is to say, the set of real zeros of the following polynomial of degree eight

$$P(X,Y,Z) = -72ADE^2 - 18DEG(C+E) + 3BG^2(3B-4D) + 4G^4,$$

where

$$\begin{cases} A = Z(1-Z), \quad B = 2(X^2+Y^2), \quad C = 2(Y^2-X^2), \\ D = 2Z^2, \quad E = 4XY, \quad G = 3(X^2+Y^2+Z^2) - 4Z. \end{cases}$$

By construction, this surface is generated by a one-parameter family of conics all passing through the origin and which are the images of the meridians of the sphere by the parametrization

$$f(\vartheta,\eta)=\frac{\cos\eta}{1-\sin\eta\cos\eta\,(\sin 2\vartheta-\cos 2\vartheta)}\left|\begin{array}{c}\frac{1}{\sqrt{3}}(\cos\eta+2\sin\eta)\cos\vartheta\\ \frac{1}{\sqrt{3}}(\cos\eta-2\sin\eta)\sin\vartheta\\ \cos\eta\end{array}\right.,$$

where ϑ and η respectively denote the longitude and the latitude on the sphere. This formulation fits our present purposes.

Description of the wire model. Each meridian on the surface (in other words, the image of a meridian of the sphere by f) is an ellipse tangent to the horizontal plane at the origin. In order to fix an ellipse it suffices to impose five independent linear conditions. Two conditions are given by the meridians passing through the origin with a horizontal tangent line. Two additional conditions are fixed by imposing the intersection points with the horizontal plane at height 1. The corresponding level curve decomposes into a circle and an elongated astroid, that is to say a curve of order two and a curve of order six, which amounts to eight as expected (Fig. 7).

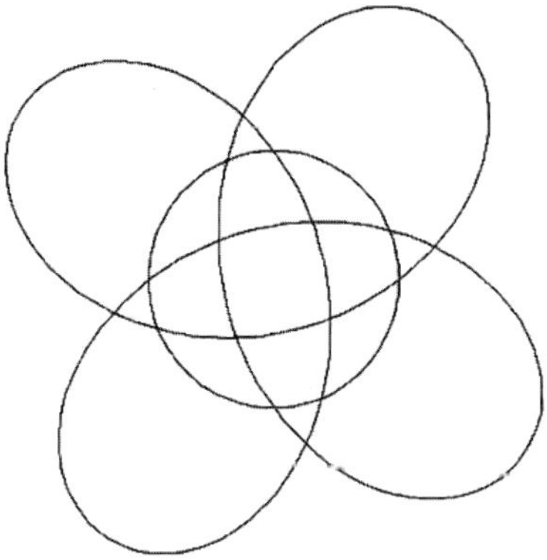

Fig. 7. The level curve of the closed halfway model at the height 1

The last condition (which is not linear itself) is given by the length of the ellipse, since we want the wire used to realize it to be inextensible.

The wire, chosen for its elastic properties, is a nickel-chrome stainless steel of section 0.9 mm in diameter and endowed with a large resistance to traction (about 2000 MPa). The wire is fastened to a reel and, in absence of constraints, bends naturally into the shape of a semi circle. The framework on which we plan to fix the wires is made up of two parts: a brass hub located at the origin where all the meridians converge, and a brass plate taking the shape of a tubular neighbourhood of the level curve at height 1 in the surface.

The hub as well as the plate are perforated by 116 holes through which we thread the 58 meridians of the model.

For symmetry reasons (the model has a fourfold symmetry around the vertical axis), the number of meridians should be a multiple of four, 60 for instance. But there are two double meridians, which explains the $58 \times 2 = 116$ apertures. In addition, we have used wires with slightly larger diameter (1 mm in diameter) for the two double meridians which happen to be circles.

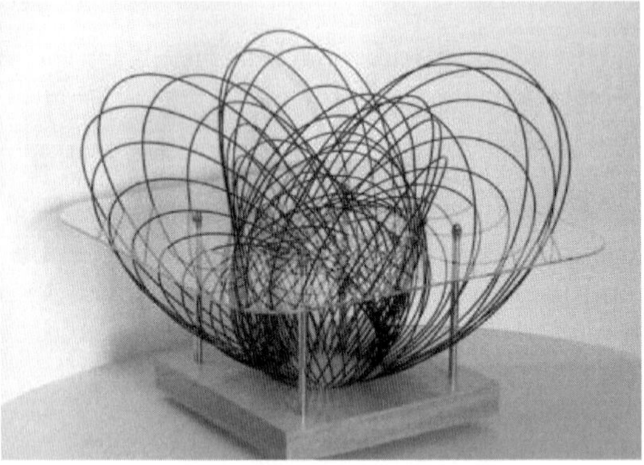

Fig. 8. [FA]2: Halway model in Teflon and Plexiglass realized by G. Franzoni

In the first model constructed under my direction by Gregorio Franzoni in Cagliari University in 1998 ([4]), the meridians are in Teflon. They are fixed on a horizontal plane in wood (the base of the model) of slightly positive height, and go through the height one horizontal plane in Plexiglas. Both planes are held in position by four metallic rods (see the figure (8) and the photograph [FA]2 in the appendix). The base being a little bit above the height 0 hides the neighbourhood of the pole.

Since it is essential to see the behavior at the pole clearly, the present model has a hub at the pole in order to fix all the meridians and to get a better representation of the surface in its neighbourhood. In addition, by using the wire specified above, the tensions balance so that it is no longer necessary to hold both horizontal planes in position by a set of rods. The base is now unnecessary. And, for the height-1 horizontal plane, it is held in position during the construction of the meridians by a central rod, which is removed at the end.

All that remains is a one-parameter of conics gathered at the polar hub, and the neighbourhood of the height one level curve which appears to be levitating. Thus, our wire model displays its apparent contour clearly, just as thread models of ruled surfaces do. What is remarkable is that the model stays in position by itself. There are neither screws nor welds. It can be completely dismantled. These are the only forces at work. On the one hand, they hold each wire in position as well as in shape, and, on the other hand, maintain the piece of brass at height one that the wires must go through along the level curve (see the photograph [FA]3 in the appendix).

4.2 The gastrulation

The amphioxus's egg. We shall now discuss a deformable model that illustrates, at the same time, a mathematical object called homotopy, and an embryological phenomenon, called **gastrulation**. A description of this has appeared in [3], but to make the present discussion self-contained, I shall repeat the essentials here. The reader may recall the school science lesson about the gastrulation stage in the development of the frog embryo. In fact, it is not the frog but the amphioxus embryo which displays the sort of gastrulation discussed here. The amphioxus is a more primitive but also aquatic vertebrate animal with sharp points at each end, which accounts for its name.

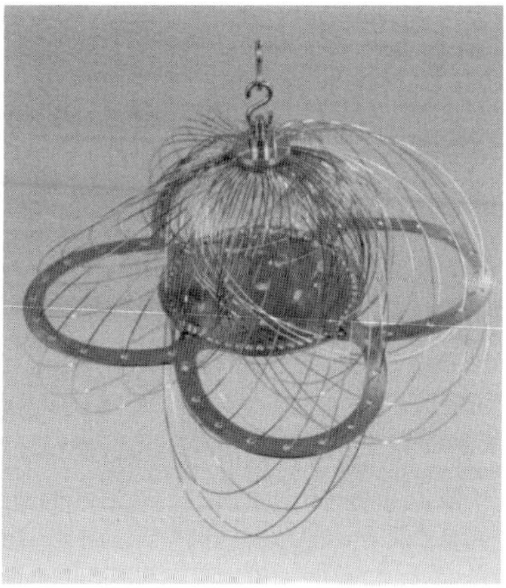

Fig. 9. A view of the wire halway model realized by F. Apéry. The bug pied of brass has been milled in the Laboratoire de Génie mécanique de l'IUT de Mulhouse. The hub has been turned in the Officina della Università degli Studi di Calgari (Italy).

At the time of laying, the egg has a diameter less than 2mm and admits roughly speaking an axis of revolution passing through the vegetative and the animal poles. After fertilization there begins the segmentation, that is, a cleavage of the egg into cells called blastomeres. It starts by a vertical cleavage along a meridian plane, that is to say a plane containing the polar axis, something which yields two symmetric blastomeres. The next cleavage, which is again vertical and orthogonal to the first one, yields four blastomeres in quarters of an orange. Then appears an equatorial cleavage, that is to say defined by a horizontal plane, which generates eight blastomeres. The more the cleavage proceeds, the more the egg looks like a mulberry, hence the name of morula.

Then, geometrically speaking, things becomes a little messy, for the cleavages no longer occur with the same geometrical and chronological regularity. Furthermore, the blastomeres move aside from the center in order to create a cavity filled up with liquid, the so-called blastocœl. We get roughly a sphere whose interior is liquid: the blastula. Afterwards starts the morphological stage called the gastrulation, for it eventually results in the stomach of the being, namely the archenteron. This is precisely the stage we are interested in.

This stomach, contrary to what a naive interpretation suggests, does not take place inside the blastula. For if it did, there would be a problem with the anal aperture, the blastopore. If the blastopore were made in the surface of the blastula, then the blastocœlian liquid would leak out. That is why the gastrulation gives rise to the simultaneous formation of the archenteron and the blastopore by a geometric deformation with neither tearing nor adding handles. The mathematical designation of this kind of continuous deformation preserving the topology is the isotopy.

The process creates a cavity in the blastula by pushing down the animal pole, in an invagination move, to a point close to the vegetative pole, in the way of a deflated balloon whose valve is pushed down (Fig. 10).

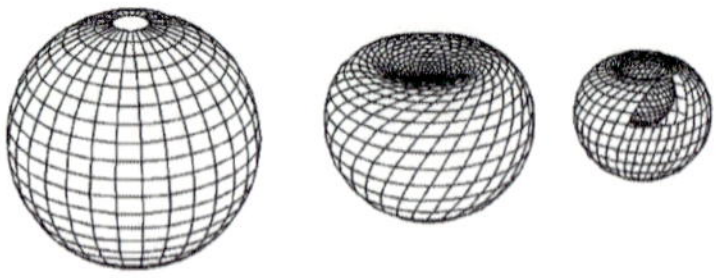

Fig. 10. Gastrulation of the blastula by embolism

As a result of this embolism, the blastula takes the shape of a bowl. It is the circular boundary of this bowl which becomes the anus by contraction and delimits the volume of the archenteron (The matter of how the oral aperture appears later does not concern us here). However, if we realize this deformation on a deflated balloon, the contraction of the blastopore involves the shrinking of the archenteron volume and eventually we are led to a more and more narrow sock where the stomach as well as the intestine have no adequate place to be.

It is clear that the foregoing is an extremely skeletal view of the gastrulation just given in order to capture the geometric nature of the process.

Quasi-eversion of the sphere. The geometric model of the gastrulation defined above shows a way to approach the sphere eversion. Indeed, the gastrulation describes a process for continuously joining the sphere to a double covered sphere (the so-called gastrula). During this deformation, the equatorial circle of the sphere collapses into a point (the blastopore), and one of the two hemispheres, say the northern hemisphere, goes inside the interior of the double sphere while the southern hemisphere stays on the outside. This way, exchanging the hemispheres with each other and performing the reciprocal of the gastrulation will yield an everted sphere.

The exchange of the hemispheres introduces a singularity along the rim. Every parametrization of this motion would cease to be an immersion along the rim because the rank of its Jacobian is no longer maximal there. If we allow such a singularity, we could do it more easily by flattening a sphere to a doubly covered equatorial disk. Let the two sheets of the disk pass through each other and inflate the everted sphere. In that case, the rank of the map fall on each point in the boundary of the double disk which besides come up from the equatorial circle of the source sphere as well, therefore presumably nothing more singular happens than with the gastrulation, but the simplicity.

However, a geometric argument pleads in favor of the gastrulation as the preferred model. Here is why. During an eversion, the tangent plane at a given point must rotate one full turn. This phenomenon does not happen in the double disk process. This suggests that no small perturbation can desingularize this process, and so produce an eversion. On the other hand, during the gastrulation the tangent planes at the poles are submitted to a relative rotation of π, and therefore to a rotation of 2π during a full eversion. As a matter of fact, the gastrulation is a quasi-eversion of the sphere, in other words the limit of a sequence of eversions. This leads us to look for a mathematically reasonable eversion of the sphere in the neighbourhood of the gastrulation.

Description of the model. Granting this, we want to use the properties of the special wire mentionned above in order to illustrate the homotopy we have called gastrulation. The object that we are going to construct is a sphere made variable by the springiness of the wires materializing the meridians. The model is made of four kinds of pieces:

1. sixteen meridians in a wire of section 1mm in diameter, similar to the one used for the closed halfway model.
2. two brass hubs: each hub is a solid cylinder 3cm in diameter traversed along its vertical axis by a hole 6mm in diameter. A mortise, 4mm in depth and 12mm in width, is hewed out of each brass hub. In addition, sixteen horizontal holes 1.1mm in diameter regularly spaced out are drilled through , in order to receive the meridians. The nonhewed hub base is drilled out a hole 1.5cm in diameter with a depth large enough to make the sixteen apertures visible (Fig. 11).
3. A stainless steel rod 34cm in length and 5.5mm in section diameter, threaded at one end, and welded at the other end to a tenon just adapted to the mortise of the hub.

4. A handle in three pieces: one aluminium disk 5cm in diameter and 8mm thick, drilled and threaded in its center in order to catch the threaded rod; one hexagonal nut for interlocking the rod and the disk; one aluminium cylinder 9cm in length and 3cm in diameter whose one end is equipped with a tenon well-matched with one of the hubs, and the other end is a disk similar to the former (Fig. 11). The axis of this cylinder is drilled by a hole of 6mm in which the rod must be able to slide. It is appropriate to corrugate both aluminium disks on the edge so as to improve the grip.

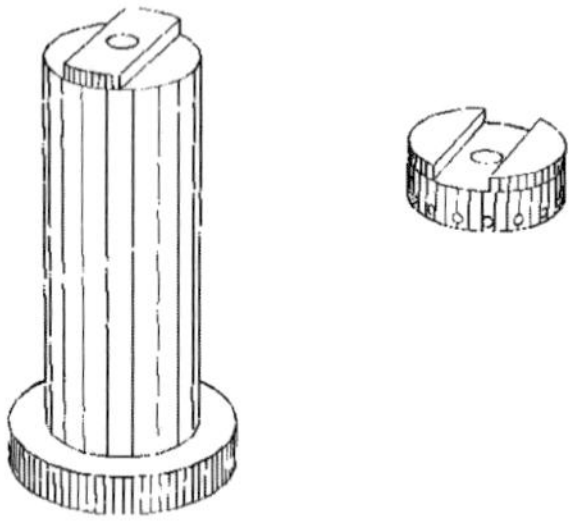

Fig. 11. The handle and one hub of the gastrulation model

In order to fit the pieces together, we begin by setting up the meridians on the hub. Each wire is slipped into one of the apertures in the hub just far enough to come out the other side. With a blowtorch we melt the ends of the emerging wires to form a small bead which prevents the wires from slipping back out of the hubs. When the sixteen spokes are set on the first hub, we carry out the same procedure with the second hub in such a way that both mortises are turned towards the outside of the sphere generated by the sixteen meridians. Then we slip the rod through both hubs and tenon in the mortise of the upper hub. We slip the other end of the rod, what sticks out of the sphere, into the aluminium cylinder which tenons in the mortise of the lower hub. Afterwards we screw the aluminium disk at the threaded end of the rod and lock it with the hexagonal nut. The model so constructed looks like a sphere set on a stand.

Handling. The model keeps a symmetry of rotation during the handling. Each meridian is submitted to a similar deformation. The handling of the model decomposes in two moves: first of all, a movement of translation of the south pole of the wire sphere towards the north pole obtained by pushing the lower hub towards the upper one along the rod by means of the handle, and secondly, a rotation made possible by the tenon-mortise system which allows carrying away the upper hub with the disk tightly screwed on the rod, while the lower hub is hold in position by means of the handle. The translation tends to flat the sphere on a disk by forcing the meridians to fold in their middle. The potential energy of the wire, and therefore the stress to be supplied, become too considerable, and we might exceed the limit of elasticity beyond which the steel keeps a permanent plastic deformation. However, thanks to the universal principle of least action recalled in introduction, the wire tends to minimize its potential energy, so that it is getting twisted and forces a movement of rotation, as we can see by releasing the handle and just pushing it with a finger. The sphere lies down flat by itself on a double disk, and each meridian takes the rough shape of a horizontal half figure eight (Fig. 6) with an energy not far from that of a wire devoid of constraints. That is why the degree of freedom in rotation of the meridians in the apertures of the hubs is so crucial. Without this degree of freedom the flattening becomes impossible.

During this flattening the lower hub turns about 90° with respect to the upper hub. By means of the handle, we can prescribe an additional rotation in the same orientation in order to reach the required 180°. Then occurs a phenomenon unexpected at first sight: the disk suddenly snaps into a doubly covered sphere half as large as the original.

Explanation of the phenomenon. The total rotation of 180° tends to give each meridian the shape of a circle, except that the two ends are not exactly aligned since the hubs cannot be merged into each other. Actually the curve is the arc of circular helix described in paragraph (2.8). Each meridian tends to minimize its potential energy, as mentionned there. Although the arc of circular helix is an extremal, it is not a minimum of the potential energy. The minimum of the potential energy with the initial conditions prescribed by the model is an arc of the plane curve depicted in figure (Fig. 4), so that the meridian seeks a plane figure, and such a figure has to be vertical, since such is the plane of both apertures receiving the ends of the meridian.

Notice that if a curve minimizes the potential energy, so is the symmetric curve with respect to the horizontal plane passing through the origin. This corresponds to two events of equal probability in the experiment: either the disk closes up on the hand of the experimenter, or it closes up outside. In order to promote the second case, more appropriate for the observation, it suffices, at the time as the additional rotation is prescribed, to slightly push up the boundary of the disk generated by the flattened meridians.

Let us reconsider the deformation of a meridian. It goes from a quasi-horizontal position to a vertical one by the movement described in figure (2). As soon as a loop starts standing up, it carries away its neighbour in the same move and step by step all follow and come locking in the vertical position (see the photograph [FA]4 in the appendix).

5 Conclusion

We have shown that wire models continue to be useful for mathematical purposes, especially when they move under the laws of elasticity. We hope that

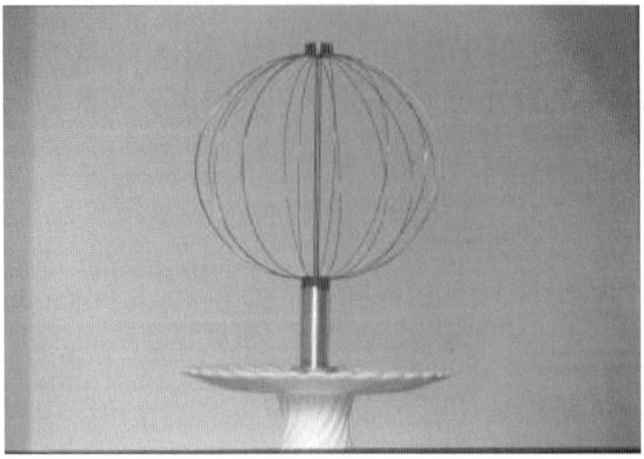

Fig. 12. [FA]4: A view of the gastrulation model realized by F. Apéry in Kyushu University (Japan).

the examples described here will inspire similar productions in the future. Especially science museums that encourage hands-on interaction between visitor and model, should take note. Interactive computer simulations are the current vogue. Yet some place should also be reserved for physical models, which provide so much more tactile sensation than the computer screen. The ingenuity and elegance of physical models makes them works of art, especially since they are not as easily made and controlled as the images produced by computer graphics.

References

1. F. Apéry, Models of the real projective plane, Vieweg Verlag, Wiesbaden, 1987.
2. F. Apéry, An algebraic halfway model for the eversion of the sphere, Tôhoku Math. J., 44, 1992, 103–150.
3. F. Apéry, L'oeuf et la sphère, Pour la Science, n° 276, octobre 2000.
4. F. Apéry, G. Franzoni, Il rovesciamento della sfera: un modello materiale della fase centrale, Rendiconti del Seminario della Facoltà di Scienze dell'Università di Caglari, 1999, Cagliari.
5. H. Cartan, Formes différentielles, Hermann, Paris 1967.
6. G. Fischer, Mathematical Models, Vieweg-Verlag, Wiesbaden, 1986.
7. G.K. Francis, A Topological Picturebook, Springer-Verlag, New York, 1987.
8. B. Morin, J.-P. Petit, Le retournement de la sphère, Pour la Science, n° 15, janvier 1979.

Sphere Eversions: from Smale through "The Optiverse"

John M. Sullivan

Mathematics Dept., Univ. of Illinois, Urbana IL, USA 61801
<jms@math.uiuc.edu>

Abstract. For decades, the sphere eversion has been a favorite subject for mathematical visualization. The 1998 video *The Optiverse* shows minimax eversions, computed automatically by minimizing elastic bending energy using Brakke's Evolver. We contrast these geometrically optimal eversions with earlier ones, including those by Morin, Phillips, Max, and Thurston.

1 A Short History of Sphere Eversions

To evert a sphere is to turn it inside-out by means of a regular homotopy: this is a continuous deformation, in which we allow the surface to pass through itself, but forbid puncturing, ripping, creasing, or pinching the surface. An abstract theorem proved by Smale in the late 1950s showed that spheres could be everted [22], but it remained a challenge for many years to exhibit an explicit eversion. Because the self-intersecting surfaces involved are complicated and nonintuitive, communicating an eversion is yet another challenge: one in mathematical visualization. More detailed histories of this problem can be found in [15] and in Chapter 6 of [7].

The earliest sphere eversions were designed by hand, and made use of the idea of a halfway-model. This is an immersed spherical surface which is "halfway inside-out", meaning that it has a symmetry interchanging the two sides of the surface. If we can find a way to simplify the halfway-model to a round sphere, we get an eversion: first unsimplify the round sphere to the halfway-model, then apply the symmetry, and finally simplify back down to the round sphere. The eversions of Arnold Shapiro (see [8]), Tony Phillips [21], and Bernard Morin [19] can all be understood in this way.

In practice, two kinds of halfway-models have been used, shown in Figure 1. The first, used by Shapiro and by Phillips (see Figure 2, left), is a Boy's surface, which is an immersed projective plane. In other words it is a way of immersing a sphere in space such that antipodal points always map to the same place. Thus there are two opposite sheets of the surface lying on top of each other. If we can succeed in pulling these sheets apart and simplifying the surface to a round sphere, then pulling them apart the other way will lead to the inside-out sphere.

The other common kind of halfway-model is a Morin surface, having four lobes: two show the inside and two the outside. The symmetry interchanging

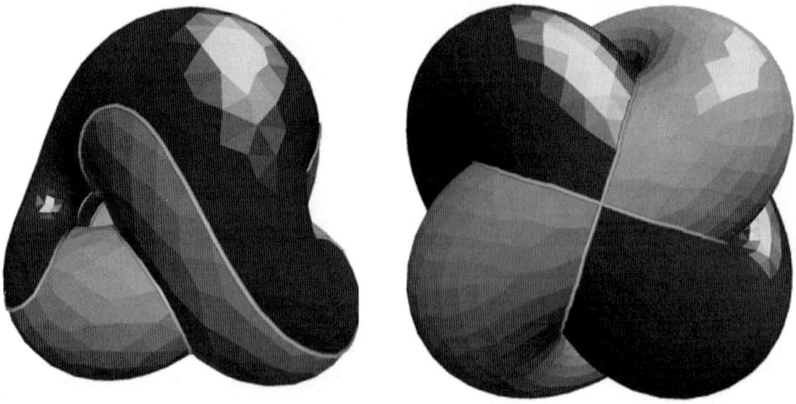

Fig. 1. These are the halfway-models for the first two minimax eversions. Boy's surface (left) is an immersed projective plane with three-fold symmetry and a single triple point; the particular shape shown here minimizes Willmore's elastic bending energy. The figure actually shows an immersed sphere, double covering Boy's surface, with the two (oppositely-oriented) sheets pulled apart slightly. A Morin surface (right) has a four-fold rotational symmetry which reverses orientation, exchanging the lighter and darker sides of the surface. The Morin surface shown also minimizes Willmore energy.

the sides is a ninety-degree rotation. Morin (whose blindness incidentally shows that mathematical visualization goes well beyond any physical senses) and Apéry [2] have shown that an eversion based on a Morin surface halfway-model has the minimum possible number of topological events.

Models of this eversion were made by Charles Pugh; then Nelson Max digitized these models and interpolated between them for his famous 1977 computer-graphics movie "Turning a Sphere Inside Out" [18], a frame of which is shown in Fig. 2. Morin eversions have also been implemented on computers by Robert Grzeszczuk and by John Hughes (see Fig. 3), among others. These eversions have usually been laboriously created by hand, by splining or interpolating between key frames, or by looking for algebraic or trigonometric equations which capture the topological motions desired.

Francis and Morin realized that the Morin surface and Boy's surface are just the first in an infinite sequence of possible halfway-models for what they call the tobacco-pouch eversions (see Fig. 4, taken from [7]).

A new proof of Smale's original theorem, providing more geometric insight, was developed by Bill Thurston (see [15]). His idea was to take any homotopy (continuous motion) between two surfaces, and, when possible, make it regular (taking out any pinching or creasing) by adding corrugations in a controlled way. The corrugations seem to make the surface more flexible, eliminating kinks. This kind of sphere eversion was beautifully illustrated in

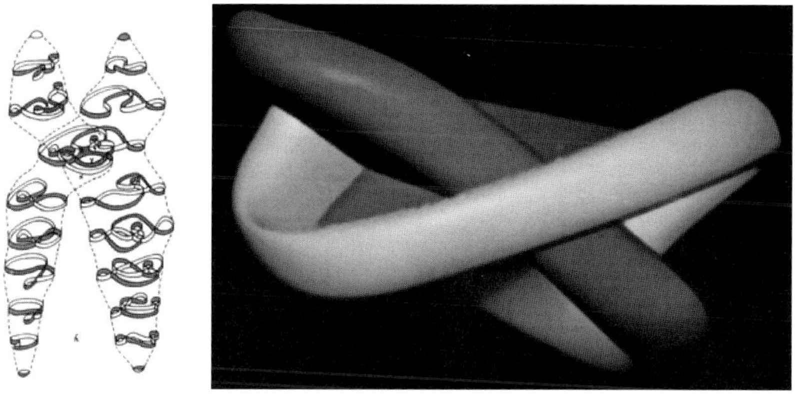

Fig. 2. This drawing (left) by Tony Phillips [21] shows one stage of his sphere eversion based on a Boy's surface halfway-model. This frame (right) from Nelson Max's classic computer animation of a Morin eversion shows a stage near the halfway-model.

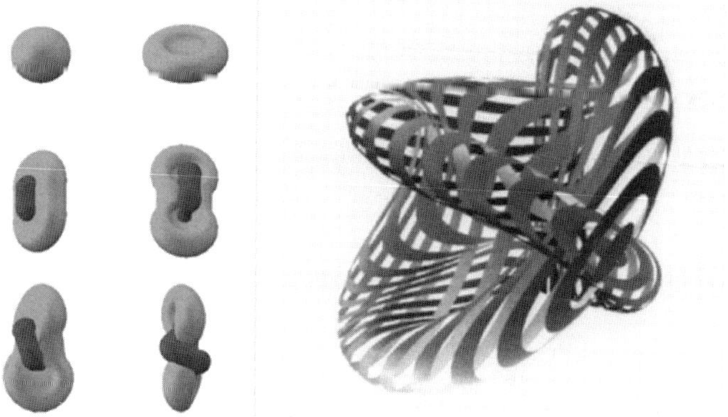

Fig. 3. Still pictures from the Morin eversion implementions by Robert Grzeszczuk (left) and John Hughes (right).

the 1994 computer-graphics video "Outside In" [16] produced at the Geometry Center (see Fig. 5).

The corrugation idea is quite natural, and provides a way to understand all regular homotopies, not just the sphere eversion. However, the resulting eversion is quite elaborate, and has many more topological events than necessary. Like the tobacco-pouch eversions described below, the Thurston eversion has rotational symmetry of essentially arbitrary order around a vertical axis. But instead of the temporal symmetry seen in all sphere eversions based on halfway-models, it has an unusual top-bottom symmetry.

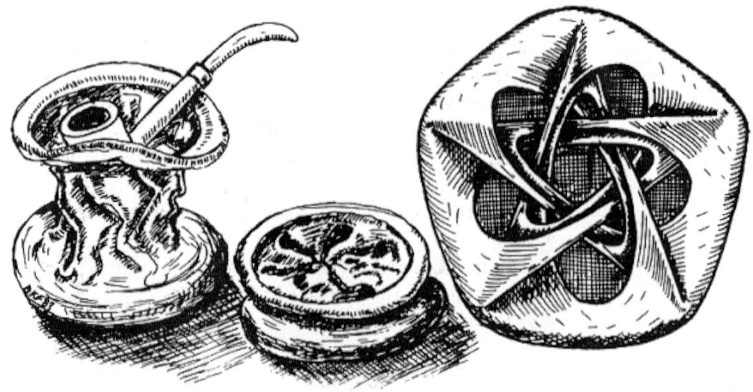

Fig. 4. These sketches by George Francis show a French tobacco pouch, and a cutaway view of the halfway-model for the order-five tobacco-pouch eversion.

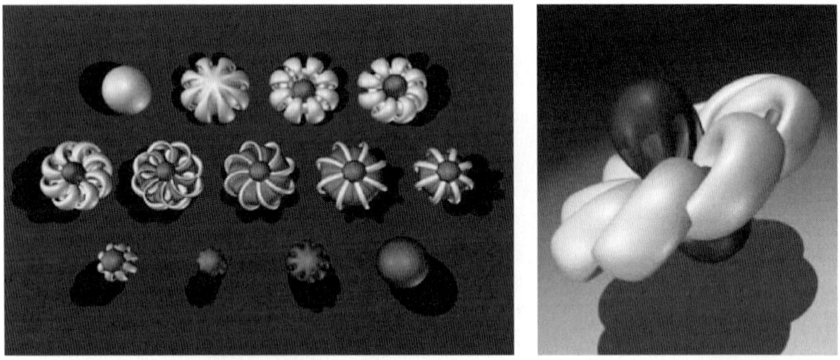

Fig. 5. These pictures, from "Outside In", show Thurston's sphere eversion, implemented through corrugations.

2 Bending Energy and the Minimax Eversions

Our minimax sphere eversions differ from the earlier ones mentioned above in that they are computed automatically by a process of energy minimization. They are the first eversions whose geometry, rather than merely topology, is part of the design.

The elastic bending energy for a stiff wire is the integral of curvature squared. A surface in space has, at each point, two principal curvatures; their average, the mean curvature, shows how much the surface deviates from being minimal. The integral of mean curvature squared is thus a bending energy for surfaces, often called the Willmore energy [24]. (Mathematically, by the Gauss-Bonnet theorem, this is equivalent to many other formulations.)

Among all closed surfaces, the round sphere minimizes this bending energy. (The energy is scale-invariant; we normalize so the sphere has energy 1.) It is also known [17,13] that any self-intersecting surface with a k-tuple point has energy at least k. To evert a sphere, it is necessary to pass through some stage which has a quadruple point [3,12], and hence has energy at least 4.

In the 1980's, Robert Bryant [6] classified all critical points for the Willmore energy among spheres; they have integer energy values which (except for the round sphere) are at least 4. Rob Kusner, being familiar with this result and with the tobacco-pouch eversions, realized that he could find, among Bryant's critical spheres, ones with the tobacco-pouch symmetries. Among surfaces with those symmetries, Kusner's have the least possible Willmore energy.

In particular, Kusner's Morin surface with four-fold orientation-reversing symmetry has energy exactly 4, the minimum possible for any surface with a quadruple point, and thus the minimax energy for a sphere eversion. If we don't enforce the symmetry, then presumably this surface is no longer a minimum, but now an unstable critical point (or saddle point) for the energy.

Pushing off from this saddle in one downhill direction, and flowing down by gradient descent, we should arrive at the round sphere, since it is the only critical point with lower energy. As we saw before, such a homotopy, when repeated with a twist, will give us a sphere eversion. This eversion starts at the round sphere (which minimizes energy) and goes up over the lowest energy saddle point; it then comes back down on the other side, arriving at the inside-out round sphere.

Of course, there is not enough theory for fourth-order partial differential equations for us to know in advance that the surface will remain smooth under this gradient descent. Perhaps some neck may pinch off, while staying close enough to a catenoid to have negligible Willmore energy. Recently, Kuwert and Schätzle [14] have proved short-term existence for the Willmore gradient flow. But the flow does not have long-term existence for all initial surfaces.

Around 1995, in collaboration with Kusner and Francis, I computed this minimax eversion [10], using Ken Brakke's Evolver [4]. This is a software package designed for solving variational problems, like finding the shape of soap films or (see [11]) minimizing Willmore energy. The computations give good evidence that indeed the Morin surface is an unstable critical point, and that the Willmore gradient flow stays smooth all the way down to the round sphere.

This eversion is, by design, geometrically optimal in the sense of requiring the least bending at any stage. We were pleased that the computed eversion is also topologically optimal, in that it is one of the Morin eversions with the fewest topological events.

Computations of the higher-order minimax sphere eversions [9] (like the one with a Boy's surface halfway-model) had to wait until Brakke and I implemented some new symmetry features in the Evolver (see [5]). The computation of the order-k minimax tobacco-pouch eversion starts from its halfway-model, which has $2k$-fold symmetry. The eversion maintains k-fold rotational symmetry throughout; the Evolver works with only a single fundamental domain for this symmetry. We can find the initial halfway-models by minimizing bending energy while enforcing the full $2k$-fold symmetry.

Alternatively, we can compute them directly. Bryant's classification says that all critical spheres are obtained as conformal (Möbius) transformations of minimal surfaces. Kusner gave explicit Weierstrass parameterizations for the minimal surfaces he needed to generate his halfway models. In Fig. 6 we see the minimal surface with four flat ends which transforms into our Morin halfway-model.

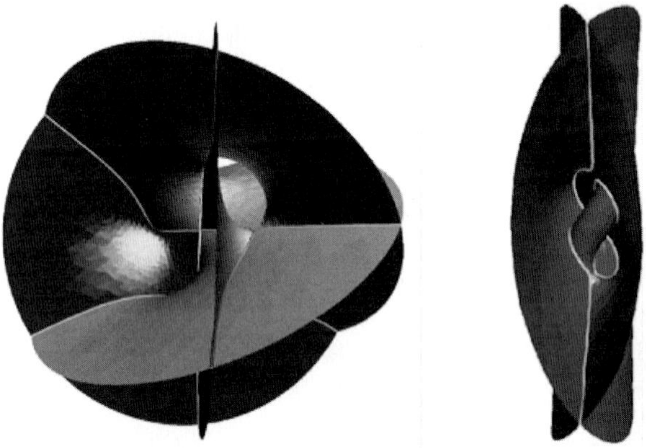

Fig. 6. This minimal surface (left), with four flat ends, gives rise to Kusner's Morin surface of least Willmore bending energy, when a conformal (Möbius) transformation is applied to compactify it. If the transformation sends the double-tangent point to infinity, we get an interesting picture (right).

The Evolver works with triangulated approximations (with a few thousand triangles) to the true smooth surfaces, and we update the triangulation as needed to maintain a good approximation. Initially, to push off the saddle point and find our way downhill, we must use the second-order Hessian methods implemented in the Evolver.

In 1998, working with Francis and Stuart Levy, I produced a computer graphics video, "The Optiverse" [23], which shows the first four minimax eversions.[1] It premiered at the International Congress of Mathematicians in Berlin, was also shown at SIGGRAPH and 3ecm, and was written up nicely in Science News [20].

Many scenes in the video (as in Fig. 7) capture views of the eversion also available in our real-time interactive computer animation, which runs

[1] This video was produced at the NCSA at the University of Illinois. More information about it is available on our website at http://new.math.uiuc.edu/optiverse/.

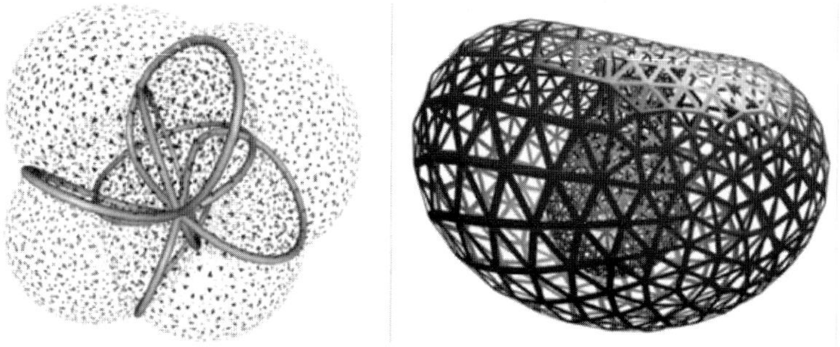

Fig. 7. The Morin halfway model (left), shown with all triangles shrunk, has an elaborate set of double curves where the surface crosses itself. An late stage in the eversion is like a gastrula (right), shown here as a triangular framework.

on desktop workstations or in immersive virtual-reality environments. These scenes show a solid, colored surface, with white tubes around the double-curves of self-intersection. Here we are aware of the triangulation used for the computation, especially if we leave gaps between (or in the middle of) the triangles.

Most stages in a sphere eversion involve a highly convoluted surface that passes through itself many times. In real life, we are accustomed to looking only at the surfaces of objects, not at internal structure. Our good sense of spatial relationships comes from walking around objects, not from seeing through them. It is therefore a significant challenge to visually present the self-intersecting surfaces found in sphere eversions.

Thus in "The Optiverse", we used several rendering techniques, since no one method can give all useful visual information for the nonintuitive phenomenon of self-intersecting surfaces. Parts of the video (like Fig. 8) were rendered as transparent soap films, with the soap bubble shader I wrote for Renderman (see [1]).

3 Topological Stages in the Eversion

A generic regular homotopy has topological events occuring only at isolated times, when the combinatorics of the self-intersection locus changes. These events happen at times when the surface normals at a point of intersection are not linearly independent. That means either when there are two sheets of the surface tangent to each other (and a double curve is created, annihilated, or reconnected), or when three intersecting sheets share a tangent line (creating or destroying a pair of triple points), or when four sheets come together at a quadruple point.

The double tangency events come in three flavors. These can be modeled with a rising water level across a fixed landscape. As the water rises, we can

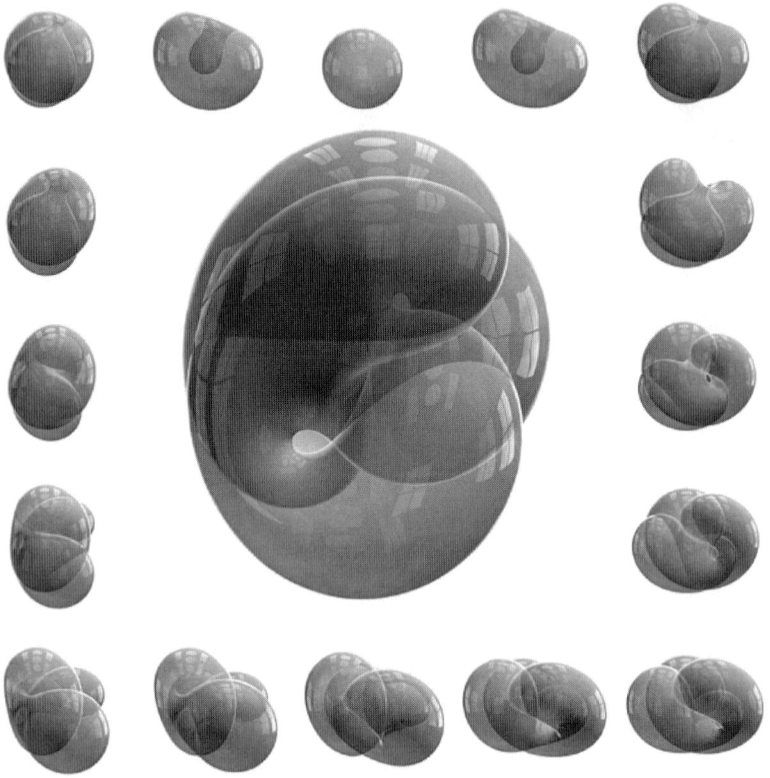

Fig. 8. This minimax sphere eversion is a geometrically optimal way to turn a sphere inside out, minimizing the elastic bending energy needed in the middle of the eversion. Starting from the round sphere (top, moving clockwise), we push the north pole down, then push it through the south pole (upper right) to create the first double-curve of self-intersection. Two sides of the neck then bulge up, and these bulges push through each other (right) to give the second double-curve. The two double-curves approach each other, and when they cross (lower right) pairs of triple points are created. In the halfway-model (bottom) all four triple points merge at the quadruple point, and five isthmus events happen simultaneously. This halfway-model is a symmetric critical point for the Willmore bending energy for surfaces. Its four-fold rotation symmetry interchanges its inside and outside surfaces. Therefore, the second half of the eversion (left) can proceed through exactly the same stages in reverse order, after making the ninety-degree twist. The large central image belongs between the two lowest ones on the right, slightly before the birth of the triple points.

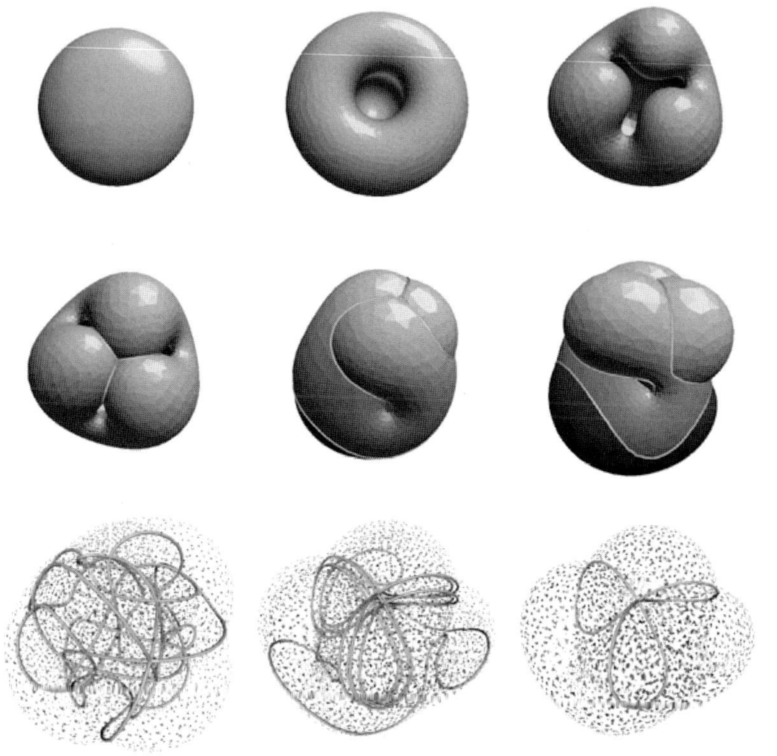

Fig. 9. This three-fold minimax eversion starts (top row, left-to-right) with a gastrula stage like that of the two-fold eversion, but the three-fold symmetry means that three fingers reach up from the neck instead of two. They intersect each other (middle row, left-to-right) and then twist around, while complicated things are happening inside. The bottom left image is at the same stage as the middle right image, but with triangles shrunk to show the elaborate double-curves. These double-curves separate into two pieces (bottom row, left-to-right), one of which is a four-fold cover of the propeller-shaped double-curve seen in the Boy's surface halfway-model (lower right).

observe the creation of a new lake, the conversion from a isthmus of land to a channel of water, or the submersion of an island. These correspond to the creation, reconnection, or annihilation of a double-curve, which is seen here as the shoreline.

In the minimax eversion with two-fold symmetry, seen in Fig. 8 and Fig. 9, the first two events create the two double-curves. The next event happens when these twist around enough to intersect each other: then two pairs of triple points are created. At the halfway stage, six events happen all at the same time : Along the symmetry axis, at one end we have a quadruple point, while at the other end we have a double tangency creating an isthmus event.

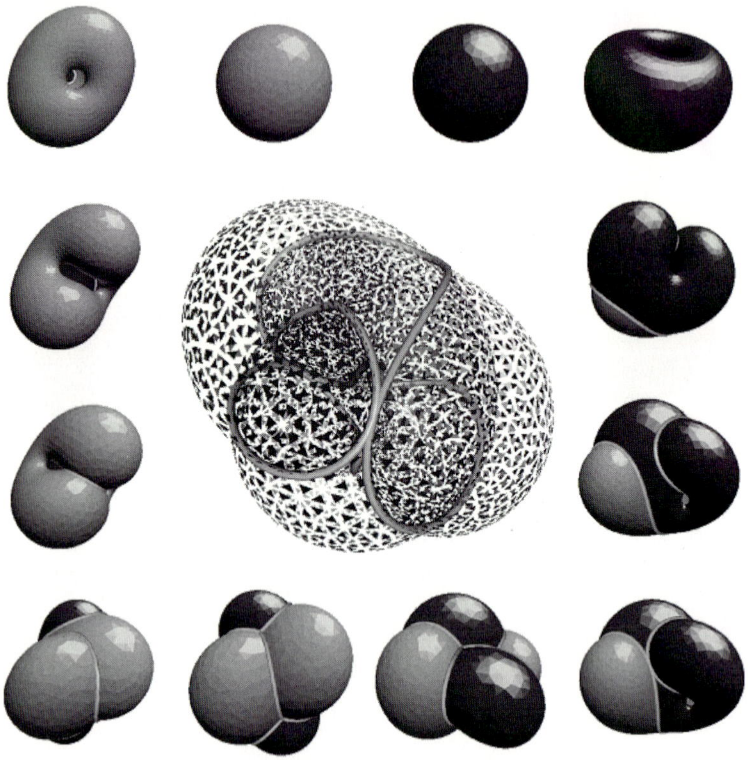

Fig. 10. This is the same eversion shown in Fig. 8, but rendered with solid surfaces. Again, we start at the top with a round sphere, and proceed clockwise. Down the right-hand side we see the creation first of two double-curves, and then of a pair of triple points. (Another pair is created at the same time in back; the eversion always has two-fold rotational symmetry.) Across the bottom, we go through the halfway-model, interchanging the roles of the dark and light sides of the surface. Up the left column, we see the double-curves disappear one after the other. (In this montage, the figures on the left are not exactly the same stages as the corresponding ones on the right.) In the center, we examine the double locus just when pairs of triple points are being created, by shrinking each triangle of the surface to a quarter of its normal size.

Finally, at the four "ears" (at the inside edge of the large lobes) we have additional isthmus events: two ears open as the other two close. (See [10] for more details on these events.)

The three-fold minimax eversion, using the Boy's surface of Fig. 1 as a halfway-model, has too many topological events to describe easily one-by-one, and its three-fold symmetry means that the events are no longer all generic. But we can still follow the basic outline of the eversion from Fig. 10.

Acknowledgments

The minimax sphere eversions described here are joint work done in collaboration with Rob Kusner, Ken Brakke, George Francis, and Stuart Levy, to all of whom I owe a great debt. I would also like to thank Francis, Robert Grzeszczuk, John Hughes, AK Peters, Nelson Max and Tony Phillips for permission to reproduce images of earlier sphere eversions. This paper is a revised and updated version of my article *"The Optiverse" and Other Sphere Eversions*, ArXiv eprint math.DG/9903101, which appeared in the proceedings of two 1999 conferences on mathematics and art: ISAMA 99 (June, San Sebastián, Spain), and Bridges (July, Winfield, Kansas). My research was supported in part by NSF grant DMS-97-27859.

References

1. Frederick J. Almgren, Jr. and John M. Sullivan (1992) Visualization of soap bubble geometries. *Leonardo* **24**:267–271. Reprinted in *The Visual Mind.*
2. François Apéry (1992) An algebraic halfway model for the eversion of the sphere. *Tohoku Math. J.* **44**:103–150. With an appendix by Bernard Morin.
3. Thomas Banchoff and Nelson L. Max (1981) Every sphere eversion has a quadruple point. In D. N. Clark, G. Pecelli, and R. Sacksteder, editors, *Contributions to Analysis and Geometry*, Johns Hopkins Univ. Press, 191–209.
4. Kenneth A. Brakke (1992) The Surface Evolver. *Experimental Math.* **1**:141–165.
5. Kenneth A. Brakke and John M. Sullivan (1997) Using symmetry features of the surface evolver to study foams. In Hege and Polthier, editors, *Visualization and Mathematics*, Springer, 95–117
6. Robert Bryant (1984) A duality theorem for Willmore surfaces. *J. Differential Geometry* **20**:23–53
7. George Francis (1987) *A Topological Picturebook.* Springer, New York
8. George Francis and Bernard Morin (1979) Arnold Shapiro's eversion of the sphere. *Math. Intelligencer* **2**:200–203
9. George Francis, John M. Sullivan and Chris Hartman (1998) Computing sphere eversions. In Hege and Polthier, editors, *Mathematical Visualization*, Springer, 237–255
10. George Francis, John M. Sullivan, Robert B. Kusner, Kenneth A. Brakke, Chris Hartman, and Glenn Chappell (1997) The minimax sphere eversion. In Hege and Polthier, editors, *Visualization and Mathematics*, Springer, 3–20
11. Lucas Hsu, Rob Kusner, and John M. Sullivan (1992) Minimizing the squared mean curvature integral for surfaces in space forms. *Exper. Math.* **1**:191–207
12. John F. Hughes (1985) Another proof that every eversion of the sphere has a quadruple point. *Amer. J. Math.* **107**:501–505
13. Rob Kusner (1987) Conformal geometry and complete minimal surfaces. *Bull. Amer. Math. Soc.* **17**:291–295

14. Ernst Kuwert and Reiner Schätzle (2001) The Willmore flow with small initial energy. *J. Differential Geom.* **57**:409–441
15. Silvio Levy (1995) *Making Waves: A Guide to the Ideas Behind Outside In.* AK Peters, Wellesley, MA
16. Silvio Levy, Delle Maxwell and Tamara Munzner (1994) *Outside In.* AK Peters, Wellesley, MA, narrated video (21 min) from the Geometry Center
17. Peter Li and S. T. Yau (1982) A new conformal invariant and its applications to the Willmore conjecture and the first eigenvalue of compact surfaces. *Invent. Math.* **69**:269–291
18. Nelson L. Max (1977) *Turning a Sphere Inside Out.* Narrated film (21 min)
19. Bernard Morin and Jean-Pierre Petit (1980) Le retournement de la sphère. In *Les Progrès des Mathématiques*, 32–45. Pour la Science/Belin, Paris
20. Ivars Peterson (1998) Contemplating the Optiverse: Surreal films. *Science News* **154**:232–234.
21. Anthony Phillips (1966) Turning a sphere inside out. *Sci. Amer.* **214**:112–120.
22. Stephen Smale (1959) A classification of immersions of the two-sphere. *Trans. Amer. Math. Soc.* **90**:281–290.
23. John M. Sullivan, George Francis and Stuart Levy (1998) The Optiverse. In Hege and Polthier, editors, *VideoMath Festival at ICM'98.* Springer, narrated video (7 min).
24. Thomas J. Willmore (1992) A survey on Willmore immersions. In *Geometry and Topology of Submanifolds, IV (Leuven, 1991)*, World Sci. Pub., 11–16

Tactile Mathematics

Stewart Dickson

MathArt, 23115 Bluebird Drive, Calabasas, CA 91302 USA,
<Stewart.Dickson@disney.com>

Abstract. Visual Computing has led directly to Visual Mathematics – this by virtue of the ability to directly compile 'natural' mathematical language into machine-executable and graphical form. Direct 3-D computer printers allow fully concrete, 3-D representations for mathematical systems, directly from the numerical representation. The author references his work to date in the field of mathematical sculpture. The author has begun work on integrating text information in Braille into three-dimensional models of mathematical surfaces. Future work, including manipulating computer-specified tactile surface texture in Computer-Aided Design, presents challenges to the technical interfaces in common practise in the Mechanical Prototyping industry. This paper will outline some proposed solutions. This paper proposes the thesis that a richer, synergistic tactile experience can be afforded by combining the abstract information on a mathematical surface with the surface itself in-the-round in physical form.

1 Background: Visual Computing/Visual Mathematics

The Mathematica [1] system for doing mathematics by computer is a compiler of "natural" mathematical language with integrated graphical methods. Mathematica supports full mathematical typesetting while retaining the ability for the computer to execute typeset statements without modification. Rendering the results of mathematical statements is the direct concretization of a human language, whose evolution predated computers.

The advantage of visual computing is that an image condenses voluminous data to a single representation which can be quickly assimilated. [2] Visual computing affords the scientist a view of the problem which is different from considering raw data or abstract statements. To view an image, the scientist makes broader use of his/her sense of vision.

Visual computing has resulted in unexpected discoveries and breakthroughs in the sciences. In 1975, Benoit Mandelbrot used a computer at IBM to make a graph of a dynamical procedure which was known to be chaotic. [3] Complex dynamics was a "monstrous" topic of inquiry, because there were no known methods for dealing with such systems – least of which was simply graphing the behavior, because so many – billions – of calculations are required. What Benoit found was a graph with symmetry, hierarchical self-similarity and infinite detail, which was only apparent visually. No graph of this kind had ever been seen before. Mandelbrot had to invent a new branch of mathematics – fractal mathematics – to describe what he found.

In 1983 a graduate student in Rio de Janeiro named Celsoe Costa wrote down an equation for what he thought might be a new minimal surface, but the equations were so complex, they obscured the underlying geometry. [4]

David Hoffman at the University of Massachusetts at Amherst enlisted programmer James Hoffman to make computer-generated pictures of Costa's surface. The pictures they made suggested first, that the surface was probably embedded (non-self-intersecting) in three-space – which gave them definite clues as to the approach they should take toward proving this assertion mathematically – and second, that the surface contained straight lines, hence symmetry by reflection through the lines.

The symmetry led Hoffman and William Meeks, III to extrapolate that the surface was radially periodic and that new surfaces of the same class could be achieved by increasing the periodicity. They did so by altering the mathematical description of the surface to be the solution to a boundary-value problem constrained by the behavior of a minimal surface at the periodic lines of symmetry. The result: Hoffman and Meeks proved that Costa's surface was the first example of an infinitely large class of new minimal surfaces which are embedded in three-space.

The technique Hoffman and Meeks used was to make a picture which caused them to modify their mathematical theory and discover something totally unexpected about that theory. They later extended their techniques to find minimal surfaces of more complex geometry and they also created pictures of them. This is a new kind of experimental mathematics and a procedure not far from creative visual art.

2 Computer 3-D Printing – Concrete Mathematics

Breakthroughs in science and mathematics due to visual computing may now be brought back from the vacuum of the cathode-ray tube by so-called Rapid-Prototyping (R-P) printers which can render a CAD model in physical materials in three dimensions. [5] [6] A growing list of similar Automated Fabrication technologies are appearing, among them: Stereolithography (3D Systems) [7], Selective Laser Sintering (DTM Corp.) [8] Fused Deposition Modeling (Stratasys) [9] Laminated Object Manufacturing (Helisys) [10] and the several licensees of the MIT 3-D Printing Consortium [11] Direct Shell Production Casting (Soligen) [12] and Z-Corporation [13].

These various technologies all build three-dimensional objects via the common principle of dividing the object in software into a sequence of horizontal slices, from bottom to top, which the machine constructs in a physical material and binds together. Thus, these technologies may also be termed various means of Layer-manufacturing [14] each of which uses some kind of three-dimensional computer object slicing software.

Compared to a two-dimensional computer rendering, a physical model in three full dimensions restores a dimension of information which was abstracted away in the perspective rendering of the image [15] Of course, the

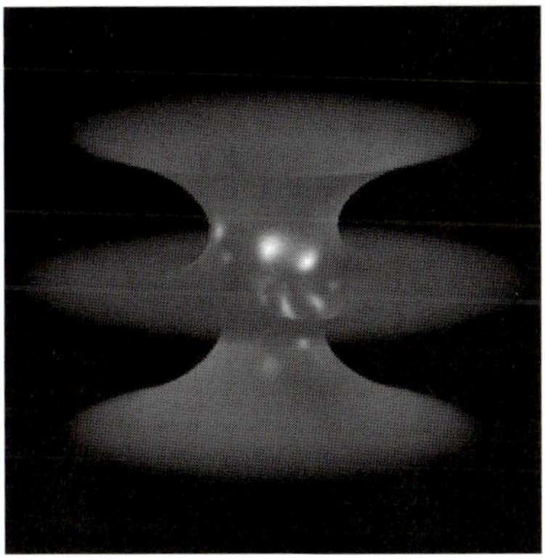

Fig. 1. [SD]1: *Costa's Three-Ended Minimal Surface, Image by David Hoffman and James Hoffman.*

actual, 3-D object gives access to a computer model to someone who might not otherwise be able to benefit from computer modeling because of a visual impairment [16]. But, as viewing an image exercises the scientist's visual cortex and integrates this additional processing path into the mental experience of evaluating an abstract hypothesis, might not viewing a sculpture be a more corporeal experience than viewing a two-dimensional image? [17]

I would like to make the claim, to be tested later, that viewing a sculpture is a different sense of apprehension compared to viewing a two-dimensional image. Certainly the tactile experience of an object enhances our understanding of the object. Is the integration of abstract information on an object with the physical object itself a synergistic experience? Is the total effect greater than the sum of the component elements?

3 4th, 5th, ... Dimensions

There is precedent for integrating additional dimensions of information into three-dimensional representations of computer data. Typically, this is done by mapping a varying color to the surface of the object. [18]

In the work of David Hoffman and James Hoffman in the Hoffman-Meeks extrapolations of Costa's three-ended minimal surface, the computer-generated images show the Gauss Map as a color which varies across the surface according to the point-wise orientation of the normal vector to the surface (see the figure [SD]1 in the Appendix).

Likewise, the Sullivan/Francis/Levy "Optiverse" shows the orientation of the sphere as a color map during the eversion metamorphosis. [19]

The 'inside' of the sphere has a different mapping of the normal vector to color from the 'outside', such that the two sides may be identified during the metamorphosis.

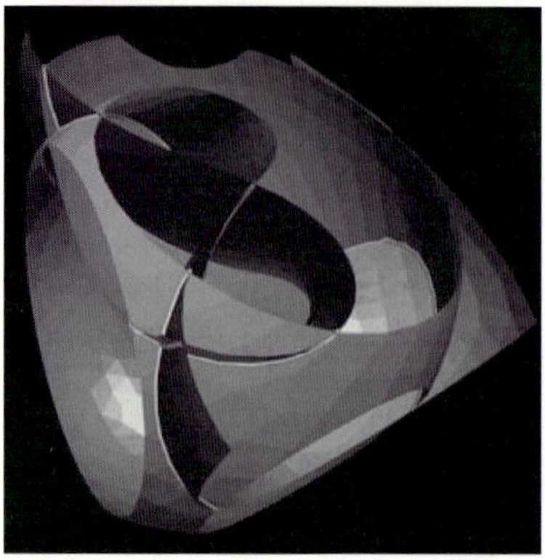

Fig. 2. [SD]2: *Optiverse Sphere Eversion, image by George Francis, John Sullivan and Stuart Levy.*

Hanson et al. have used color to encode 4-D "Depth" and the complex phase of their so-called "Fermat" equations. [20]

In many usual cases, a parametric surface maps a two-dimensional domain (u, v) to a three-dimensional range (x, y, z) in a one-to-one fashion. That is, every point (x, y, z) in the parametric surface in three-space corresponds to a unique (u, v).

Computer rendering systems typically employ a two-dimensional parameterization of a three-dimensional surface in order to depict a detailed colored texture on the surface. To achieve photorealism in a computer rendering, the image used for the texture map may be derived from a photograph of a real-world object's surface.

In communicating abstract information on a system represented in three-space in a computer visualization system, I propose mapping text to the surface to convey the connection between information-space and real three-space. This figure ([SD]3) depicts visually-readable captions on a mathematical 3-D surface in-the-round. [21] In particular, depicted in the captions are the rearranged equations and curves one achieves by holding one of the variables in the equation for the hyperbolic paraboloid to a constant value.

I.e., the parabolas in the X = 0 and Y = 0 planes, the hyperbolas parallel to the X-Y plane and the degenerate hyperbola (two straight lines) in the X-Y plane at Z = 0.

I believe this is a form of high-level 3-D Information Integration. It restores abstract information on the surface to the physical representation. Typically, when a mathematical surface is cast into physical form – concretized – the abstract information which brought about the three-dimensional object is left behind in the computer. There is typically a good deal of verbal explanation required to supplement the object itself. The object does not stand on its own.

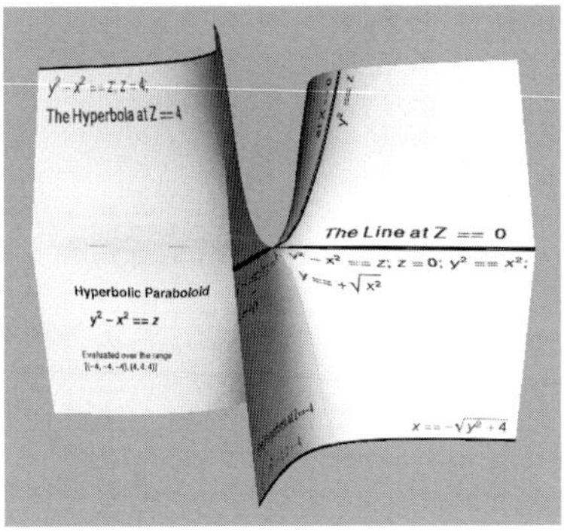

Fig. 3. [SD]3: *Annotated Hyperbolic Paraboloid – Computer Rendering by Stewart Dickson.*

Attaching captions to the surface might be a way of restoring the integrity of the abstract system which the object is intended to represent.

4 Tactile Fonts, 3-D Textures

There seems to be a synergism inherent in the idea of attaching captions, intended to be "read" by touch – by a blind person – to the surface of a three-dimensional, in-the-round representation of an abstract system.

In the following figure ([SD]4) I have attached captions printed in the DotsPlus proposed Braille mathematical typesetting standard to the surface of a Hyperbolic Paraboloid, rendered in Stereolithography. [22] [23]

Other work has been done on creating in a computer system tactile textures to represent abstract information, as a stand-in for color. Examples have been produced including two-dimensional, tactile geographical maps utilizing multiple, distinct textures to denote political boundaries. [24]

However, there are limitations in today's commonly used computer-aided manufacturing infrastructure which impede what can be done.

Computer representation of tactile fonts and textures today almost always takes place in a text editing system, such as Microsoft Word. It is a strictly two-dimensional view of the world. The height or depth of Braille dots or textures are not represented in any explicit way until they are presented to the embosser or to the thermal-swell paper. [25] [26]

So, the first step toward modeling Braille text directly into three-dimensional models in CAD is to create an explicit, 3-D representation of the Braille font in CAD, along with the system for kerning the type. [27]

Computer rendering systems typically represent surface quality as a hierarchy of 'macrostructure', 'mesostructure' and 'microstructure'. [28] Macrostructure is the gross surface geometry, typically expressed as a polygon hull or parametric spline surface. Microstructure is typically detailed reflected

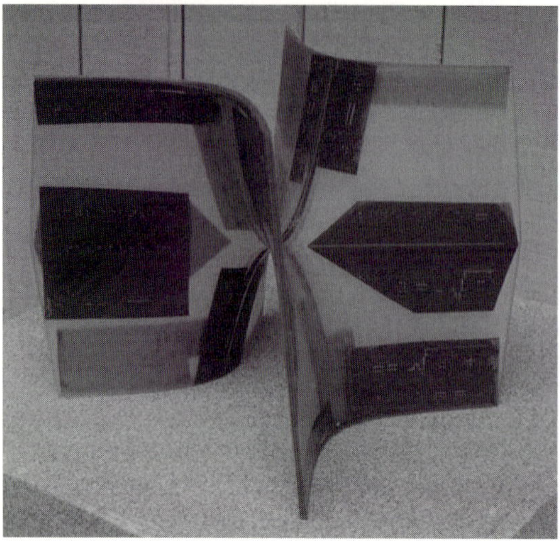

Fig. 4. [SD]4: *Hyperbolic Paraboloid 3-D Model (Stereolithograph) Annotated with Self-Adhesive Braille Captions, image by Stewart Dickson.*

color, expressed as a 'texture' map. Mesostructure is surface 'bump' or displacement information – more detailed than comfortably expressed in explicit polygons – and is also typically represented as a binary image, parametrically mapped to the geometry. [29]

Tactile texture, and possibly also tactile (Braille) captions fall into the class of 'mesostructure', which we would like to express more compactly than in an explicit geometrical representation (e.g., in polygons). However, the R-P Industry standards do not support displacement maps nor do the file exchange standards support parametric surfaces. [30]

Parametric surfaces are more convenient than polygon models for applying image-mapped information, because the surface parameterization (u, v) can correspond identically (or to within scale factor and translation) to the dimensions of the rectangular image map. Polygon models, even when derived from parametric equations, may not carry with them useful information on the parametric domain over which they were generated.

However, the geometric object exchange standard generally accepted by the Automated Fabrication industry at present is the so-called '.STL' file – which is composed of nothing but triangular polygons, without additional (U, V) coordinates.

Proposed solutions include the following: i) A geometric object exchange standard for the Rapid Prototyping industry accepting Topologically Cognizant parametric Patch models which can be sliced. A Layer-Manufacturing Slicing program needs precise knowledge of what is inside the computer-represented 3-D model, separate from the outside – so that the printer may correctly fill the solid (inside) portions and leave empty the rest of the build space (envelope). [31]

Models composed of a collection of parametric patches (such as Non-Uniform, Rational B-Spline – NURBS), in which patch edges are meant to be coincident are generally not explicitly closed. Additional information is usually required in order to fully 'stitch' patch edges together, so that the object can be known to be 'closed' by a computer program.

This topological problem is an extension of Andrew Glassner's "Winged Edge" model for polygons – extended to parametric surfaces, in which an edge will contain at least four control vertices, instead of only two endpoints. [32]

ii) Alternately, we may map 'mesostructure' detail information to polygon meshes by devising ways of "Parameterizing" the surfaces. This problem is related to "Parameterization" of implicit surfaces – those expressed as a function f(x, y, z) = 0. Pedersen describes a method of obtaining piece-wise parametric 'patchinos' for interactive placement of texture (two-dimensional color 'microstructure') on an implicit surface. [33]

Mapping explicitly-modeled 3-D Braille text into surface parametric space is a similar problem to texture mapping. The geometry of each dot is modeled as a polygon mesh. The dots are composed in 3-D by a program for generating Braille characters from ASCII text. The height of each dot is oriented to the normal vector to the surface at the U-V parametric coordinate to which the text string is mapped. We would like to be able to use the features Pedersen has demonstrated with color 'microstructure' for interactively repositioning 3-D mapped tactile text and texture ('mesostructure') on an arbitrarily-formed, curved surface in a 3-D CAD system. [30]

Braille text could also be represented as an image-based surface displacement map. What is required in this case is An R-P slicer which can evaluate surface displacement maps. Such a program or capability does not currently exist in the Rapid Prototyping industry. This will also be required to resolve highly-detailed 3-D texture which one would not want to represent any less compactly in other than an image-based displacement map [30].

5 Conclusion / Future Work

After overcoming the few technical limitations outlined above, there remains the task of testing tactile 3-D models representing high-level abstraction. The appropriate venue for this is probably an educational curriculum using mathematical and scientific tactile pedagogical models in a classroom laboratory.

So, what does this have to do with art? Well, the problem I have stated so far is the "hard" problem. This is the obvious problem which can be logically formulated – a solution to an obvious particular need: Access to 3-D computer graphics by those who cannot see a computer video screen.

But this is only the beginning. Beyond this, I would like to test the following thesis: Is the tactile integration of multi-dimensional abstract-space a

Synergistic experience? Is the total experience of tactile mathematics greater than the sum of its constituent elements?

I imagine that the experience of reading tactile captions on a mathematical surface in physical 3-D would be as follows: Assuming the caption describes a curve, and follows that curve across the surface, (as in Figs. 3 and 4) then the fingers doing the reading will be following the instantaneous tangent vector to the curve through space as the reader is reading the abstract, mathematical description of that curve.

Furthermore, the hand is also constrained to be oriented according to the tangent vector to the curve and the instantaneous normal vector to the surface as the reader is reading the information on the mathematics of the surface.

I see the potential synergism here. However, I have found that blind people tend to be hand-centric rather than object-centric as they read my objects in 3-D. That is, they will rotate the frame of reference of the object in space while keeping the frame-of-reference of their hands stationary. I don't know whether this relative "reversal" is significant or not. Again, this will have to be tested.

The emphasis in computing is on the Virtual – to immerse the imagination in this plastic, abstract space – to further disembody the mind. This is simply the continued tendency of writing and literature.

Art and sculpture have been in trouble for a long time. Consider Brancusi's "Sculpture for the Blind".

But this not the way you will see it in the art museum! What you will see is a plexiglas box completely surrounding and covering the sculpture – protecting the material surface of the sculpture. The philosophical value of

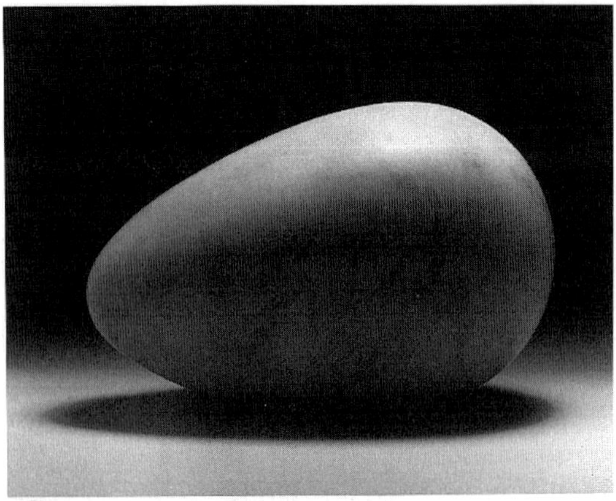

Fig. 5. *Constantin Brancusi, "Sculpture for the Blind" (Beginning of the World).*

this sculpture is lost to its intended audience. That is the tragedy of the plastic arts – that they are exclusively visual.

For those whose eyes do not work well enough to use a video screen or head-mounted display, the value of visual computing is totally lost. Modern tools like automated fabrication, on the other hand, now enable us to project the internal world outward, into physical space.

And, for the rest of us, casting the virtual into physicality forces the illusion to withstand the light of day – to test its honesty.

Experiencing a physical object – which occupies the same space we occupy – is a different sense of apprehension of the object than seeing a flat, even a moving picture, of the object. Viewing a film clip of an object rotating, the brain has forgotten what the front looks like by the time the back rotates into view. Viewing the physical object, we have a more integrated idea of the whole object.

The composer Harry Partch established the need for a Corporeal Music which encompasses sound, vision and performed ritual. It is a theory of synergy, synaesthesia and a reaction against abstraction in music. It can be said that making physical sculpture from computer-generated designs is a similar reaction against the sterility of abstract data space. Only the physical object relates to us as physical beings. Only the physical object has life. Only the physical object has the power to resonate with our lives.

I would like to propose that in tactile, Corporeal Mathematics, one can achieve a true Integration of mind and body.

References

1. Wolfram S. (1996) (The Mathematica System for Doing Mathematics by Computer `<http://www.wolfram.com>`) The Mathematica Book, 3rd edn. Cambridge University Press, Cambridge
2. DeFanti T.A., Brown M.D. (1991) Visualization in Scientific Computing, Advances in Computers Vol. 33, Academic Press, New York, 247–305
3. Mandelbrot B. B. (1983) The Fractal Geometry of Nature, Freeman and Company
4. Hoffman D. (1987) New Embedded Minimal Surfaces The Mathematical Intelligencer 9 (3)
5. Dickson S. Mathematica Visualizations, <http://emsh.calarts.edu/~mathart/portfolio/SPD_Math_portfolio.html>
6. Razdan A., Mayer J.W., Steinberg B. (1997) Scientific Visualization using Rapid Prototyping Technologies, Proceedings of the Sixth European Conference on Rapid Prototyping, Nottingham, U.K., 171–175
7. 3D Systems Inc., Solid Imaging and Solid Object Printing, <http://www.3dsystems.com>
8. DTM Corporation, Advanced Rapid Prototyping and Manufacturing Solutions, <http://www.dtm-corp.com/>
9. Stratasys, <http://www.stratasys.com>
10. Helisys, Layered Material Technology, <http://www.helisys.com>

11. MIT 3-D Printing Laboratory <http://me.mit.edu/groups/tdp/>
12. Soligen, Parts Now, <http://www.partsnow.com>
13. Z Corporation, Office Compatible 3D Printers, <http://www.zcorp.com>
14. Hodgson E., (1998) Creating Art with Layer Manufacture (CALM), report to TASC, University of Central Lancashire, UK <http://www.uclan.ac.uk/clt/calm/overview.htm>
15. Peterson I. (1991) Plastic Math, Science News, 140 (5) 65–80
16. Skawinski W. J., Venanzi C. A., Ofsievich A. D., REAL VIRTUALITY: The Use of Laser Stereolithography for the Construction of Accurate Molecular Models, Department of Chemical Engineering, Chemistry, and Environmental Science, New Jersey Institute of Technology, <http://www-ec.njit.edu/~skawinsk/nano/nano.html>
17. Partch H., (1974) Genesis of a Music, 2nd edn., Da Capo Press, New York
18. Hoffman J., The Gauss Map, Mathematical Sciences Research Institute (MSRI), Berkeley <http://www.msri.org/publications/sgp/jim/geom/surface/maps/gauss/mainc.html>
19. Sullivan J. M., Francis G., Levy S., (1998) The Optiverse, University of Illinois <http://new.math.uiuc.edu/optiverse>
20. Hanson A. J., Munzner T., Francis G., (1994) Interactive Methods for Visualizable Geometry, IEEE Computer, 27 (4) 73–83 <http://www.geom.umn.edu/docs/research/ieee94/node8.html>
21. Dickson S., Braille-Annotated Tactile Models In-The-Round of Three-Dimensional Mathematical Figures, <http://emsh.calarts.edu/~mathart/Annotated_HyperPara.html>
22. Ibid.
23. Gardner J.A., (1998) The DotsPlus Tactile Font Set, Journal of Visual Impairment and Blindness, 836–840
24. Gardner J. A., (1999) Science Access Project, Oregon State University, Personal communication <http://dots.physics.orst.edu> <http://www.physics.orst.edu/people/faculty/gardner.html>
25. ViewPlus Technologies, TIGER Advantage Tactile Graphics and Braille Embosser <http://www.viewplustech.com/products.html>
26. American Thermoform Corporation, Swell-Touch paper <http://www.atcbrleqp.com/swell.htm>
27. Army High Performance Computing Research Center, Wavefront Fonts, <http://www.arc.umn.edu/gvl-software/wavefront-fonts.html>
28. Yu Y., Dana K., Rushmeier H., Marschner S., Premoze S., Sato Y., (2000) *Image-based Surface Details*, SIGGRAPH 2000 course notes, New Orleans <http://www.cs.berkeley.edu/~yyz/publication/>
29. Upstill S., (1989) *The RenderMan Companion: A Programmer's Guide to Realistic Computer Graphics*, Addison-Wesley, NewYork
30. Razdan A., (2000) *Partnership for Research in Stereo Modeling (PRISM)*, Arizona State University, Personal communication <http://surdas.eas.asu.edu/prism/prism/>
31. Ibid.
32. Glassner A., *Maintaining Winged-Edge Models*, in James Arvo (Ed.) Graphics Gems II, Academic Press, New York, IV.6, 191–201
33. Pedersen H. K., (1996) *A Framework for Interactive Texturing on Curved Surfaces*, SIGGRAPH 96 Conference Proceedings, Addison-Wesley, New York, 295–302

Hyperseeing, Knots, and Minimal Surfaces

Nathaniel A. Friedman

Department of Mathematics, University at Albany-SUNY
<artmath@math.albany.edu>

Abstract. We wish to highlight the fact that seeing is of basic importance in mathematics just as in art. We will introduce hyperseeing which is a more complete all-around seeing from multiple viewpoints. Hyperseeing is then applied to study knot forms and their corresponding minimal surfaces.

1 Hyperseeing

One can say that the operative word that unifies art and mathematics is SEEING. More precisely, art and mathematics are both about SEEING RELATIONSHIPS. One can see certain mathematical forms as art forms and creativity is about seeing from a new viewpoint. Thus it is all about seeing. As the Basque sculptor Eduardo Chillida states "to look is one thing, to see is another thing", "to look is to try to see", "to see is very difficult, normally" [1]. I would like to add that from my own experience as a research mathematician and sculptor, it can take a lot of looking before one finally sees what has been there all the time. Seeing better is a lifetime endeavor. An excellent related article is See-Duction by Howard Levine [2].

We will now discuss a more complete way of seeing a three-dimensional object that is called hyperseeing. First we note that to see a two-dimensional painting on a wall, we step back from the wall in a third dimension. We then see the shape of the painting (generally rectangular) as well as every point in the painting. Thus we see the painting completely from one viewpoint. Now theoretically, in order to see a three-dimensional object completely from one viewpoint, we would need to step back in a fourth spatial dimension. From one viewpoint, we could then (theoretically) see every point on the object, as well as every point within the object. This type of all-around seeing, as well as a type of x-ray seeing, was known to the Cubist painters such as Braque, Duchamp, and Picasso, as discussed in [3]. In particular, Cubists were led to showing multiple views of an object in the same painting. In mathematics four-dimensional space is referred to as hyperspace and I refer to seeing in hyperspace as hyperseeing ([4]–[7]). Thus in hyperspace one could hypersee a three-dimensional object completely from one viewpoint. The Cubists were approximating hyperseeing in their paintings with multiple views. In general, we can regard hyperseeing in our own three-dimensional world as a more complete all-around seeing from multiple viewpoints. We will apply hyperseeing to study knots and their soapfilm minimal surfaces.

2 Knots

Knots are usually presented by two-dimensional knot diagrams that indicate where the knot crosses itself. Three diagrams of a trefoil knot are shown in Figure 1.

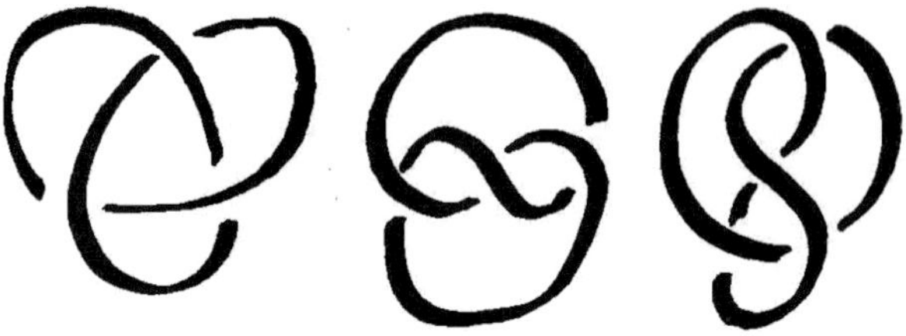

Fig. 1. *Trefoil Diagrams.*

Knots are usually presented as two-dimensional knot diagrams. However, a knot is actually a three-dimensional object and a knot really comes alive in a three-dimensional model of the knot which can be constructed from wire, folded aluminium foil, folded aluminium foil with a wire insert, copper tubing, or other material. Models of knots are ideal mathematical forms that can generate ideas for sculptures. Examples of models of knots made from folded aluminium foil are shown in Figure 2 ([NF]1). A knot made from copper tubing is shown in Figure 3 ([NF]2).

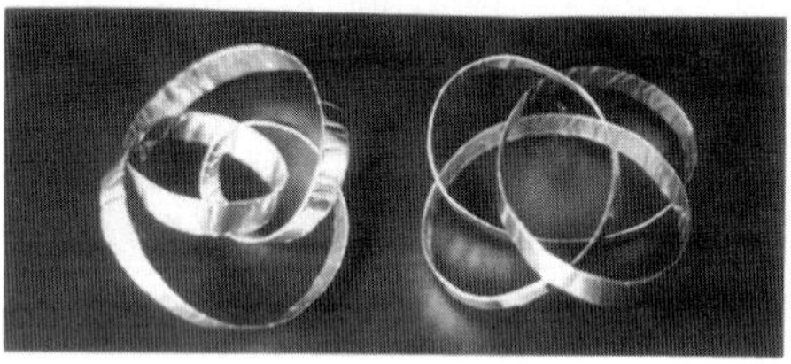

Fig. 2. [NF]1: *Aluminum Foil Knots.*

Models of knots are perfect mathematical forms for generating ideas for sculptures. They are completely three-dimensional with no preferred top, bottom, front, or back. Furthermore, a knot can look completely different when viewed from different directions. John Robinson has made beautiful bronze sculptures based on knot forms (see Sculpture in the Directory at `<www.isama.org>`).

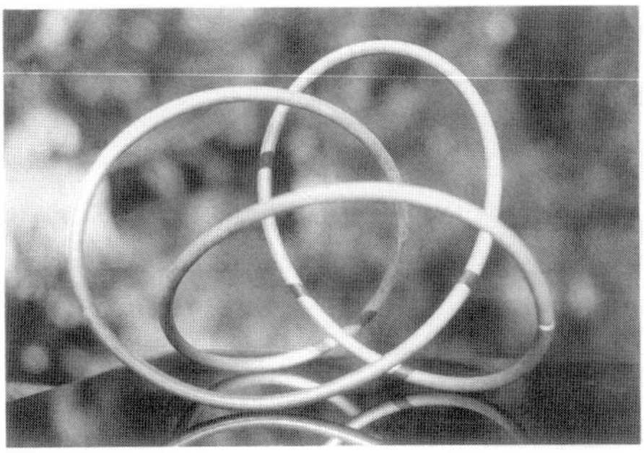

Fig. 3. [NF]2: *Copper Tubing Knot.*

As mentioned above, we can regard hyperseeing in our three-dimensional world as a more complete all-around seeing from multiple viewpoints. Since knots can look so different from different viewpoints, knots are excellent examples of interesting forms on which to practice hyperseeing. Knots are also open forms that one can actually see through. This is another reason that knots are ideal forms for hyperseeing.

For example, if one makes a wire model of a trefoil knot with front view as in Figure 1(a), then the top view will appear similar to Figure 1(b). A right side view of the model with a little tip forward will appear similar to Figure 1(c). In general, a model of a knot yields an infinite number of diagrams of the knot depending on your viewpoint of the knot.

There are several interesting mathematical exercises related to hyperseeing a model of a knot. For example, every knot admits a quadra secant. This is a viewpoint where one sees a diagram containing a point where four arcs of the knot coincide. Normally a crossing is where two arcs coincide. Colin Adams has introduced an exercise where the model is placed inside a plastic sphere and one maps the regions on the sphere corresponding to how many crossings one sees viewing the knot from a point in the region. For a trefoil knot there will be regions corresponding to at least three crossings. Adams has conjectured that a model of a trefoil knot always admits a viewpoint with a diagram having six crossings. In particular, this will imply that a trefoil model will admit viewpoints with diagrams having n crossings, n = 3,4,5, and 6. A projection of a knot is a diagram of a knot with the crossings not indicated as under/over. Thus a projection corresponds to replacing crossings by just intersections of two arcs. That is, a projection is just a "shadow" of a knot. The crossing number c(K) of a knot K is the least number of crossings that occur in a diagram of K. For example, if K is a trefoil knot, then c(K) = 3. If K is a loop, then c(K) = 0. Knots are listed in knot tables according to their crossing numbers.

When we hypersee a knot model from different viewpoints, we obtain a variety of diagrams. Given a diagram, we can consider the corresponding projection. Given a projection with n crossings, one can assign under/over at each crossing. This will yield a variety of knots with crossing numbers between 0 and n. One can always assign crossings so that one obtains a loop with crossing number 0. In the reduced case ([8], page 68), one can also assign the crossings to obtain an alternating knot with crossing number n. Therefore given a knot model, one can choose a viewpoint and obtain a diagram with n crossings. By changing the crossings, one obtains knot diagrams with a variety of crossing numbers between 0 and n. Thus one viewpoint yields a variety of knots. In particular, a trefoil model can have viewpoints yielding diagrams of knots with crossing number between 0 and 6.

An interesting property that results from the open structure of a knot concerns looking at a model of a knot from opposite directions. Suppose we regard one viewpoint as the front viewpoint. The viewpoint directly opposite will be the rear view. In general, the rear view of an object cannot be determined at all from the front view. However, for a knot, the rear view will be the reflection of the front view with the crossings reversed. For example, if one considers the views in Figure 1 as front views, then the corresponding rear views can be obtained by holding the turned page to the light and reversing the crossings.

Another property of knots is that one sees all points on a knot in any one view except for a finite number of crossing points. This implies that front and top views determine the side views. To see why this is true, let us fix one point P on the knot in an x,y,z coordinate system. A view along the x-axis determines the y and z coordinates of P. A view along the y-axis determines the x and z coordinates of P. Thus two orthogonal views determine the x,y,z coordinates of P in space. Applying this to points P on a knot, it follows that two orthogonal views of a knot will determine the position of the knot in space. In particular, the front and top views determine the side view. This is an interesting exercise in hyperseeing the knot. A suggested first exercise is to consider a straight line in space with endpoints P = (a,b,c) and Q= (d,e,f). A view along the x-axis shows P at (b,c) and Q at (e,f). A view along the z-axis shows P at (a,b) and Q at (d,e). The view along the y-axis is therefore determined with P at (a,c) and Q at (d,f). One can now connect (a,c) and

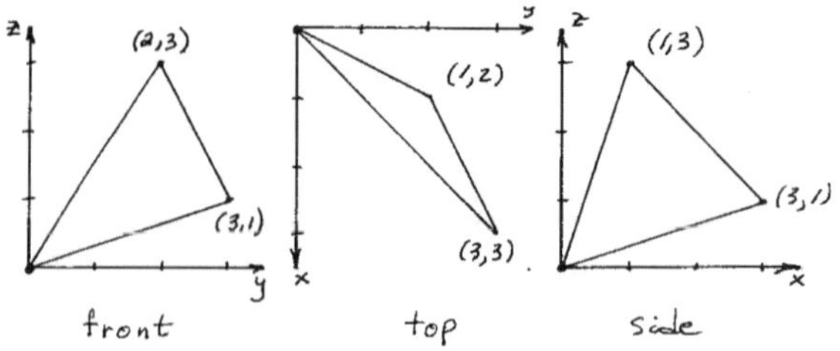

Fig. 4. *Front, Top, and Side Views.*

(d,f) by a straight line to obtain the side view of the line. An example of a triangle in space is shown in Figure 4, where the side view is from the left.

The next exercise is to consider a knot formed by straight line segments. One can then obtain the side view from front and top views using coordinates of the endpoints of the lines. No line can be parallel to an x,y, or z axis.

3 Kenneth Snelson Sculptures

Kenneth Snelson sculptures consist of line elements suspended in space. For example, Free Ride Home is shown in Figure 5. The discussion above implies that a front and top view of a Snelson sculpture would determine the side view.

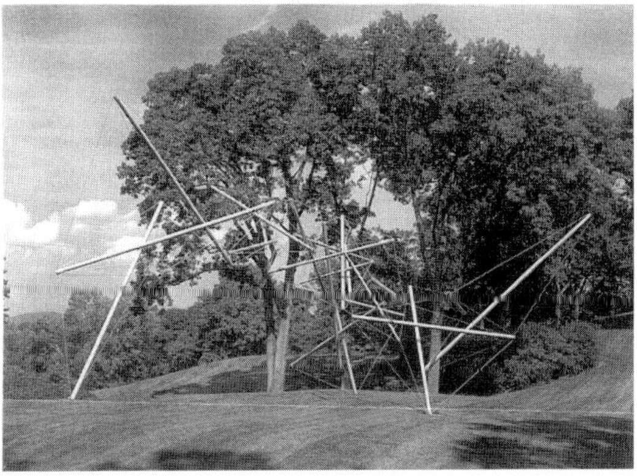

Fig. 5. [NF]3: *Free Ride Home, Kenneth Snelson, 1974, Stormking Art Center, Mountainville, New York.*

4 Soapfilm Minimal Surfaces of Knots

If one dips a wire model of a knot in a solution of liquid soap and water, one obtains a soapfilm minimal surface with the knot as the single edge of the surface. For example, the edge of a half-twist Möbius band is a simple loop as shown in Figure 6(a). If a wire is bent in this shape and dipped in a soap solution, one obtains a minimal surface with a central disk. If this disk is puntcured, then one obtains the Möbius band minimal surface in Figure 6(b).

The corresponding soapfilm minimal surfaces for the trefoil knots in Figure 1 are shown in Figure 7. In (a) we have a triple twist Möbius band with one side. In (b) we have a two-sided minimal surface. In (c) we have a one-sided minimal surface. The exercise students of all ages really enjoy is dipping

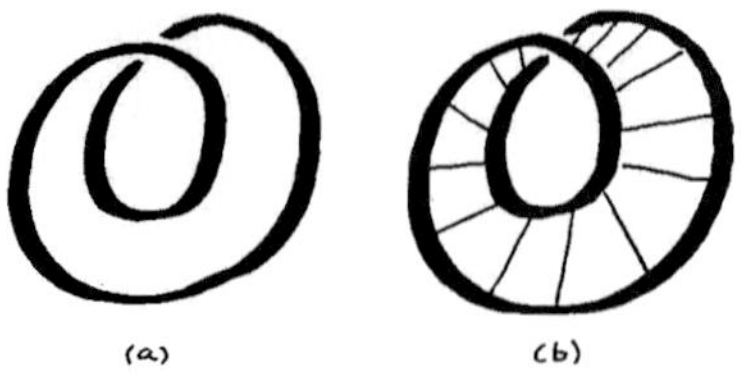

Fig. 6. *Möbius Band.*

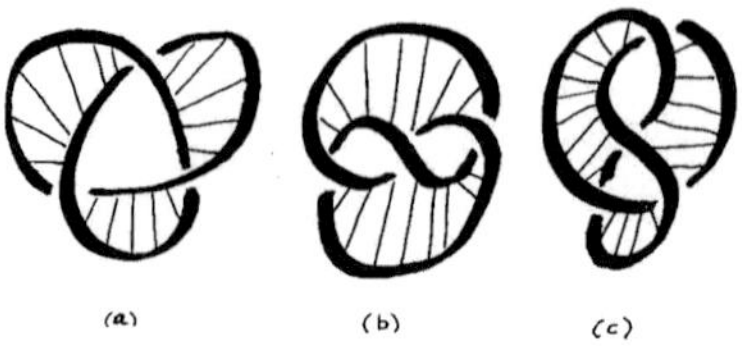

Fig. 7. *Trefoil Minimal Surfaces.*

wire models of knots in order to obtain the corresponding soapfilm minimal surfaces. In order that students can learn to anticipate the shape of the surface, it is helpful to use masking tape in order to approximate the surface. An example is shown in Figure 8 for a trefoil knot.

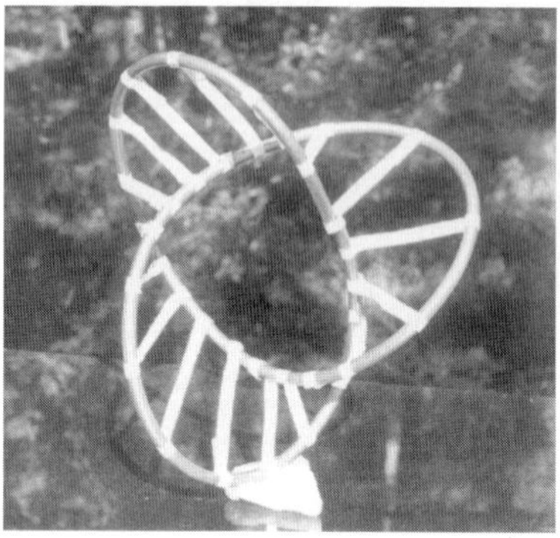

Fig. 8. [NF]4: *Trefoil Taped Surface.*

5 Framed Knots

In Figure 7(a) we have a minimal surface that could be described as a form consisting of three leaves with a central space. We will now modify the knot so that the leaves become space and the center is form. The modified knot

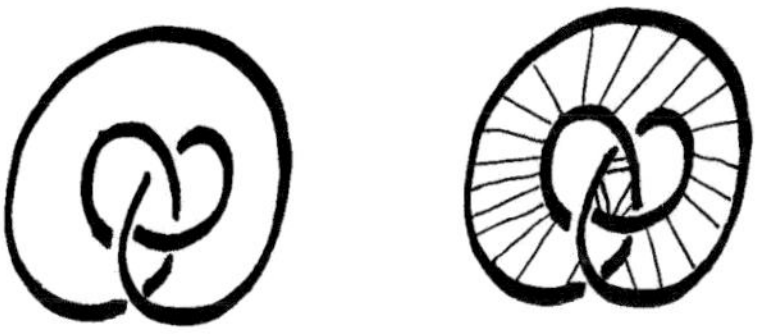

Fig. 9. *Framed Trefoil.*

is shown in Figure 9(a) and the corresponding minimal surface is shown in (b). We refer to the knot in (a) as a framed knot. The original idea is to form a link obtained by placing the knot inside a circle. This was first shown to me by the sculptor Charles Perry. Later I saw this link in several books on knots. This led me to the idea of a framed knot. From a mathematical point of view the framed knot is not much different from the original knot since it is easy to see that one can deform the framed knot by lowering the added "frame" to obtain the original knot. However, from a sculptural viewpoint, the framed knot is interesting since it has a minimal surface that reverses form and space in the minimal surface of the original knot. One can also suspend a wire knot in a wire circle using an extra piece of wire to suspend the knot. This is the form that Charles Perry introduced in order to obtain a sculpture where form and space are switched. Perry forms the surface using flexible metal screen and then applies automobile body putty on the screen. A mold is made and then the minimal surface is cast in bronze. The idea of forming a sculpture from the minimal surface of a framed knot or a knot in a circle is a very recent development.

6 Multiple Möbius Bands

In Figure 6(b) we have a minimal surface for a loop that consists of one Möbius band. It will now be shown that there is a configuration of a trefoil knot that has a minimal surface consisting of two Möbius bands that share part of their edges as in Figure 10(a) and also alternately cross over each other as in Figure 10(b). The knot is shown in Figure 11(a). In Figure 11(b) Möbius band 1 is shown and in Figure 11(c) Möbius band 2 is shown. The intersection lines on the cross overs are indicated by dotted lines.

If Figures 11 (b) and (c) are combined, then we obtain the complete minimal surface shown in Figure 12. The two bands share parts of their edges and alternately cross over each other. Band 1 crosses over band 2 at the top and band 2 crosses over band 1 at the bottom. It is interesting to see each soapfilm Möbius band twist as it crosses over the other Möbius band. The bands share edges until they twist.

The knot in Figures 10 and 11 is the representation of a trefoil knot as the 2-3 torus knot. In general, given p and q mutually prime, the p-q torus

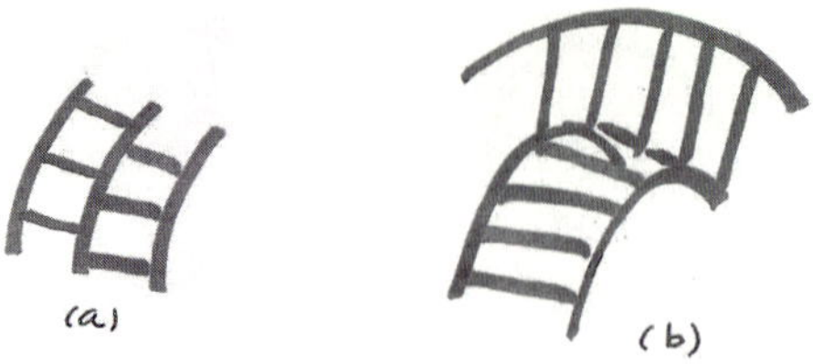

Fig. 10. *Sharing edges and crossing.*

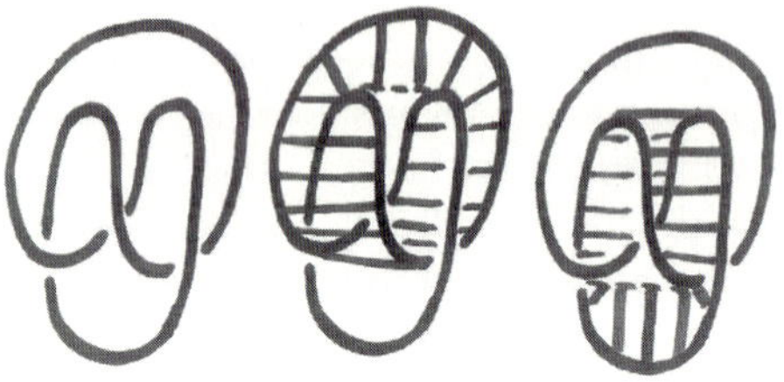

Fig. 11. *Möbius band on a TrefoilKnot.*

knot wraps meridionally around the torus p times and wraps longitudinally around the torus q times. The p-q torus knot is equivalent to the q-p torus knot. That is, one is deformable into the other (see [8], page 111). Thus from a mathematical viewpoint, the p-q and q-p torus knots are the same knot. However, from a sculptural viewpoint, they look quite different and they have configurations with very different minimal surfaces. In particular, the 3-2 torus knot is shown in Figure 1(a) and it has a minimal surface as in Figure 7(a), which is a triple twist Möbius band. In Figure 11 we have the 2-3 torus knot with a configuration having a minimal surface consisting of two single twist Möbius bands. In general, the n-(n+1) torus knot has a configuration with a minimal surface consisting of n single twist Möbius bands that partly share edges and alternately cross over each other.

For example, the 3-4 torus knot has a minimal surface consisting of three single twist Möbius bands. Suppose we color these three bands red, white, and blue. Part of the surface will have the red band and white band sharing an edge with the red band to the left of the white band. The blue band will be crossing over the red and white bands. The blue band will then twist and share an edge with the white band, where the white band is now to the left of the blue band. The red band will now twist and cross over the white and blue bands. This structure goes around a central space so that each band twists once to cross the other two bands. In order to picture this, we will first draw the 3-4 torus knot as in Figure 13. We begin with three points as in (a). We then draw arcs as in (b). The knot is completed as in (c). The corresponding minimal surface is shown in Figure 14.

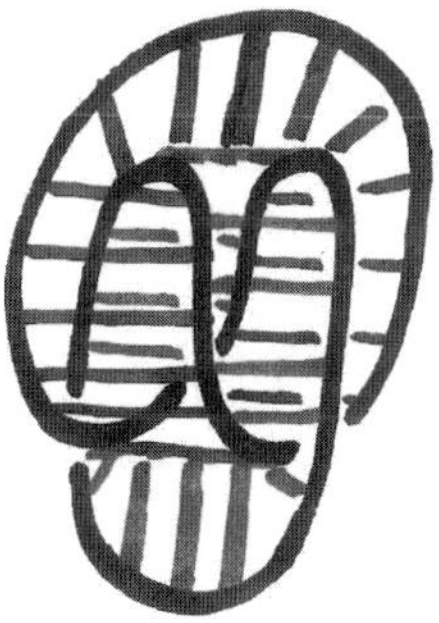

Fig. 12. *Double Möbius Band Minimal Surface.*

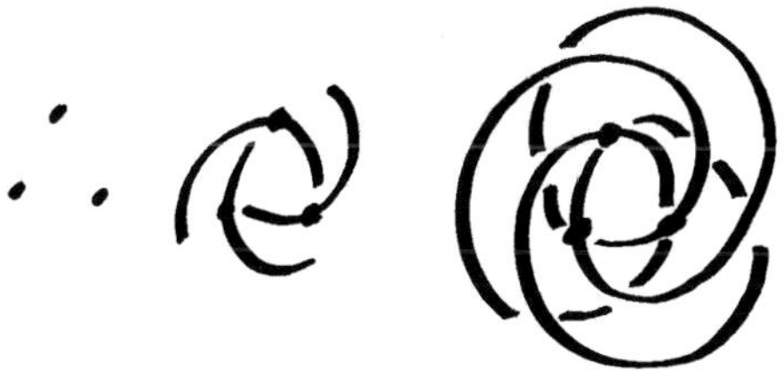

Fig. 13. *3-4 Torus Knot.*

In general, the behavior of minimal surfaces for configurations of (n+1)-n torus knots may consist of one or more bands with one or more half twists. In Figure 7(a) for the 3-2 torus knot we have one band with three half twists. The 4-3 torus knot has a minimal surface consisting of two bands that partly share edges and cross over each other twice, hence each band has two half twists. The 5-4 torus knot has a minimal surface that consists of one band that crosses itself five times as it wraps around four times. Part of the surface will appear as described above for the 3-4 torus knot except there is only one band alternately crossing over itself as it wraps around.

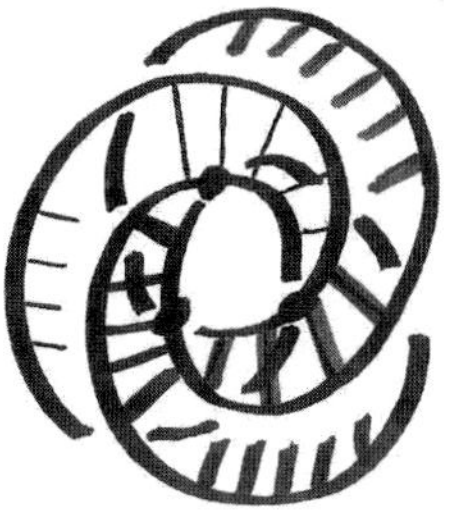

Fig. 14. *Triple Möbius Band Minimal Surface.*

References

1. Chillida E., (1985) Basque Sculptor, Video, Home Vision, 24
2. Levine H., (1997) See-Duction, Humanistic Mathematics Network Journal #15
3. Henderson L.D., (1983) The Fourth Dimension and Non-Euclidean Geometry in Modern Art, Princeton University Press, Princeton
4. N.A. Friedman, (1998) Hyperspace, Hyperseeing, Hypersculptures, in: Conference Proceedings, Mathematics and Design 98, Javier Barrallo Ed., San Sebastian, Spain.
5. N.A. Friedman, (1998) Hyperseeing, Hypersculptures, and Space Curves, in Conference Proceedings, 1998 Bridges: Mathematical Connections in Art, Music, and science, Reza Sarhangi Ed., Winfield, Kansas, USA.
6. N.A. Friedman, (1998) Hyperspace, Hyperseeing, Hypersculptures (with figures), Hyperspace, vol. 7, Japan Institute of Hyperspace Science, Kyoto, Japan.
7. N.A. Friedman, (1999) Geometric Sculptures for K-12: in Geos, Hyperseeing, Hypersculptures, Conference Proceedings, 1999 Bridges, Mathematical Connections in Art, Music, and Science, Reza Sarhangi Ed., Winfield, Kansas, USA.
8. Adams C., (1994) The Knot Book, W.H. Freeman, New York,

Ruled Sculptures

Philippe Charbonneau

24, rue Coriolis, 75012 Paris

Designer – projector in architecture, retired, I devote a great part of my time in plastic researches oriented into two main directions:

1 3rd degree ruled surfaces

Of the same type as the Möbius band, these surfaces are somewhat paradoxal. They are all in curves, but in fact they are generated by straight lines only. Their principle is simple: it determines complex forms and volumes that destabilize and enrich our sense of space. Notions as trivial as faces, boundaries, above, underneath, interior, exterior may loose their exact meaning.

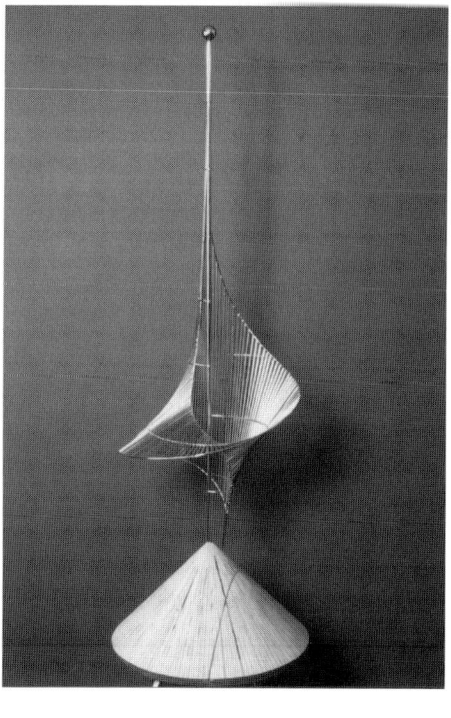

Fig. 1. [PC]1: Biconique IV

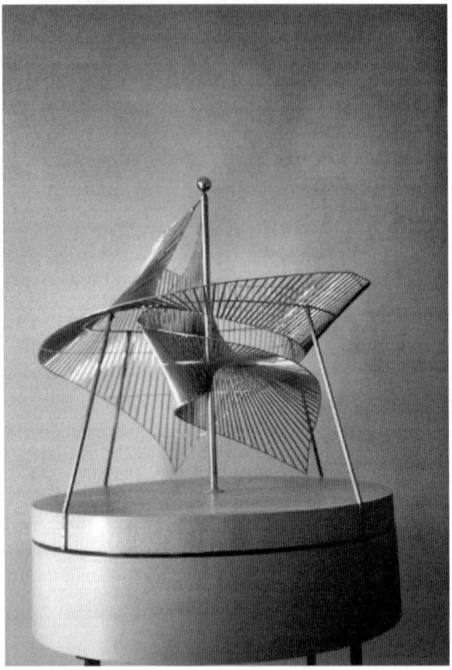

Fig. 2. [PC]2: Biconique V

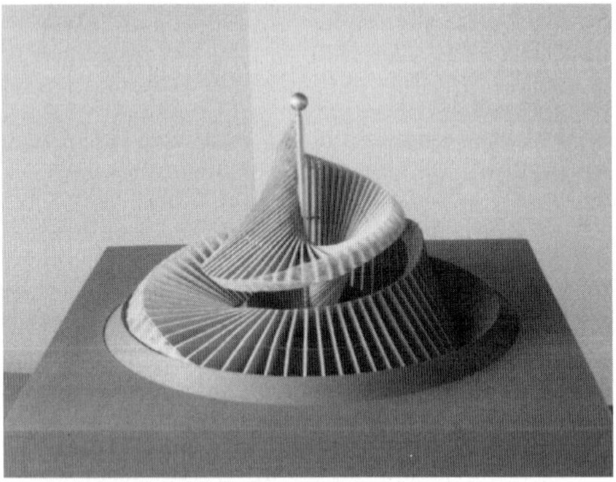

Fig. 3. [PC]3: Biconique I

To my knowledge, such surfaces do not exist in nature nor in the living life, and, until now, men would not have found any technical use for them. They have been well studied by mathematicians, and they have inspired some artists, but there still remains a huge field to explore.

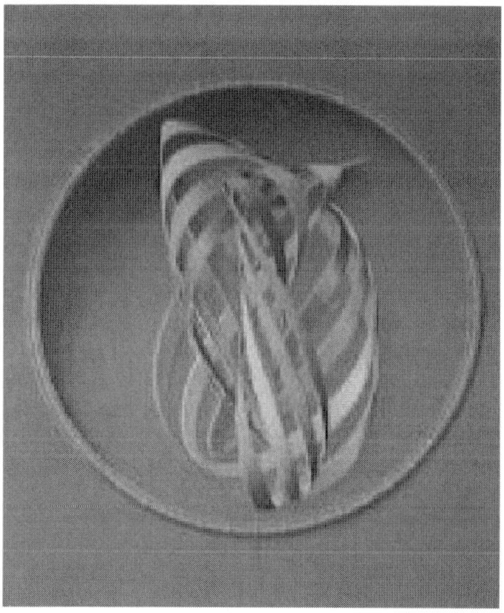

Fig. 4. [PC]4: Entrelac quatre rubans

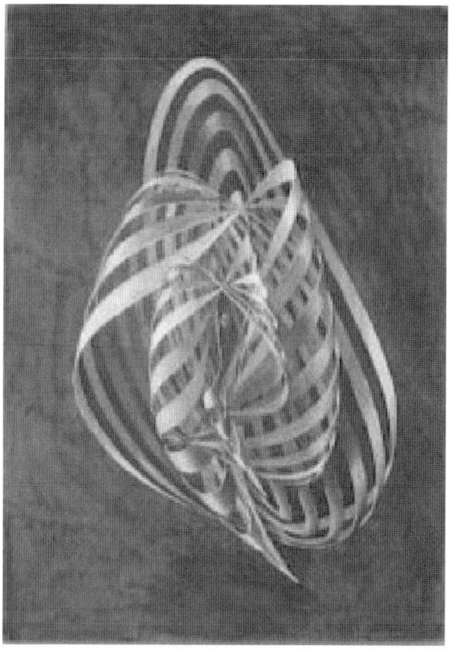

Fig. 5. [PC]5: Entrelac I 2 dix rubans

I began to materialize these surfaces using different structures, materials, colours and limits. Then I undertook research on the assembling of surfaces of the same type through plays of symmetries, of transformations and movements. Their combinatorics is infinite and the visual effects are often unexpected.

Sculptures of this type, realized in a monumental scale, would enlarge the apprehension of these spaces and enhance an aesthetic emotion lived by the body as a whole.

2 Drawings through pendular plotting

Pendular plotting first are drawings automatically realized within 2 or 3 minutes through a handover thrown pendulum. It is a game for children. It results in Lissajous curves with damping, well known from the physicists who studied the oscillatory phenomena.

From this original plotting, where the lines are often very mixed up, I let emerge a third dimension, I give the illusion of relief, and by colours, small threads, bands, I try to render sensible a lasting harmony and to transpose the dynamics that spurred on it.

The same plotting may give rise to several very different elaborations. The hazard and my subjectivity, coupled with a certain logic, end at the creation of these shapes which would be my contribution to the theme of knot-works, treated by numerous artists from the prehistoric time.

A Gallery of Algebraic Surfaces

Bruce Hunt

Max-Planck-Institut für Mathematik in den Wissenschaften, Leipzig
and
Helaba, OB 555000, Strahlenbergerstr. 11, 63067 Offenbach, Germany
<Bruce.Hunt@helaba.de>

1 Introduction

The notion of a surface is a very classical one in technology, art and the natural sciences. Just to name a few examples, the roof of a building, the body of a string instrument and the front of a wave are all, at least in idealized form, surfaces. In mathematics their use is very old and very well developed. A very special class of (mathematical) surfaces, given by particularly nice equations, are the *algebraic surfaces*, the topic of this lecture. With modern software, one can make beautiful images of algebraic surfaces, which allow us to visualize important mathematical notions; explaining this is the object of this talk.

Several years ago I started to make pictures of algebraic surfaces. Originally I was involved in an open day at the University of Kaiserslautern, and the pictures were an attempt to motivate youngsters in getting interested in mathematics. Later I made a gallery of these pictures for the Web, which is located at:

`http://www.mathematik.uni-kl.de/~wwwagag/Galerie.html`.

For my talk, which I was invited to give by Claude Bruter who had seen the gallery, I again took up to making some new images, so that the presentation here contains many images which are not presently available on the net. In addition, several colleagues and I created in Kaiserslautern some movies, which I showed at Maubeuge, some frames of which are also presented here.

I have attempted to show how these computer images can be used to visualize non-trivial mathematical concepts. After a brief introduction to the notion (definition) of algebraic surface in the first section, I consider successively three mathematical aspects to which the pictures yield a vivid visualization. The first is the notion of *symmetry*, certainly one of the most profound and important ones in all of mathematics. Often times a mathematical problem can only be treated for objects with a sufficient amount of symmetry. Of course, mathematicians have an abstract understanding of what symmetry is, but the pictures enable non-mathematicians to "see" it (mathematical symmetry, more than just the naive notion in our everyday vocabulary). Next we consider singularities of algebraic surfaces. Quite generally in mathematics, singularities are often of central interest, creating also most of the problems in the consideration of examples, as the "general theory" does not hold at these points. With the images of the algebraic surfaces, one can "see" what these singularities look like. Again, mathematicians have a kind of abstract understanding of singularities, but often do not really know what they look like, so real pictures of them are quite exciting. As a final topic, one rather specific to algebraic geometry (as opposed to differential geometry or topology), we consider some easy problems of *enumerative geometry.* In particular, with the computer images one can see the 27 lines on a smooth cubic surface.

Six of the pictures here in the text may be viewed in full living color in the colored signature at the end of the book. 34 coloured pictures can be viewed from the Appendix.

Acknowledgements: I have made all images presented here and in the gallery with public domain software, a ray tracing system called VORT (Very Ordinary Ray Tracing), available via anonymous FTP from

`ftp://gondwana.ecr.mu.oz.au/pub/vort.tar.gz.`

For the introduction to this system and help with creating the movies, it is my pleasure to thank Rüdiger Stobbe (at that time also at the University of Kaiserslautern). More help with creating movies was provided by Christoff Lossen, and in addition help with equations was provided by Duco von Straten (now University Mainz). The present system administrator at Kaiserslautern, David Ilsen, as well as his predecessor Hans Schönemann, were very helpful in getting VORT installed on my present computer and getting the equations I used in Kaiserslautern to me. My thanks to them all. Finally, my present employer, a state bank in Germany, was gracious enough to allow me to visit Maubeuge as a business trip. For that and more support, I thank them also.

2 Algebraic surfaces

To start this talk, I would like to motivate the notion of algebraic surface for non-mathematicians. For this, recall from your school courses the *graph of a function.*

Typically one starts with the function $f(x) = x^2$, the graph of which is plotted in the (x, y)-plane by setting $f(x) = y$, i.e., $y = x^2$. You may remember that the graph of this function is a parabola, see Figure 1. The special point of the graph $(x, y) = (0, 0)$ is the *minimum* of this function. As a next example, consider $f(x) = x^3$, i.e., the graph given in the (x, y)-plane by $y = x^3$. You might recall that this graph is as in Figure 2, and here the special point at $(0, 0)$ is now not a minimum or maximum, but an *inflection point.* Finally, a more typical example is given by taking the slightly different function $f(x) = x^3 - 3x$ which now has a local minimum at $x = 1$ and a

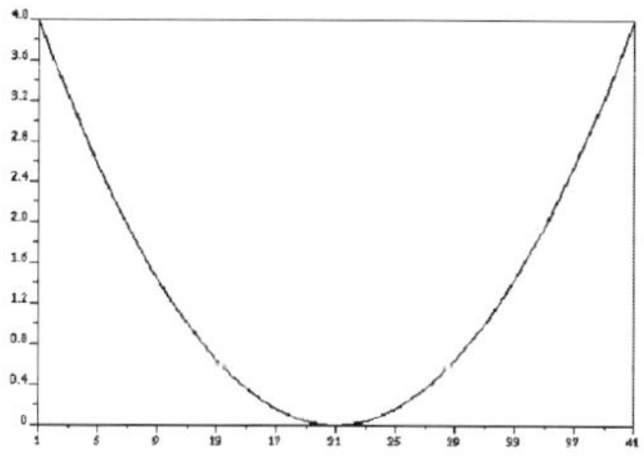

Fig. 1. The graph of the function $f(x) = x^2$

local maximum at $x = -1$ [1]. Now, for any such graph, given by an equation $f(x) = y$ as above, we can form the *equation for a curve as a zero locus:*

$$g(x, y) := y - f(x) = 0. \tag{1}$$

The function g of two variables is defined by the expression (1). Thus, the *graph of a function* is a *special case* of a *curve* in the plane.

If, moreover, $f(x)$ is a polynomial in x (a sum of *powers of* x) (as opposed to more complicated functions of x like sin, cos or tan) with certain coefficients, then the graph is said to be *algebraic*, and again, more generally, a curve defined by an equation (1) is said to be an *algebraic curve*, if $g(x, y)$ is a polynomial in two variables, i.e., $g(x, y) = \sum a_{ij} x^i y^j$.

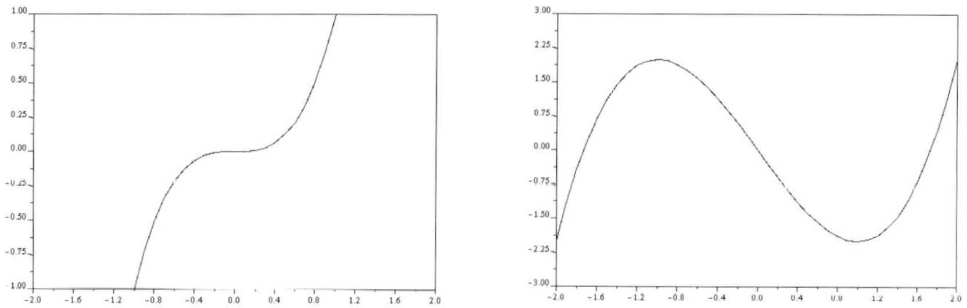

Fig. 2. The graphs of the functions $f(x) = x^3$ and $f(x) = x^3 - 3x$

Perhaps the simplest example of a curve which is not a graph is the so-called *Niel parabola* or *cuspidal cubic*, given by the equation $g(x, y) = x^3 - y^2 = 0$, which is depicted in Figure 3. Note that the point (0,0) for this curve is even more important than for the functions above: here we have an example of an algebraic curve with a *singular point* at (0,0). Another example

[1] This is obtained by considering the derivatives of f, $f'(x) = 3x^2 - 3 = 3(x^2 - 1) = 3(x + 1)(x - 1)$, $f''(x) = 6x$, hence $f''(1) = 6 > 0$ and $f''(-1) = -6 < 0$.

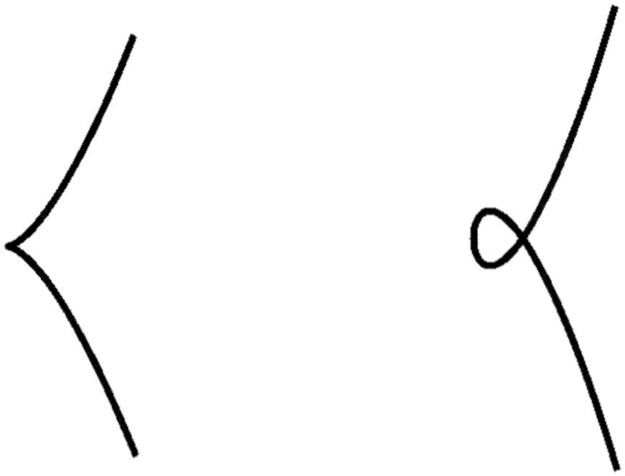

Fig. 3. Two singular curves which are not graphs of functions. Left: The cuspidal cubic curve $x^3 - y^2 = 0$. Right: The cubic curve with a double point $x^3 - x^2 - y^2 = 0$.

of this kind is given by the *cubic with a double point*, given by the equation $x^3 - x^2 - y^2 = 0$, depicted in the second picture in Figure 3. There are two branches at $(0,0)$, determined by the two *tangents* which one gets by taking the derivatives.

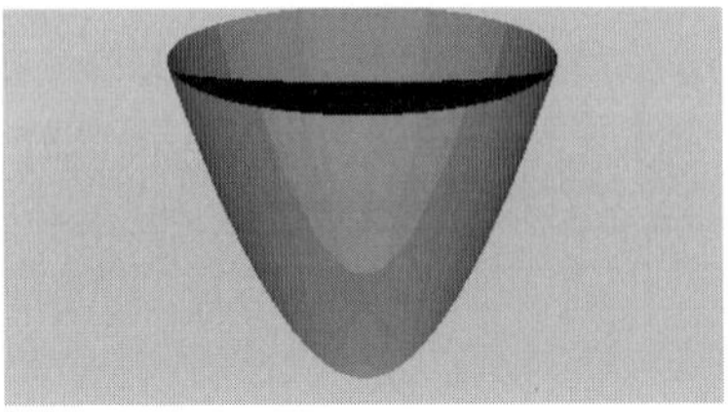

Fig. 4. The graph of the function $g(x, y) = x^2 + y^2$

Now a statement that mathematicians love is: "What we can do once we can do again." So we now start with the function $g(x, y)$ and consider *its* graph, that is $z = g(x, y)$ in three-space. Plotting the graph now results in a landscape (which is referred to in mathematics as a *surface*). For $g(x, y) = x^2 + y^2$, we get the three-dimensional analog of the first example above, see Figure 4. This surface has a special property: it has a rotational *symmetry*, i.e., it is like a top which can spin on its vertex at $(0, 0, 0)$. A more interesting example is obtained by starting with our function $g(x, y) = x^2 - y^3$, which we explained above is the simplest example of a curve which is not a graph. In this case we get a surface $x^2 - y^3 - z = 0$, which is pictured in Figure 5.

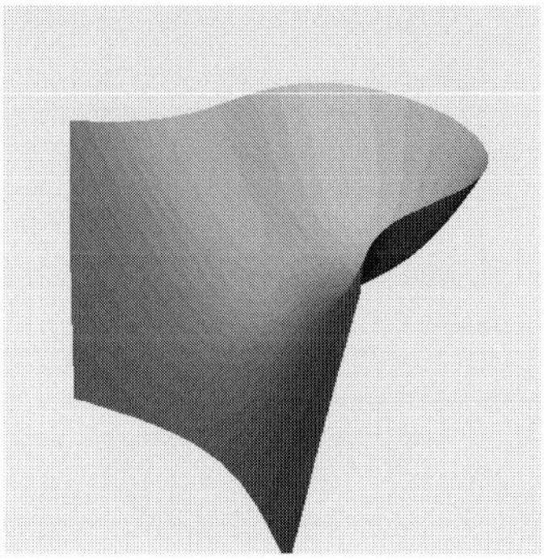

Fig. 5. The algebraic surface with the equation $x^2 - y^3 - z = 0$.

As opposed to the previous surface which is a *quadric*, i.e., its equation has degree 2, this is an example of a surface of degree 3, a cubic.

At any rate, we do again as we did in the discussion of curves and consider the surfaces given by equations $\{h(x, y, z) = 0\}$, with the case of a graph being characterized by the condition $h(x, y, z) = z - g(x, y)$. This surface is said to be *algebraic*, if $h(x, y, z)$ is a polynomial in x, y, z: $h(x, y, z) = \sum a_{ijk}x^i y^j z^k$. Again, the simplest example of h which is not the graph of a function is $\{x^2 + y^2 - z^2 = 0\}$, shown in Figure 6. Note that this surface has a *singular point* at $(x, y, z) = (0, 0, 0)$.

For the mathematicians in the audience which I have been boring up to now let me briefly explain in more detail what an algebraic surface is. What we have been considering above belongs in the realm of *real algebraic geometry*, which is not the beautiful theory studied by algebraic geometers. Instead of working over the reals $\mathbb{R}$, classical algebraic geometry works over the complex numbers $\mathbb{C}$, and an algebraic surface is given by an equation $\{(x, y, z) \in \mathbb{C} | h(x, y, z) = 0\}$ for a polynomial h with complex coefficients. Consequently, this object is actually a four-dimensional space (naturally embedded in a six-dimensional one). Hence, viewing an algebraic surface as a topological space, it is four-dimensional topology which is relevant.

Furthermore, what we have done above is *affine* geometry, and what is more interesting (in the wonderful theorems one gets) is *projective* geometry, so that an algebraic surface is a *compact* four-dimensional topological space, and the affine pictures above are real slices, locally around some point on the surface. The affine surface is the complement of a "curve at infinity" on the projective surface. For the non-mathematicians which might be surprised that starting with something compact (which can be held in your hand) and taking away some curve, one gets something which is infinite, consider the following. If you take a sphere in your hand and delete the north pole, then a so-called *stereographic projection* maps it to the plane, which is infinite. This projection maps a point P (see Figure 7) to the point Q in the (x, y)-

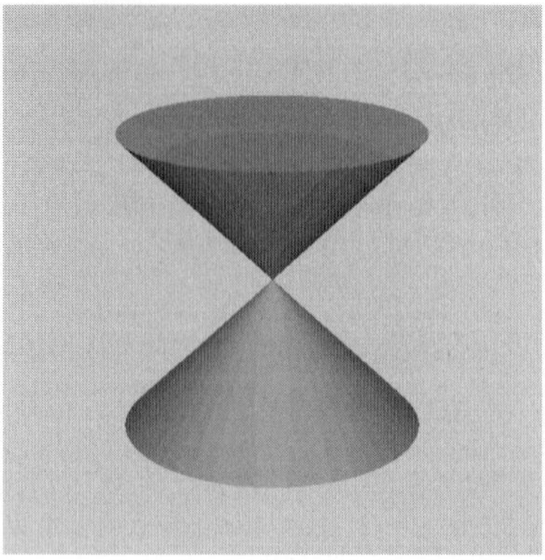

Fig. 6. The surface $x^2 + y^2 - z^2 = 0$, which is not the graph of a function

plane which is where the ray $\boldsymbol{OP}$, emanating at the north pole O and passing through P, intersects the plane.

The passage to projective geometry is affected by *homogenizing* the polynomial h, which is done by setting:

$$x = \frac{x_1}{x_0}, \quad y = \frac{x_2}{x_0}, \quad z = \frac{x_3}{x_0},$$

then multiplying $g\left(\frac{x_1}{x_0}, \frac{x_2}{x_0}, \frac{x_3}{x_0}\right)$ by the smallest power of x_0 necessary to clear denominators. The result is a homogenous polynomial $g_{hom}(x_0, x_1, x_2, x_3)$, and the projective surface is

$$\{g_{hom}(x_0, x_1, x_2, x_3) = 0\} \subset \mathbb{P}^3,$$

where $\mathbb{P}^3$ denotes the three-dimensional projective space over $\mathbb{C}$ with homogenous coordinates (x_0, x_1, x_2, x_3).

Many properties of algebraic surfaces hold only for projective surfaces over the complex numbers. For example:

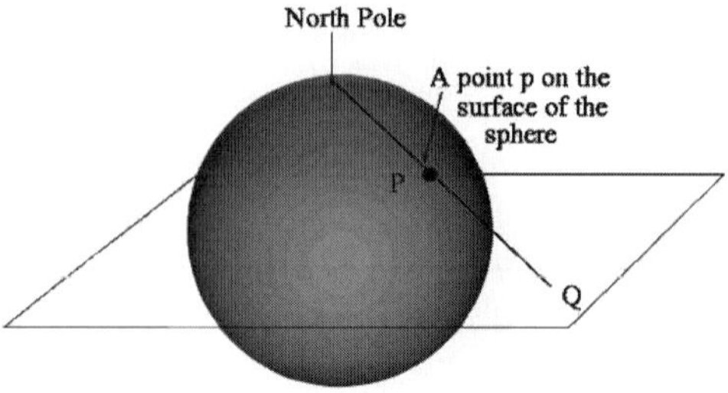

Fig. 7. The stereographic projection

- A generic line in $\mathbb{P}^3$ meets an algebraic surface S in a fixed number d of points on the surface. This number d is called the *degree* of the surface. We have met examples of degrees 2 and 3 above.
- Any two smooth quadric surface are isomorphic (complex analytically) and have the same symmetry group, which is $SO(4) \cong SO(3) \times SO(3)$. Such a quadric surface is ruled; more precisely it is isomorphic to $\mathbb{P}^1 \times \mathbb{P}^1$, where $SO(3)$ is the symmetry group of $\mathbb{P}^1$.
- Any two smooth surfaces of the same degree are *diffeomorphic* to one another, thus in particular have the same topological type. There is a finite number of *moduli* of such surfaces, i.e., parameters giving rise to complex analytically distinct surfaces.
- The notion of *duality* holds in projective space. This states that to every point there is a unique plane associated with it (the dual), and conversely, to every plane there is a unique point associated with it. An example of an application of this notion, the dual variety, is considered below.

3 Symmetry

A short dialog between a mathematician and an artist. The mathematician tries to explain to him the meaning of some mathematical concepts, first that of symmetry.

> ARTIST: "I have often heard that mathematicians have a keen sense of esthetics. Can you explain this to me?"
> MATHEMATICIAN: "As a first point I'll try to elucidate the notion of symmetry. Look at this ball I have in my hand. In your opinion, is it symmetric?"
> ARTIST: "It's the most symmetrical object there is. It has a certain symbolic importance because of that."
> MATHEMATICIAN: "Is it only the shape, or do you associate with it some other property which gives rise to a characterization of 'ball'?"
> ARTIST: (After thinking a moment) "The shading, the coloring gives a unique pattern."
> MATHEMATICIAN: "Now let me describe another property, which we mathematicians see in this object. Instead of the shape of the ball, consider the fact that it *rolls*. Phrased differently, we consider the symmetry of the object as its set of motions; in the case of the sphere these are the rotations you get by rolling the ball around on the table. Note that this is different than, for example, the case of an egg, which wobbles – it has less symmetry."

The "Erlanger Programm", put forth by Felix Klein in 1872, is the basis for a modern mathematical treatment of symmetry. It states that a property should be considered as part of some geometry, if it is *invariant* under the set of all automorphisms (motions) of the geometry. We have seen this above

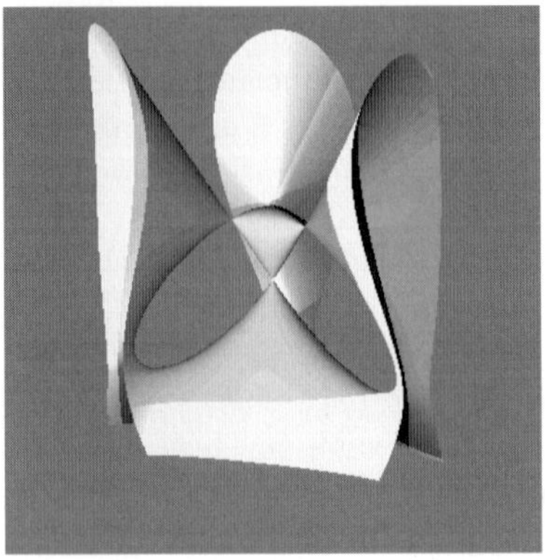

Fig. 8. The Cayley cubic surface

in the case of *continuous* groups of symmetry. However, we are actually more interested in *discrete* groups of symmetry, and even more specifically, in *finite* symmetries, i.e., objects which are invariant under finitely many motions of some kind. Think of, instead of a ball, a cube. It does not roll, but by turning it on an edge, it flops back down onto the table. You can do this in two different directions, leading in fact to a group of symmetries of order 24 (i.e., containing 24 elements). The fact that there are so many is based on the fact that these motions are not *commutative*, i.e., the order in which they are carried out is important. This group is known to mathematicians as the *symmetric group on four letters*, and can be described as a group of permutations of four objects (usually referred to as letters) denoted Σ_4.

We now present some surfaces with nice symmetry. First of all, a special surface with the group Σ_4 as symmetry group: the *Cayley cubic*. This surface was first studied by Arthur Cayley around 1850, and Felix Klein had a plaster model of it prepared for the Chicago World Fair in 1893, thus initiating the creation of plaster models to visualize mathematical objects. This surface is depicted in Figure 8.

The fact that this surface has the permutation group of four letters as its symmetry group is easily seen from its projective equation, which is:

$$\frac{1}{x_0} + \frac{1}{x_1} + \frac{1}{x_2} + \frac{1}{x_3} = 0,$$

which is the equation of an algebraic surface perhaps contrary to appearances: just multiply through by $x_0x_1x_2x_3$.

This surface is, by the way, also of particular interest as it is the unique cubic surface which has four ordinary double points. It is not difficult to visualize the symmetries of this surface which consist of the *rotations* of the tetrahedron whose vertices are the singular points. It is more difficult to envision the symmetries which arise from a *simple permutation* of two of the

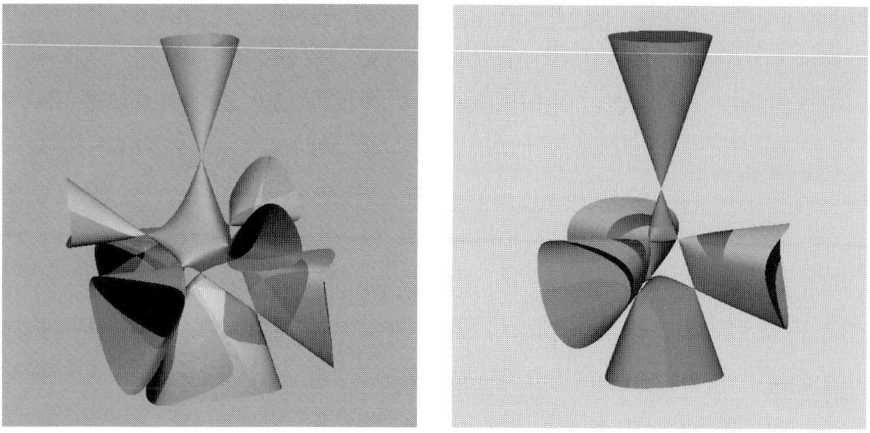

Fig. 9. Hessian varieties of cubics surfaces ; these are quartic surfaces. Left: [BH]1 The Hessian of the Cayley Cubic. Right: The Hessian of the Clebsch cubic

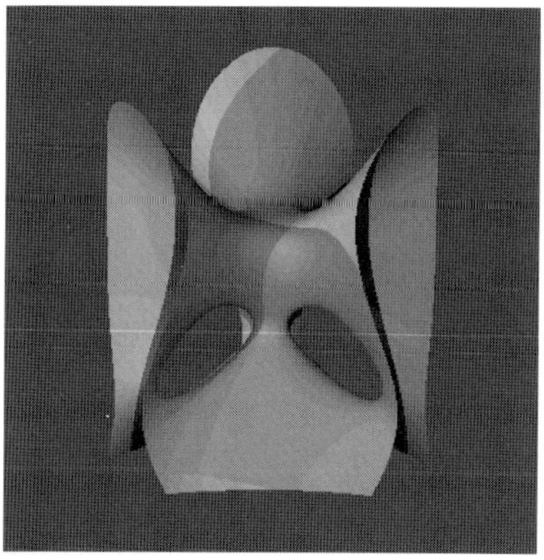

Fig. 10. The Clebsch "diagonal" cubic surface

(homogenous) coordinates, which is a *reflection* on a certain plane. These planes are those which contain two of the vertices and pass midway between the other two. These symmetries can be described as follows: suppose that we place a mirror exactly at the location of this plane, and you look at the mirror from one side. Then what you see is *exactly* the same thing you would see if we had put, instead of the mirror, a transparent pane of glass.

Another very beautiful cubic surface, the unique such which has the symmetric group on *five* letters as its symmetry group, is the *Clebsch cubic*, also known as the *diagonal cubic surface*. In this case the surface is smooth, and one does not "see" the symmetry just looking at the surface, which is depicted in Figure 10. Again, the symmetry can be seen by looking at the projective equation, this time with a twist. One describes the surface in $\mathbb{P}^4$ instead of

$\mathbb{P}^3$, in which it lies in a particular *hyperplane*, which cuts out of $\mathbb{P}^4$ a $\mathbb{P}^3$. The equation is then

$$C = \{y_1^3 + \cdots + y_5^3 = 0 = \sum y_i\}; \tag{2}$$

the equation clearly stays the same (is *invariant*) under an arbitrary permutation of the y_i.

Some other surfaces of interest for their symmetry groups follow. There is an interesting notion of *covariant* of hypersurfaces, in particular also of surfaces. These are surfaces whose equation is obtained in some manner from a given equation in such a way as to preserve all symmetries. A prime example is that of the *Hessian variety*, whose equation is easily written down: let $f(x_0, x_1, x_2, x_3)$ be the defining equation of surface, let $\partial_i f$ denote its partial derivative with respect to x_i, and form the *Hessian matrix*:

$$\{\mathrm{Hess}(f)\}_{i,j} := \partial_i \partial_j f.$$

This is a four by four matrix, in the case of our cubic surfaces above for example with entries which are linear forms in the variables x_i. Then one forms from this matrix the *determinant*, which is then a single polynomial of degree $4(d-2)$ for a surface of degree d, so for the cubic surfaces above this is a *quartic*. We have some nice pictures of the Hessian of the Cayley and Clebsch cubic, see Figure 9. As mentioned above, the Hessian is a covariant of a polynomial, so it has the same symmetry as the original surface.

A further example of covariants is the so-called *dual variety* of a hypersurface. See my book "The Geometry of some Special Arithmetic Quotients", Springer Lecture Notes in Mathematics **1637**, Springer-Verlag 1996, section B.1.1.6 for more details on this notion. The dual variety is defined to be the union of all hyperplanes of the ambient projective space (in this case, all planes in three-space) which are *tangent* to the hypersurface. Here one uses the notion that the set of *all* hyperplanes of projective space is itself a projective space (duality principle alluded to above). Thus, starting with a given surface, we get another surface, the dual. In general, even for surfaces of low degree the degree of the dual is quite high, namely $d(d-1)^2$, where d denotes the degree of the surface.

For example, the dual of a smooth cubic has the degree $3(2)^2 = 12$. However, if the surface has singularities, this reduces the degree of the dual. In the case of nodes, each node reduces the degree by two. Hence, the dual of the Cayley cubic is $3(2)^2 - 2 \cdot 4 = 4$, which is a quartic surface (surface of degree four).

Being the dual of a very unique cubic surface, this quartic is very unique. It turns out to be the unique quartic surface which has three singular lines which meet at a point (here the set of singular points is not a finite set of points, but consists of a union of lines, which is referred to by saying the "singular locus is of dimension one"). It can be shown that this quartic surface has the following property: it is the projection into three-space of the *Veronese surface* in $\mathbb{P}^5$. This is the unique surface in $\mathbb{P}^5$ whose variety of chords (a chord is a line

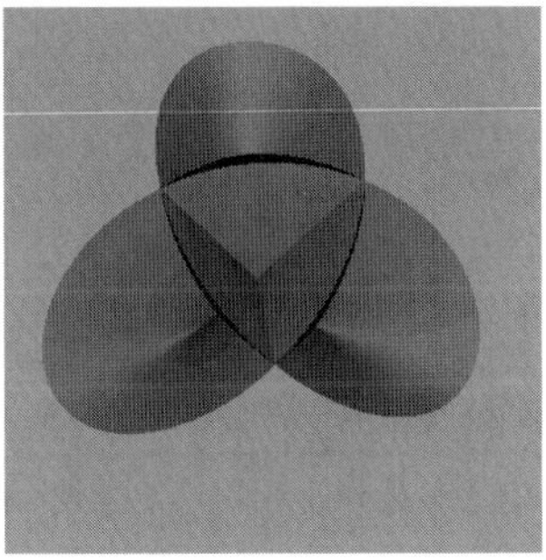

Fig. 11. The dual of the Cayley cubic, a quartic surface which is the projection of the Veronese surface in $\mathbb{P}^5$

intersecting the surface in two points; in general a line in five-space will not meet a given surface at all) is a proper subvariety of $\mathbb{P}^5$. It is in fact a quartic hypersurface. The Veronese is also (another favorite of mathematicians: give as many different descriptions of given objects as you can) the image of $\mathbb{P}^2$ under the so-called Veronese map $(s,t) \mapsto [1 : s : t : s^2 : st : t^2]$ (the former coordinates are affine coordinates on $\mathbb{P}^2$, the latter are homogenous coordinates in $\mathbb{P}^5$). The image is a surface of degree four, and this degree is preserved under projection. As the Cayley cubic has Σ_4 as symmetry group and as the dual variety is a covariant, also the dual surface has this symmetry.

So far, we have essentially discussed only two groups as symmetry groups: the symmetric groups on four and five letters, Σ_4 and Σ_5, respectively. Actually, the interesting groups which can occur are not too numerous. The reason is that if we have a surface with the given symmetry group, then this group also acts on the ambient projective space, and *these* groups are highly restricted, there just are not too many of them. In the theory of groups, one has certain kinds of building blocks, the so-called *simple* groups. There is a short list of simple groups which act on the projective three-space, and thus there is a short list of simple groups which can occur as the symmetry groups of surfaces embedded in projective three space.

In addition to those already mentioned, we just mention two further ones. As to the first, whose symmetry group has order (number of elements) 11,520, Figure 12 suggests it just also has a symmetry group which permutes four letters, in this case the four "lobes" in the picture. In fact, we *cannot* see all the symmetries at once. This is an example where the real pictures we are drawing are misleading; in this case the symmetries themselves are complex! The symmetry group is a so-called *unitary reflection group*, which is generated by *complex reflections*, rather that real ones. Even experienced mathematicians have difficulty envisioning this.

As to the second, this is a wonderful group of order 168, which acts on the projective plane and on projective three-space as well. In mathematical terms, it is the simple group $G = PSL(2, \mathbb{F}_7)$. We display pictures of the

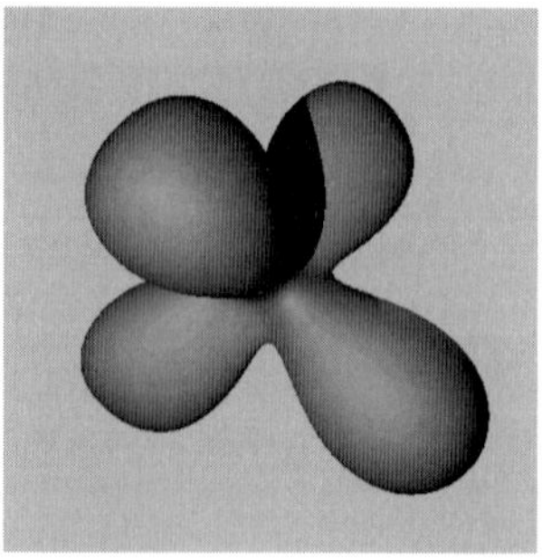

Fig. 12. The invariant of degree 8 of the simple group of order 11,520

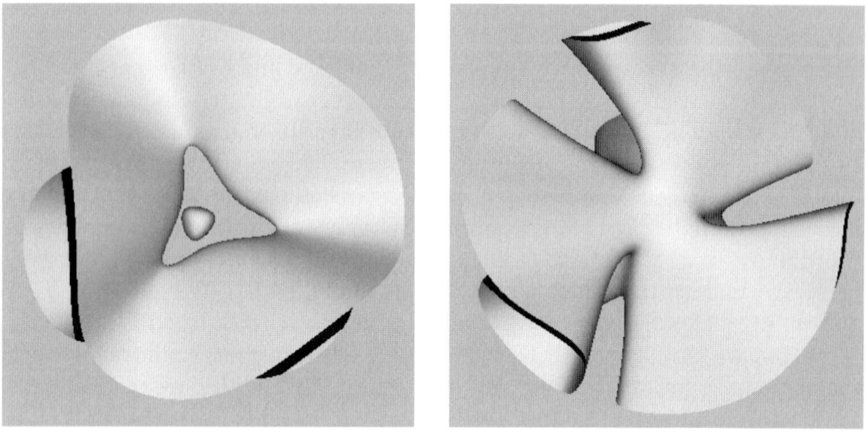

Fig. 13. Two invariants of the simple group of order 168. Left: The invariant of degree 4. Right: The invariant of degree 6

unique invariants of degrees 4 and 6 in Figure 13. The equation of the degree 4 invariant is simple enough to write down here: it is $t^4+6\sqrt{2}xyzt+2(y^3z+z^3x+x^3y)=0$, the parenthesized expression being the equation of the famous *Klein curve* in the projective plane (this is the unique invariant *curve* of degree four under the action of G_{168} on the projective plane we have already mentioned). For the mathematicians let us add a few details on this remarkable curve. It is, on the one hand, the quotient of the upper half-plane by a principal congruence subgroup in the arithmetic triangle group $(2,3,7)$. The latter group is generated by three elements of orders 2, 3 and 7, and the principal congruence subgroup arises from a certain subalgebra in a division quaternion algebra (a maximal order in this algebra is the arithmetic group $(2,3,7)$ in a different guise). On the other hand, the same curve is the compactification of the quotient of the upper half-plane by the principal congruence subgroup of level 7 in $SL(2,\mathbb{Z})$. This is an example of a *Janus-like algebraic variety* (see B. Hunt & S. Weintraub, *Janus-like algebraic varieties*, J. Diff. Geo. **39** (1994), 509).

4 Singularities

The dialog continues.

> MATHEMATICIAN: "The next thing I would like to explain to you is the notion of a singularity. Suppose you go to take a seat at the market place in some European city, and you observe people, thinking about a nice motive for your next painting. Do all people you see leave the same impression on you?"
> ARTIST: "Of course not. Some people are more interesting than others."
> MATHEMATICIAN: "And what is it that makes some people more interesting?"
> ARTIST: "Some people wear more interesting clothes, some people have a very particular and unusual way of moving about. Some people are pretty, while other are rather boring. And every once in a while somebody comes along that really catches your eye."
> MATHEMATICIAN: "Right, that is somebody which is truly unique and one-of-a-kind. Now, thinking instead of people just of objects of some sort, the same is also true. This one-of-a-kindness is what mathematicians refer to as *singular*, and an object which has this property is called a *singularity*."

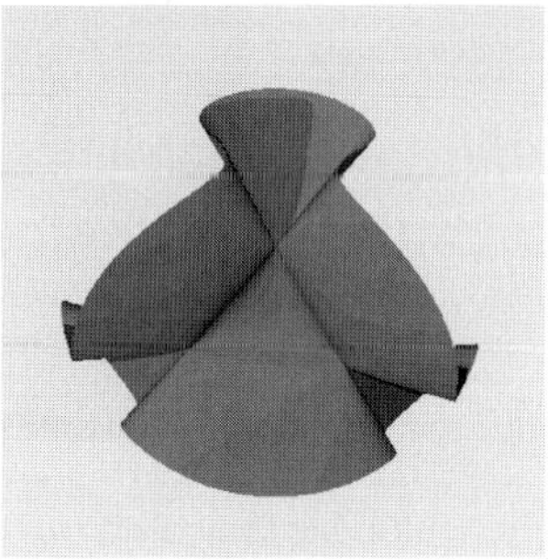

Fig. 14. The cubic surface with three singularities of type A_2

The objects which catch most of the mathematicians interest, just like I suppose most people's, are those objects which are different and more special than all others. These are the objects which are referred to as singular ones, or, in some cases, just as singularities. For example, among all surfaces of degree three, we have seen two of the most singular, in the sense of being the most special. One of them (the Cayley cubic) is also the most singular in the sense of having the most singularities, where here singularity refers to *singular points* on that surface. There are two basic kinds of questions which mathematicians are obsessed with. These are "Does it exist?" and "How many are there?". In the present context, we ask whether and how many singularities exist, and also how many *kinds* of singularities are there. This is a precise mathematical notion, which is based on the notion of *equivalence*, here equivalence of singularities. It turns out that there are in fact not so many singularities, at least not that can occur on surfaces in three-space (these are termed *two-dimensional hypersurface singularities*). In particular, there is a short list of the possible singularities which can occur on cubic surfaces (the higher the degree of a surface, the more nasty kinds of singularities which can occur). For example, there is a unique cubic surface which has four of the simplest singularities, termed A_1-singularities: this is the Cayley cubic introduced earlier. There is also a unique cubic surface which contains *three* of the next-worst singularity, called (as you might suspect) the A_2. This surface, whose projective equation is quite simply $x_0^3+x_1x_2x_3=0$, is pictured in Figure 14. A remark for those familiar with singularities: this is the most singular *semistable* cubic surface.

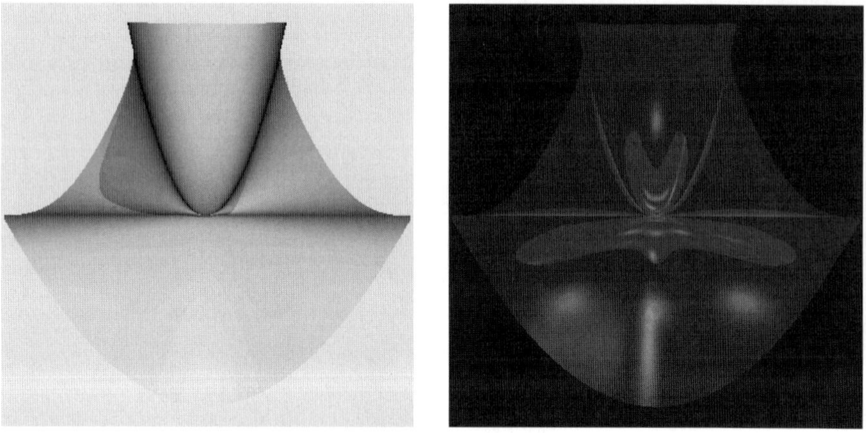

Fig. 15. The "most" singular cubic surface. Left: [BH]2 The E_6 cubic Surface. Right: The same surface, made of mirror glass, with three sources of light, one red, one green and one blue

In Cayley's famous paper "A memoir on cubic surfaces", a classification of the different types (possible singularities) of cubic surfaces was given. He came up with a list of 23 cases. The most singular of which has a beautiful singularity with the name E_6. Especially in the case of cubic surfaces this is

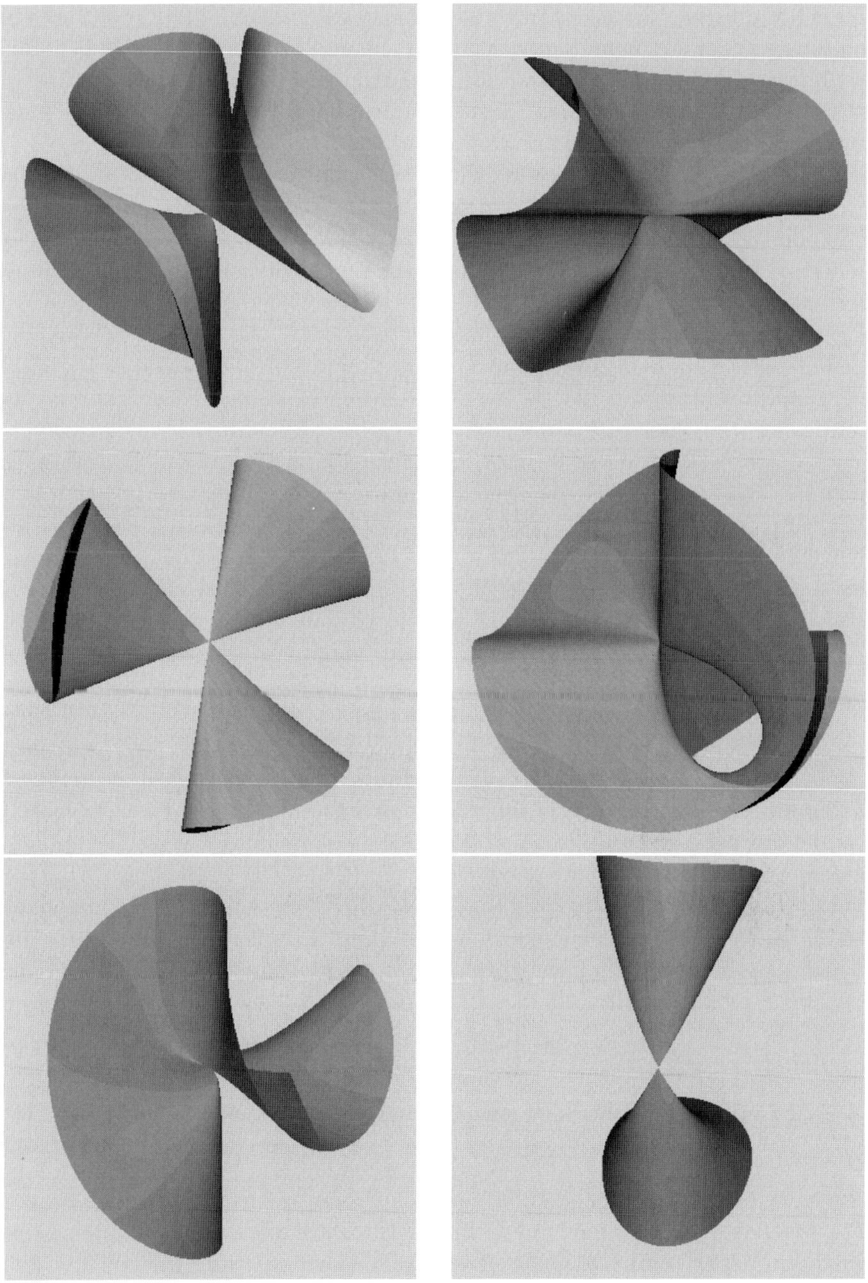

Fig. 16. Some of the singular cubic surfaces from Cayley's list. Top left: $xz + (x + z)(y^2 - x^2) = 0$. Top right: $xz + y^2z + x^3 - z^3 = 0$. Middle left: $(x + y + z)^2 + xyz = 0$. Middle right: $xz + y^2(x + y + z) = 0$. Bottom left: $xz + y^2z + yx^2 = 0$. Bottom right: $x^2 + xz^2 + y^2z = 0$

highly interesting, as the singularity of this type is related to cubic surfaces in curious ways. For the mathematicians in the audience, a brief explanation is as follows. The set of 27 lines has a very particular *combinatorial* structure, in that the set of "permutations" of the 27 lines that also preserve the incidence structures is a very special subgroup G of the permutation group Σ_{27}.

It is a group of order 51,840 (which has a normal subgroup of index two which is *simple*); this is the *Weyl group* of the Lie algebra of type E_6. The combinatorial structure of the set of Weyl chambers of the Lie algebra is identical to a similar kind of symmetry related to the singularity (versal deformation space), and a deep conjecture of Grothendieck, proved later by Brieskorn, relates the versal deformation space with the Lie algebra *directly*. Still, the relation is of mysterious kind, and is a typical example of the kind of wonderful mysteries nature has in store for future generations of mathematicians who want to explore.

We have depicted the surface in Figure 15, in the second picture as a mirror. This emphasizes the special shape of the object, and is just plain beautiful in its own right. The set of most of the interesting singularities which occur are contained in the images of Figures 16 and 17. The last one in his list is in Figure 18; this is the so-called *Whitney umbrella*, a very well-known singularity which is often used to explain and test concepts. Note that this last example is the only case in which the singular locus is one-dimensional.

In the theory of singularities there are two basic notions of how to "improve" a singularity, in the sense of making a singular point "less" singular: *deformation* and *resolution*. At least the first can be made very easy to visual using movies, showing how a singular point arises as a case of a very special set of parameters from a situation in which there are either no singular points at all, or there are singular points which are much "less singular". Another notion in algebraic geometry is that of *degeneration*, which is what happens when we start deforming something and then run into a set of parameters which are so special that the surface completely changes its structure; this again can be beautifully visualized with movies. We now give a few examples.

4.1 Maximal numbers of singularities

We have already mentioned cubic surfaces. As far as surfaces of higher degree are concerned, some famous beautiful pictures are of surfaces having a maximal number of *ordinary double points*, that is, singularities of type A_1. The following pictures from the Gallery show such for degrees $d = 4, 5, 6, 8$ and 10. We should mention here that for higher degrees there are indeed *bounds* on the number of ordinary double points they can have (and of course also for worse singularities), but it is not in general known whether this known bound is actually obtained.

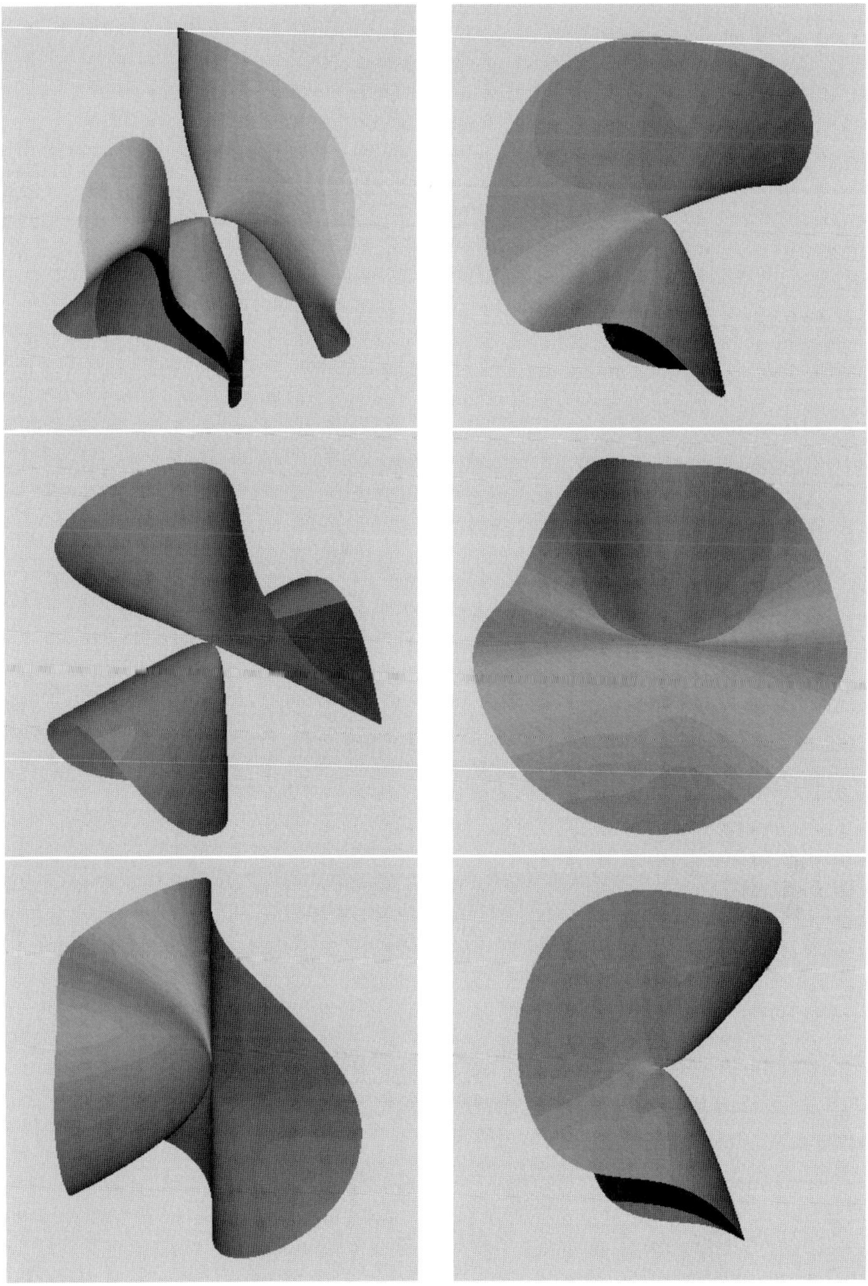

Fig. 17. Some of the singular cubic surfaces from Cayley's list. Top left: $(xy + xz + yz) + xyz = 0$. Top right: $xz + xy^2 + y^3 = 0$. Middle left: $xz + (x + z)y^2 = 0$. Middle right: $xz + y^2z + x^3 = 0$. Bottom left: $x^2 + xz^2 + y^3 = 0$. Botom right: $xz + y^3 = 0$

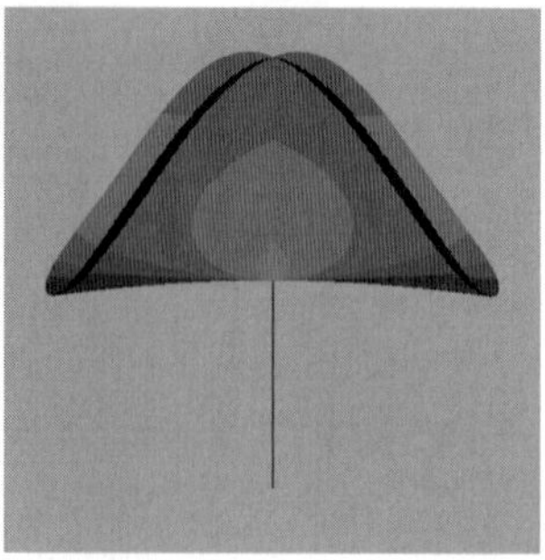

Fig. 18. The last cubic surface in Cayley's list, also known as "Whitney's Umbrella"

1. **Quartics:**
   A quartic surface (i.e., of degree 4) can have at most 16 A_1 singularities. There is in fact a three-dimensional family of such surfaces, known as *Kummer* surfaces from the research of Edward Kummer in the last century. This family of surfaces has ties to several other interesting areas of mathematics, in particular it is closely related to the family of *Jacobians* of curves of genus two, or equivalently, to *Abelian surfaces*. The latter in turn are related to many areas of topology, geometry and number theory. Each of the surfaces is beautiful to look at. There is a delightful combinatorial structure related to them, called the *Kummer* 16_6*-configuration*. This configuration consists of, in addition to the 16 ordinary double points, 16 planes, and they share the wonderful property that *in each of the 16 planes, 6 of the double points lie, and conversely, through each of the 16 double points, 6 of the 16 planes pass.* This configuration is in turn related to the even theta functions of genus two, one of the favorite topics of inquiry late in the nineteenth century.
   We show in Figure 19 one of the Kummer surfaces.

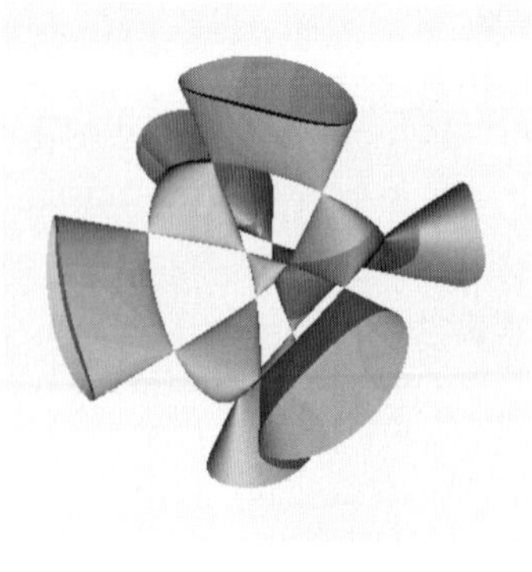

Fig. 19. A Kummer surface

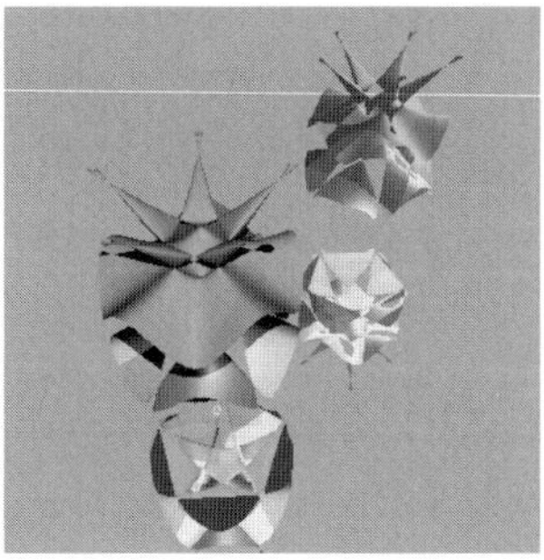

Fig. 20. [BH]3: A Togliatti surface, a quintic with 31 double points

2. **Quintics:**
In the case of quintics it is known that the maximal number of ordinary double points they admit is 31, although it was not clear for a long time whether such quintics actually exist. In fact, they do, and here is a picture of a surface which was derived by Duco v. Straten, Stephan Endrass and Wolf Barth (we will be hearing about Barth a couple of times in the sequel, as he has found many of the "records", i.e., surfaces with maximal known numbers of double points). In order to make the visualization easier, we have added mirrors in the back and below, so the viewer may see the surface from more than one side.
The equations of many surfaces with a large number of ordinary double points can be found, together with pictures, which were made with the drawing tool SURF developed by Endrass, in his article *Flächen mit vielen Doppelpunkten*, DMV Mitteilungen **4** (1995), 17–21. In many cases one finds that these surfaces not only have a large number of double points, but they also have a relatively large symmetry group, making them interesting from this point of view also. However, finding the equations of these surfaces is in general a very difficult problem, and requires not only sound knowledge in the theory of surfaces but also a bit of ingenuity.
3. **Sextics:**
Here we just present the picture of the famous sextic found by Barth, which has 65 ordinary double points, the maximal number which can occur, in Figure 21.
4. **Octics:**
There is a series of beautiful surface with many double points constructed by Čmutov. Of these, we have a picture of an octic which has 112 nodes in the left picture of Figure 22. The maximal number is 168, but we find this surface to be more interesting to look at.
5. **Surfaces of degree ten:**
As a final example we display the "most incredible of all" of the set of surfaces we are discussing: a surface of degree ten with 345 nodes! This surface was again discovered by Barth, and shows how much symmetry

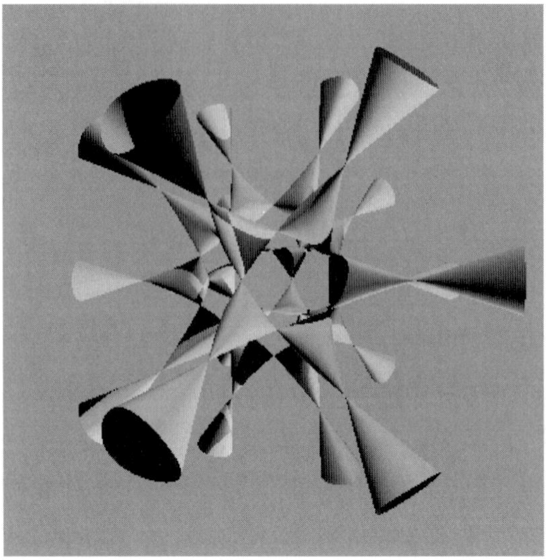

Fig. 21. Barth's sextic with 65 nodes

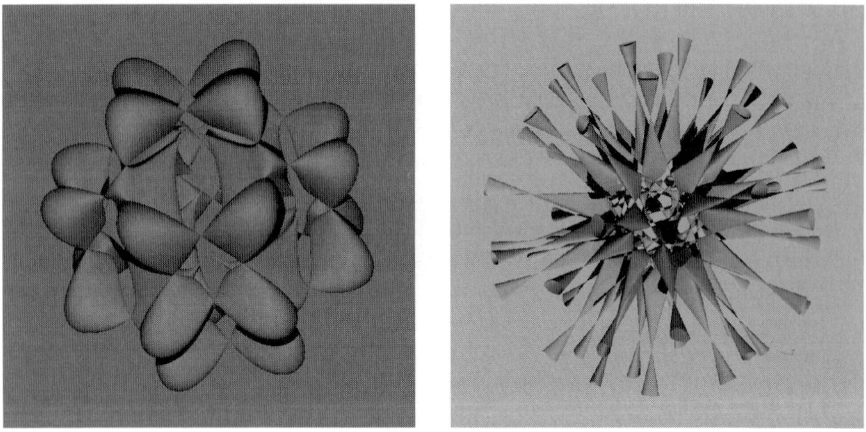

Fig. 22. [BH]4, [BH]5: A Čmutov octic with 112 nodes and the surface of degree ten with 345 nodes

a surface which is this special can have. Looking at the picture it is not difficult to imagine that it is invariant under the symmetries of an icosahedron in three-space; this group is a subgroup of index two (meaning that its order is just half of) of the symmetric group Σ_5 which was already discussed in relation with the Clebsch cubic surface.

4.2 Deformations

During the talk at the colloquium, we showed movies which impressively demonstrated the process of deforming singularities to smooth them, or at least to make them "better". Here we show a sequence of pictures which demonstrate this to some extent.

As a first example, we consider the so-called A_2-singularity. Such a singularity is given mathematically by some equation, in this case it is $x^2-y^2+z^3 = 0$. The fact that the point $(0,0,0)$ is a singular point follows from the fact that all partial derivatives of the function, which are $2x$, $-2y$ and $3z^2$, vanish at that point. In order to "smooth" the singularity, one simply perturbs this equation such that no longer all partial derivatives vanish. This is achieved for example by adding a constant term, whose precise value is indeterminant:

$$x^2 - y^2 + z^3 + t = 0,$$

which describes a family $\{S_t\}$ of smooth surfaces for which only the surface S_0 for the parameter value $t = 0$ is singular, all others are smooth.

In the upper series of three pictures in Figure 23 we show three surfaces in the family, giving an impression of what the smoothing of the singularity looks like.

On the other hand, one can deform the singularity by keeping the point singular, but lessening the degree of singularity. In this case this means we deform the A_2-singularity to a A_1. The equation for an A_1-singularity is just $x^2 - y^2 + z^2 = 0$, so we deform by the family

$$x^2 - y^2 + t \cdot z^3 + (t-1)z^2,$$

which is an A_2 singularity for $t = 1$, but to an A_1 singularity for $t < 1$. This is depicted in the bottom series of Figure 23.

Once the principal is clear, one can just observe the kind of degenerations which can be constructed. As a particularly interesting example, we show also the deformation of a E_7 singularity to four ordinary double points ($4A_1$). Both the singularity of type A_n and those of types E_N, $N = 6,7,8$ are a type of singularity which are called *rational double points* (of which the simplest, A_1, is called the *ordinary* double point).

For these singularities, the possible deformations of one type to another is clearly revealed upon inspection of the so-called *Dynkin diagram* of the singularity. In the case of A_1, this is also the simplest possible graph, consisting of just one vertex. For E_7, the diagram is:

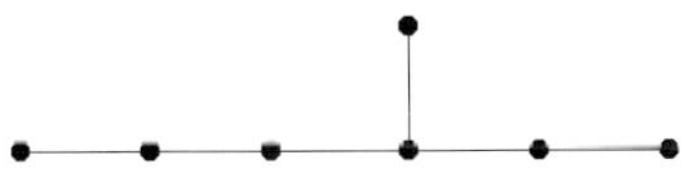

The possible ways in which E_7 can be deformed is equivalent to the possible *disjoint subgraphs* of the E_7 graph. In the case at hand, we can find a total of four disjoint vertices, indicating the pictured deformation in Figure 24:

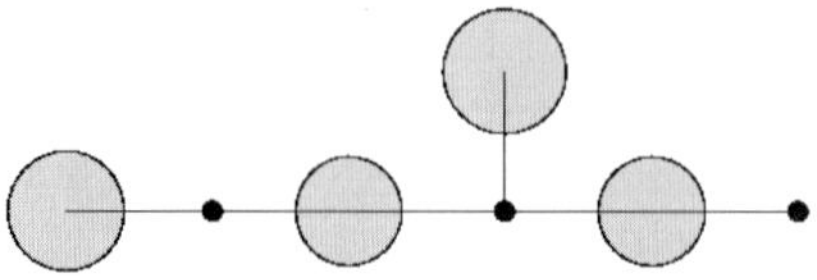

4.3 Degenerations

What we have been considering up to now is a kind of mild change. However, in mathematics also catastrophic changes are of great interest. For example, a smooth surface or a surface with *isolated* singularities of the kind we have been discussing up to now may, for particular values of the parameters, break up into several pieces. This is what is known in algebraic geometry as *degenerations*. We give some examples of these, where the families are of particular interest. During the talk we showed movies of a family of desmic surface (surfaces of degree four, i.e., quartics), which have 12 ordinary double points, which for special parameter values degenerate into *four planes* (this is also a quartic!). In Figure 26 we show a couple of the frames of this movie.

Another movie we showed is of another family of quartics, which again degenerates for special parameter values to four planes, but for which there are two different types of pictures (over the reals, but in fact these are also different over the complex numbers too). On the one hand we have a famous surface called a *Steiner surface*, on the other a famous set of quartics known as *Kummer* surfaces, which we have mentioned already above. Figure 25 shows a few frames of this movie.

As to the desmic surfaces, they are also quite interesting objects. Now more for the mathematicians in the audience, let me briefly comment on this. The Kummer surfaces which were mentioned above are the *Kummer varieties*, i.e., the quotient by an involution of Abelian surfaces which are *generic* in the sense that they are not the product of two Abelian curves (known in general as *elliptic curves*). The Kummer surfaces have 16 nodes, which are the images of the 16 so-called *2-torsion points* on the Abelian surface, that is points which, in the natural group structure of the Abelian variety[2], are 2-torsion, meaning that adding them to themselves results in zero. The desmic surfaces are also Kummer varieties, but now they are the quotients of Abelian

[2] An Abelian variety is by definition a smooth, projective variety with a group structure.

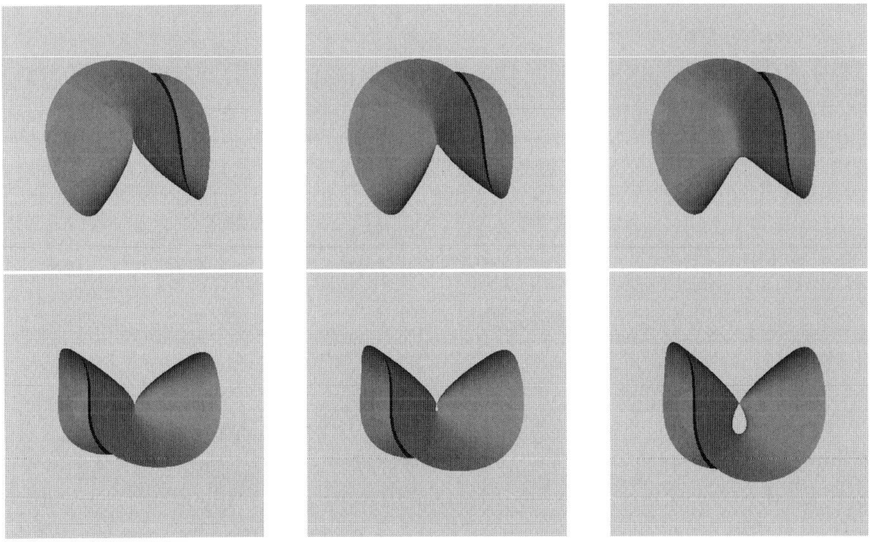

Fig. 23. The smoothing of a singularity of type A_2, and the deformation from the A_2 singularity to the A_1

surfaces which *are* products of elliptic curves, more precisely, the product of two copies of the same elliptic curve, $E_\tau \times E_\tau$. Since the moduli space of elliptic curves is just one-dimensional, this explains that there is only a one-dimensional family of desmic surface. The desmic surfaces on the other hand have 12 nodes instead of 16, but there are 16 lines which lie on them. The name desmic comes from the fact that their equations can be written

$$a\Delta_1 + b\Delta_2 + c\Delta_3,$$

where the Δ_i form a system of *desmic tetrahedra*, which means that each of the three is in perspective with respect to the remaining two (this is not easy to imagine, it means more precisely that there are four centers of perspective, and these four points are the vertices of the remaining tetrahedron). There are 16 lines through which a face of *each* of the Δ_i pass, and these 16 lines lie on each of the desmic surfaces. The twelve vertices of the Δ_i are the nodes of the desmic surfaces. The 16 lines play the role of the 16 nodes on the Kummer surfaces (points of order two), while the nodes are images of special curves on the corresponding products $E_\tau \times E_\tau$. For more details, see my book "The Geometry of some Special Arithmetic Quotients", Springer Lecture Notes **1637**, Springer-Verlag 1996, section B.5.2.3.

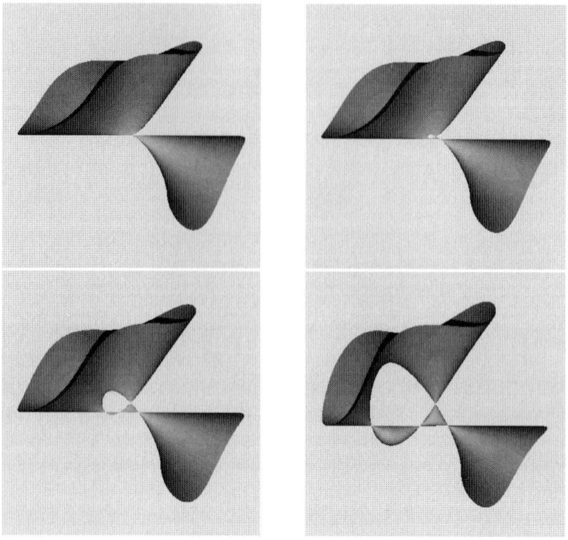

Fig. 24. A deformation of the singularity of type E_7 to four ordinary double points

5 Enumerative geometry

The dialog continues.

> MATHEMATICIAN: "The final thing I wanted to try to explain, and by far the most difficult, is what we call enumerative geometry. It is an incredibly beautiful topic, but requires a certain amount of mathematical background to really understand. But here goes. Consider a building which has, instead of a usual roof, some curved surface, as for example the Olympic stadium in Munich. This is an example of a curved surface. Can you imagine it?"
> ARTIST: Oh, of course. Almost everything I paint consists of pieces of curved surfaces like that."
> MATHEMATICIAN: "Wonderful! That is exactly what I mean. Now such a surface is not "strait", but it might still happen that certain lines could lie on such a surface."
> ARTIST: "For very special surfaces and lines, I suppose it could."
> MATHEMATICIAN: "Exactly. The word very special is important. Now what a mathematician likes to do is to find such surfaces where this can happen, and to *count* the number of such lines there are."
> ARTIST: "That reminds of a geometry course I once took."
> MATHEMATICIAN: "Think of something slightly different. You are making a perspective drawing, and you have a globe in the foreground, and want to draw lines of perspective from it. Which lines would you choose?"

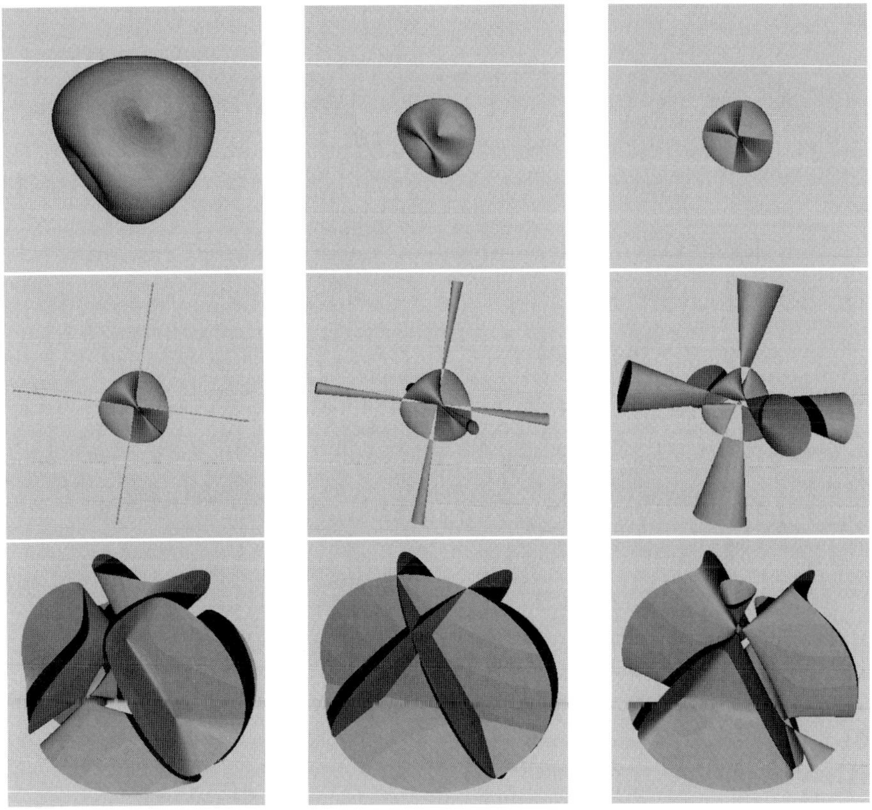

Fig. 25. A degeneration: first the Steiner surface degenerates into a Kummer surface, the latter then degenerates into four planes

ARTIST: "One which meets the globe at the top and the bottom, in just one point."
MATHEMATICIAN: "You have real mathematical talent. Those two lines are the ones we say are *tangent* to the globe, and again, we can consider objects other than the sphere and consider the same question. This is also a problem in enumerative geometry."

The subject of *enumerative geometry* is concerned with counting problems, counting objects of which there are only finitely many in some given configuration. We have already met some examples above: there are 16 planes which meet a Kummer surface six at a time in its 16 ordinary double points. There are 16 lines which lie on a desmic surface. This kind of result is definitely one which will be valid only over the complex numbers. In real algebraic geometry, there will only be certain configurations and examples for which all, say 16 planes of a Kummer surface, are also real and can be visualized.

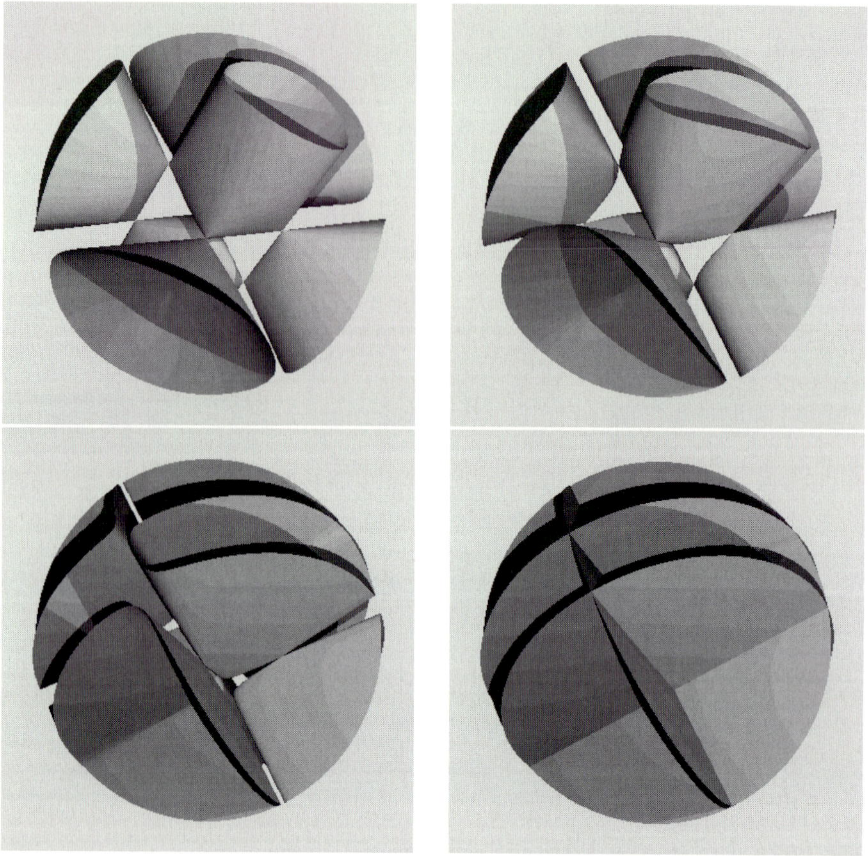

Fig. 26. A degeneration of desmic surfaces to the union of four planes

We will show some nice pictures for the easiest examples of such phenomena, which is the case of cubic surfaces. A lot has been written about this subject, but, as my experience tells me, it is still the most accessible for non-mathematicians. The magic numbers here are 27 and 45; 27 is a nice number, being the first odd number which is a cube. In the situation here, however, 27 more naturally arises as $12 + 15$. Anyhow, the statement is that:

There are exactly 27 lines on a smooth cubic surface.

This statement becomes false over the reals (there are real cubic surfaces with 27 real lines, but they are rather special). It also becomes false if we neglect the adjective *smooth* above. An example, which is ideal to begin with, is the Cayley cubic. This is the (unique) cubic surface with four ordinary double points. So first, take a close look at the picture, and see if you can see any lines which lie on the surface. This is a beautiful example because you really

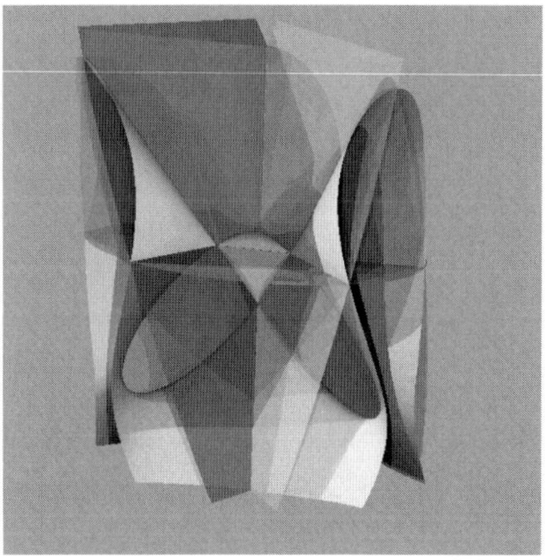

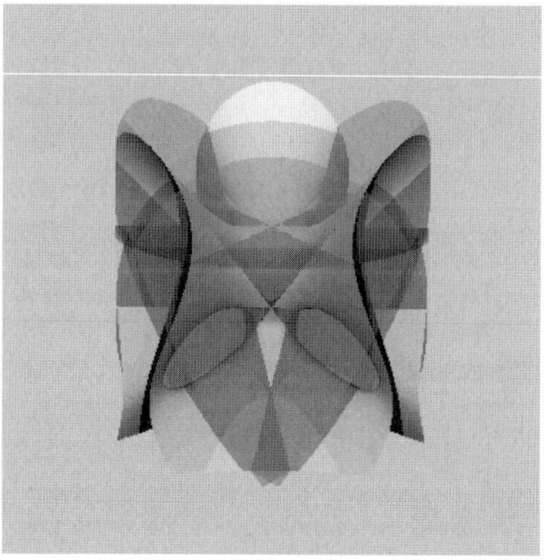

Fig. 27. Tritangents and lines on cubic surfaces. Left: The Cayley cubic with some tritangents and lines. Right: The Clebsch cubic with some tritangents and lines

can imagine them! Well, in this case, instead of 27 there are only nine lines, and six of them are just the edges of the tetrahedron whose vertices are the double points of the surface. This is easily seen in Figure 27 (a), in which we have illustrated a set of four planes which meet the surface in three lines apiece. This is the way in which the other magic number, namely 45, comes about:

> *There are 45 planes, called tritangent planes, each of which intersects the cubic surface in a set of three of the 27 lines.*

Note that "in general" (again, there is a precise mathematical definition of what this means) a plane intersects the cubic surface in a *cubic curve*, a couple of examples of which we saw in the very first section of this talk. A set of three lines is *also* a cubic curve, one in which is the cubic has degenerated into three irreducible components.

Next, let us consider a smooth cubic surface, more precisely, the Clebsch diagonal cubic, which we have already met. Recall that this cubic surface was special in having a large, in fact the largest possible, symmetry group. From our present point of view, this surface is very special in the following way. *10 of the 45 tritangent planes of the Clebsch cubic are so-called Eckard planes, which are tritangent planes with the property that the three lines they contain all meet in a single point.* We can also see this in a nice picture, Figure 27 (b).

A fascinating fact about cubic surfaces concerns their *Hessian varieties.* The Hessian is obtained in a relatively simple way from the equation of the cubic, by taking the determinant of the square matrix of second derivatives of the defining polynomial, the so-called *Hessian* matrix. In the case of cubic surfaces, this Hessian is a *quartic surface.* In general (i.e., for a smooth cubic

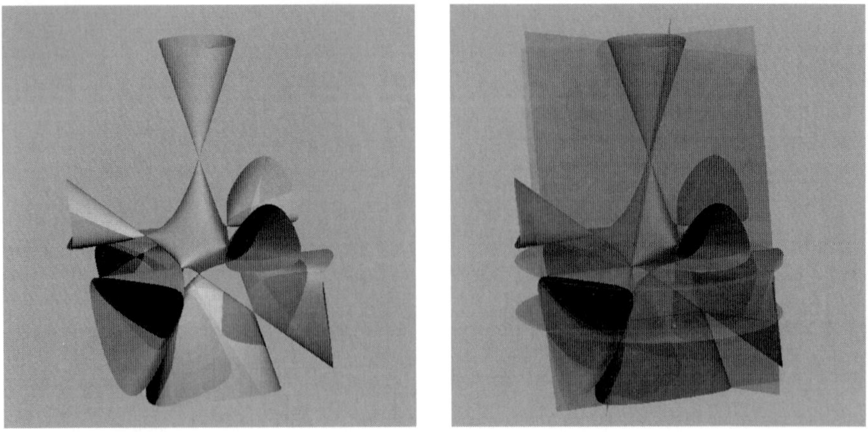

Fig. 28. The Sylvester pentahedron and lines on the Hessian of the Cayley cubic Surface. Left: The Hessian of the Cayley Cubic. Right: Lines on the Hessian

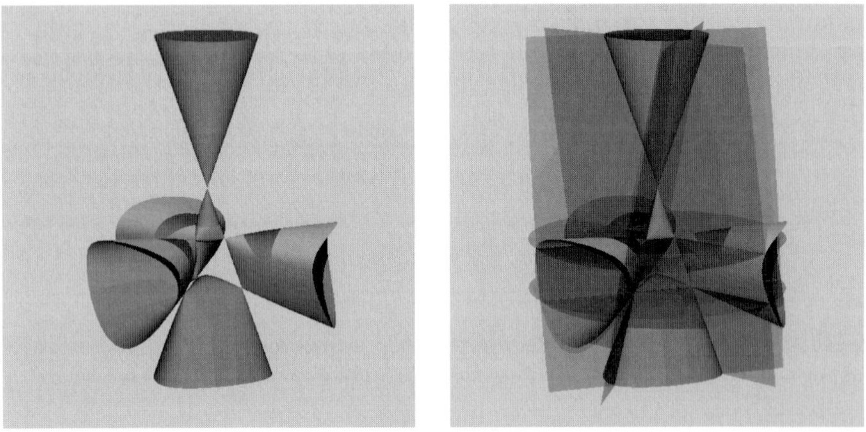

Fig. 29. The Sylvester pentahedron and lines on the Hessian of the Clebsch cubic. Left: The Hessian of the Clebsch cubic. Right: Lines on the Hessian

surface), it has 10 ordinary double points, and for every double point on the cubic, the Hessian acquires an additional node. In particular, for the Cayley cubic, the Hessian has 14 nodes instead of just 10. A nice picture of this surface is displayed in Figure 28 (a). There is a special set of five planes for a given cubic surface, a so-called *Sylvester pentahedron*, which is of use in writing down the equation. These planes are not tritangents, but it turns out they are something similar for the Hessian. In fact, these planes meet the Hessian surface in the union of four lines each (a similar situation as the case of the tritangents for the cubics). For the Hessian of the Cayley cubic, this is depicted in Figure 28 (b). The same thing for the Hessian of the Clebsch

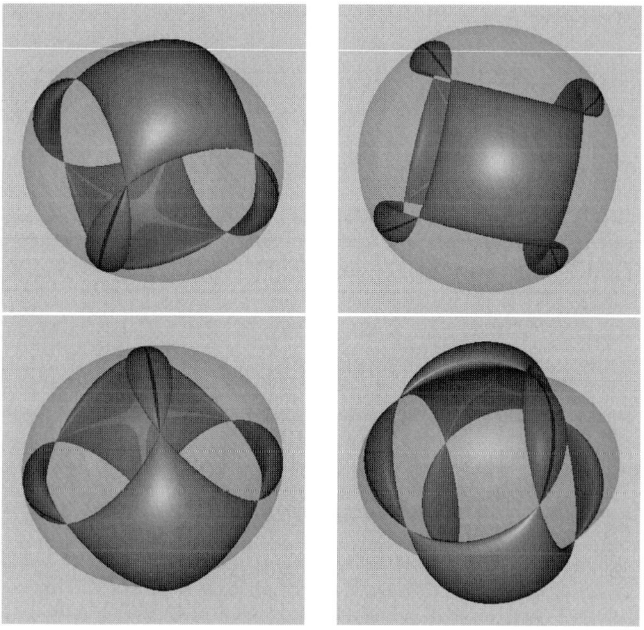

Fig. 30. [BH]6: A quartic and quadric surface in contact

cubic (since the Clebsch cubic is smooth, this only has ten nodes) is shown in Figure 29.

Needless to say, this is only the most simple aspect of the science of enumerative geometry. Much more interesting problems concern, instead of lines, curves of some degree. But these problems are much more difficult to display, for the simple reason that the equations involved get increasingly complex and difficult to derive. Another typical kind of question regards the notion of *tangency* of given objects. For example, how many lines are doubly tangent to a given plane curve? This question is for example extremely interesting for the famous *Klein curve*, a quartic curve in the plane, with particularly interesting properties (through which it is related to the problem of the solution of certain algebraic equations of degree seven). This curve (as does any quartic curve in the plane) has 28 bitangents, and the fact that 28 is 27 + 1 is no coincidence: these 28 lines are in fact closely related to the 27 lines on a cubic surface. We present an image, produced by Duco v. Straten, of a quadric surface and a quartic surface which have a high degree of contact. This is shown in Figure 30.

6 Conclusion

I hope that I have convinced the audience of the beauty and the interest in considering algebraic surfaces as pieces of art. And these pieces of art are really creations of nature, of the natural world of mathematical objects, which mathematicians just endeavor to discover.

In finishing, I would like to mention that also mathematicians have a heart. Being an algebraic geometer, of course this is an algebraic surface for me:

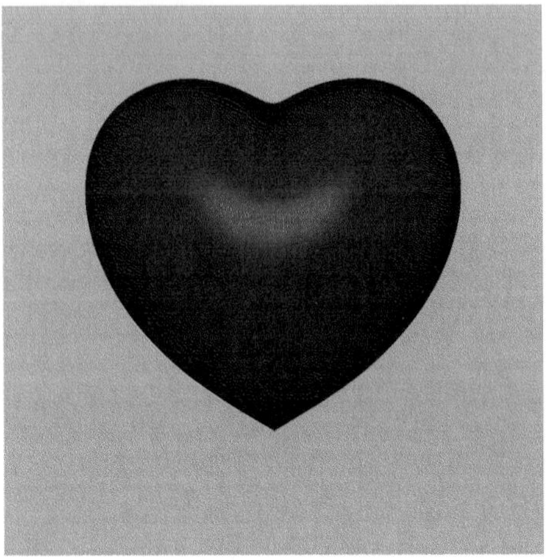

Fig. 31.

The equation for this surface is:

$$(2x^2 + y^2 + z^2 - 1)^3 - \frac{1}{10}x^2z^3 - y^2z^3 = 0.$$

I got the equation for this surface from the web gallery of Tore Nordstrand, located at `www.uib.no/People/nfytn/mathgal.htm`; thanks to him for this. His picture of the surface was made with a different ray tracing program, the picture above has been made with VORT.

The Mathematical Exploratorium

Richard S. Palais

Department of Mathematics, Brandeis University, Waltham, MA 02254-9110
<palais@brandeis.edu>

1 Introduction

In the June/July 1999 issue of the Notices of the AMS, I wrote an article in which I made a suggestion for a concept I called The Mathematical Exploratorium. This idea is closely related to the goals of the Maubeuge Colloquium, so I feel that it is fitting for me to explain here, in more detail than was possible in my Notices article, how I conceive of the Exploratorium, and how I think it might be organized and run.

2 Rationale

Every six months or so, I have been using various search engines to seek out and inventory mathematical art and visualization resources on the Web. I am sure you will not be surprised to hear that these have been increasing rapidly in quantity. But more importantly, the number of high quality sites has also been growing, and some of the material at these sites is truly outstanding. When I looked recently, in preparation for this conference, I realized that there are currently so many sites, that it is no longer feasible for me to evaluate them all with any care. In fact, a Google search for "mathematical visualization" found 1800 pages, and "mathematical art" elicited 1690.

There is something attractive about the untrammelled freedom of the Web, but it leads to a varied, unwinnowed mixture of wheat and chaff. In the real world too, we sometimes enjoy seeing art by gallery hopping, and it is exciting to occasionally discover a gold nugget amid the dross. But sometimes we are in a mood to see the very best art that has been selected and collected for us, and then we go to a fine museum.

It is a problem with Cyberspace that while there are very many galleries, there are as yet very few fine museums to complement them.

3 What is it?

The Mathematical Exploratorium will be

An Interactive Virtual Museum.

The exhibits will consist primarily of high quality:

1) items of mathematical art

and

2) visualizations of mathematical objects.

These will be selected (and perhaps commissioned) by panels of experts.

4 Who will the audience be?

My hope is that it will appeal to **everyone** with an interest in mathematics. An ideal exhibit will be designed with layers of sophistication: the outermost, visual layer should be easily accessible to young and non-expert visitors, but it should also excite their curiousity and motivate them to look further into successive layers of provenance and documentation that will gradually provide a fuller appreciation and logical comprehension of what their eyes see. And the innermost layer should have something for the expert mathematician to think about.

5 What are the Goals ?

1) To exhibit the highest quality graphic art and sculpture that is based on mathematical objects or that illustrates mathematical concepts, and to encourage the creation of such art.

2) To promote the development and use of excellent software tools to aid in the visualization of complex mathematical concepts, and to display the best visualizations produced by such software.

3) To educate non-mathematicians concerning the aesthetics of mathematics and encourage them to appreciate the deeper logical beauty that underlies "pretty" mathematical pictures.

6 Principle

I) The Exploratorium should be international.

As a web-based virtual facility, the Mathematical Exploratorium will anyway exist in Cyberspace. Like Mathematics itself it should be thought of as belonging to all mankind, and for this reason I would like to see it mirrored in many countries, and the primary site could even rotate among the mirrors.

II) The Exploratorium should be language neutral.

While pictures speak to us in a universal language, it is extremely important for the provenance of a piece of art to be clearly stated, and for a mathematical visualizations to be carefully documented. Since English has become the *de facto* common language of both mathematics and the Internet, it is likely that a majority of exhibits will be documented first in English, but all the documentation should be translated into a variety of languages, and a visitor to the Exploratorium should be able to choose a favorite language upon entering.

III) Where it is possible and useful, the exhibits in the Exploratorium should go beyond the static display of a mathematical object and use techniques of animation and virtual reality to permit the visitor to interact with and better understand the object.

IV) The Exploratorium should **not** be a mere collection of hyperlinks.

The Exploratorium will contain links to other web-sites that have galleries of mathematical art and of mathematical objects, but in order for the governing committees to maintain quality control, it is essential that all of its main exhibits be physically located on the main site and the various mirrors.

V) The contributors to the museum will retain the copyright to their works. They would only give the Exploratorium a non-exclusive license to make the item available for viewing on its primary site and mirrors.

7 Logical Organization

The Exploratorium will be divided into a number of "wings", that are further subdivided into Galleries, Alcoves, etc. Each wing will have a steering committee, consisting of experts in the relevant field, charged with selecting and commissioning exhibits.

1) The Mathematical Art Wing.

Devoted to the display of mathematical sculpture, paintings, etc. The steering committee will consist of artists who have made major contributions to the field. The Art Wing will be divided into Galleries, each devoted to a particular type of art, and these Galleries may be further subdivided into "alcoves" exhibiting the work of individual artists.

2) The Mathematical Visualization Wing.

Devoted to the visualization of mathematical objects of all types. The steering committee will consist of mathematicians interested in mathematical visualization. In addition there will be subcommittees of experts in the various specialties (surfaces, curves, ODE, PDE, fractals, conformal maps, etc.) represented by a Gallery within the wing. These Galleries may be further subdivided into "alcoves" exhibiting the work of individual mathematicians.

3) The Museum School Wing.

Devoted to the education of those who would like to learn more about mathematical art or visualization. There would be online tutorials in the use of various mathematical visualization software packages, lessons on the techniques of documenting mathematical art and visualizations, the discussion of advanced algorithms for mathematical computer graphics, etc. The steering committee will consist of mathematicians and artists with experience in programming, web-site construction, computer graphics, documentation, and mathematical visualization.

8 Administration

The Exploratorium will be governed by various committees that will operate much as do editorial boards for electronic journals. To keep expenses low, these committees will normally "meet" electronically, using email and private bulletin boards.

There will be an Executive Committee having financial responsibility and overall oversight of the Exploratorium. It will have five members, each serving five years (with a limit of two terms), with the fifth year member serving as Chair. A retiring or resigning member will be replaced by someone elected by the Executive Committtee. The steering committees of the various wings will be appointed by and report to the Executive Committee. There will be an Executive Director, appointed by and serving at the pleasure of the Executive Committee, and responsible for the day to day operation of the Exploratorium.

9 Financing

I wish I could end this description of the Mathematical Exploratorium by telling you that it will cost next to nothing to set up and maintain. Unfortunately, that is an unrealistic hope. One substantial expense that seems unavoidable to me is the salary and expenses of the Executive Director. I expect that this will require on the order of ten hours per week of effort. This is too much to expect anyone to handle on a pro bono, volunteer basis, and it will be important for the success of the project to have someone who has good administrative abilities and is also knowledgeable and respected in the worlds of art and mathematics.

Another non-trivial cost will be for web-design. It will be important to have a design that not only looks attractive, but one that is easy to maintain and that scales well—that is, the cost of adding one new exhibit should be negligeable. These days there are simple WYSIWYG web design tools that make it easy for amateurs to set up a reasonably good-looking and serviceable small web-site. But to create a top-quality complex site of the sort that the Exploratorium calls for will probably require considerable professional advice.

On the other hand I believe that those expenditures will be very highly leveraged. Most of the hard work in creating and maintaining the Exploratorium will be carried by the members of the Executive Commitee and the various steering committes discussed above, and I am hoping that, in the long tradition of editorial boards of research journals, mathematicians and artists will be happy to pitch in and contribute a couple of hours of their time each week to ensure that the Exploratorium will be a success.

Some Favorite MathArt and MathViz Links

Let us help one another to see things better. **Claude Monet**
*** = Particularly Recommended**

Mathematical Art Sculpture

Bruce Beasley, Brent Collins, Stewart Dixon, Helaman Ferguson, George Hart, John Robinson, Carlo Sequin

Graphics

Tom Banchoff, *Mike Field, Zarko Mijajlovich, *Matthias Weber, *World of Escher, Math-Art Exhibit (Dec. 9, '95)

Mathematical Visualization

Geometry: *Paul Bourke's Geometry Page, Bill Casselman's Pythagoras Gallery

Plane Curves: The MacTutor Famous Curves Index, *Xah's Visual Dictionary of Special Plane Curves

Knots and Space Curves: *Robert Scharein's The KnotPlot Site

Surfaces

*3D-Filmstrip Gallery, *The Scientific Graphics Project(MSRI), The Alfred Gray Galleries, *GANG CMC Gallery, Geometry Center Topological Zoo, *Optiverse John Sullivan et al., Ute Fuch's Gallery of Minimal Surfaces, *Jorg Hahn and Konrad Polthier Calendars, Torre Nordstrand's Gallery of ray-traced surfaces, *Robert McLachlan's Gallery of Pseudospherical Surfaces

Complex Function Theory

Hans Lundmark, Douglas Arnold

Tilings and Tesselations

Xah's Gallery of Tilings and Patterns

Chaos and Fractals

UMD Chaos Gallery, *Eric Schol Mandel Zoom, Uwe Krüger-Beauty of Chaos, *Loughborough Univ. Gallery, *Spanky Fractal Database

Polyhedra

Ha Le's Gallery of Polyhedra, Tom Getty's Polyhedral Hyperpages, George Hart's Pavilion of Polyhedrality

The Museum School

3D-Filmstrip, GRAPE, OORANGE, Maple, Matlab, Mathematica, POVRay, Surface Evolver

Maintained by Richard S. Palais: http://rsp.math.brandeis.edu
Last updated September 12, 2000

Copper Engravings

Patrice Jeener

49 Grande Rue, 26470 La Motte Chalancon

After his initial studies in a well-known high school in Paris, the "Lycée Janson de Sailly", where he discovered the universe of mathematical curves, and then at the "Ecole Nationale Supérieure des Beaux-Arts" (in particular at the engraving school), Patrice Jeener started by doing figurative engravings (landscapes).

During a visit to the "Palais de la Découverte", models of various surfaces draw his attention. He then undertook to study the surfaces, be they algebraic or minimal, and to represent them through original engravings. Two engravings of minimal surfaces, and two engravings of extensions conceived by Patrice Jeener of the original Klein bottle are here reproduced on the next page.

(Some of the engravings by Patrice Jeener have been copied by Jean-François Colonna on the computer:

http://www.lactamme.polytechnique.fr/Mosaic/images/BKLN.11.D/display.html
http://www.lactamme.polytechnique.fr/Mosaic/images/ESCA.41.d.D/display.html
http://www.lactamme.polytechnique.fr/Mosaic/images/ESCA.51.g.D/display.html
http://www.lactamme.polytechnique.fr/Mosaic/images/MINI.13.16.D/display.html
http://www.lactamme.polytechnique.fr/Mosaic/images/MINI.31.D/display.html
http://www.lactamme.polytechnique.fr/Mosaic/images/MINI.52.D/display.html

One might notice that the Jeener's extensions recall those done by Möbius of his original band. Here are the drawings by Möbius which one can find in the second volume of his complete works, page 520:

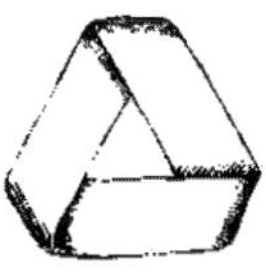

Fig. 2.

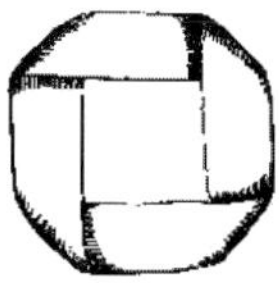

Fig. 3.

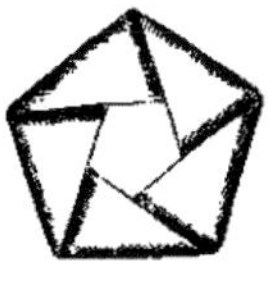

Fig. 4.

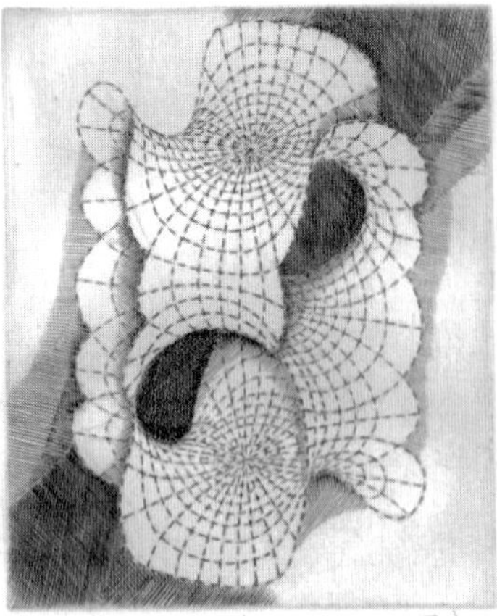

surface minimale P. Jeener

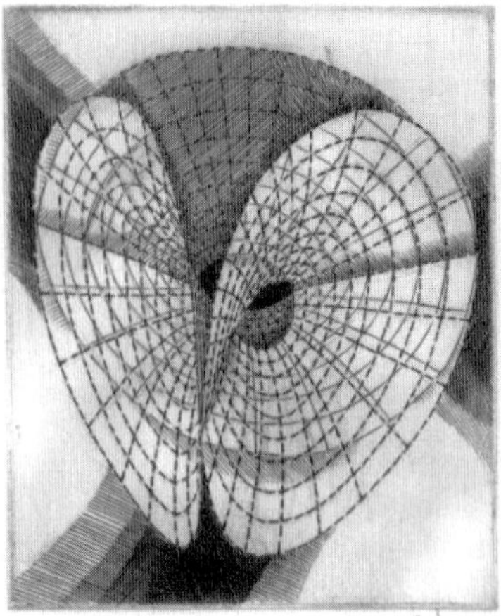

P. Jeener

$x = 2e^{2u}\cos 2v - e^{4u}\cos 4v$
$y = 2e^{2u}\sin 2v + e^{4u}\sin 4v$
$z = \frac{8}{3}e^{3u}\cos 3v$

surface minimale

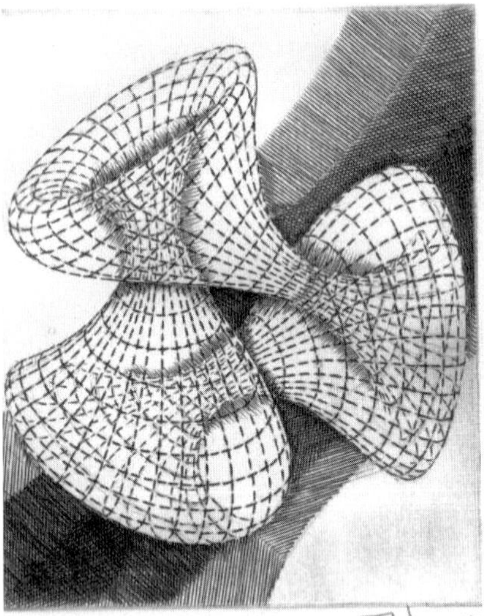

Triple bouteille unilatère P. Jeener

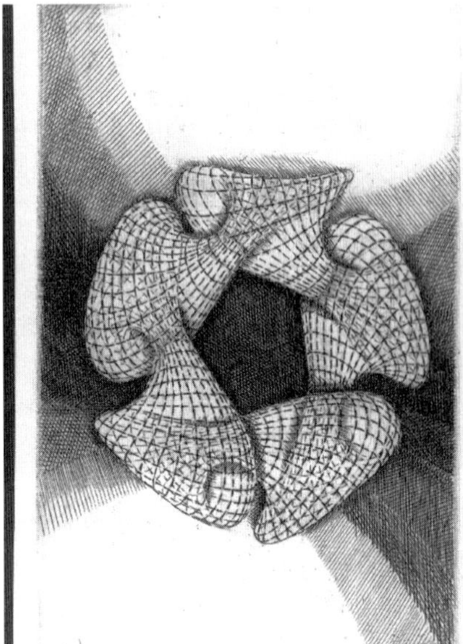

P. Jeener

Appendix

Color Plates

GEORGE HART

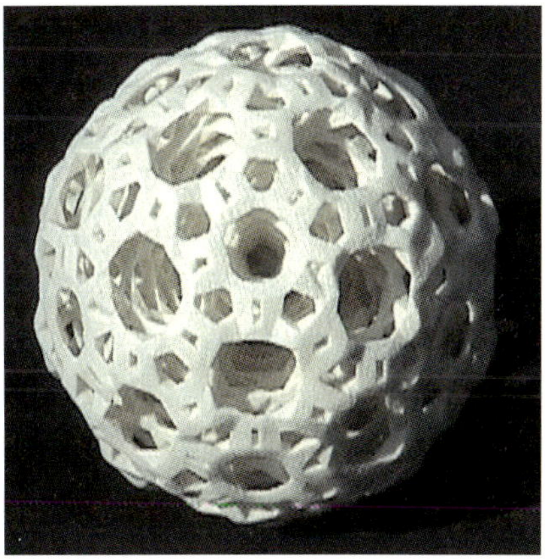

Plate 1. [GH]1: Deep Structure

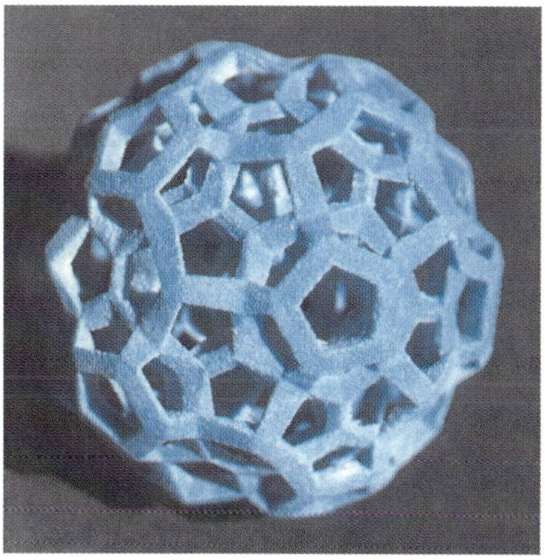

Plate 2. [GH]2: Five-Legged-Bee Hive

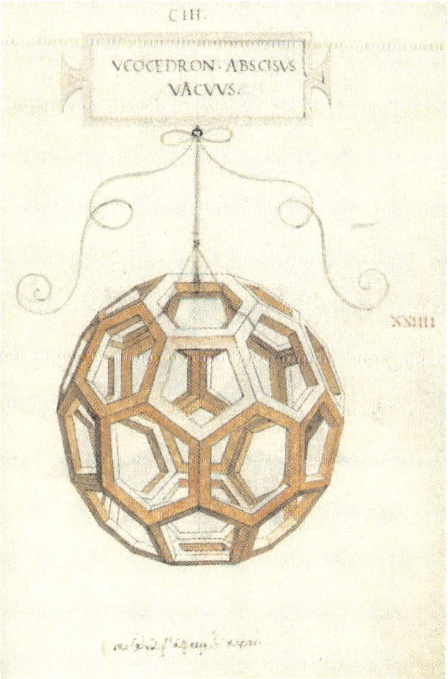

Plate 3. [GH]3: Leonardo "solid segment" presentation of truncated isocahedron

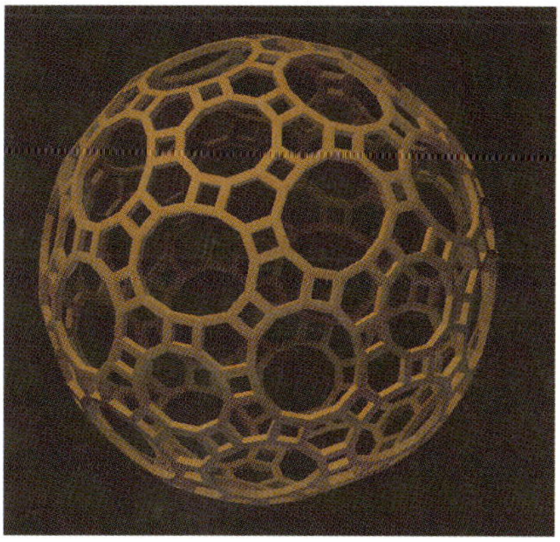

Plate 4. [GH]9: 31-zone zonhedron that underlies each layer of Fig. 8

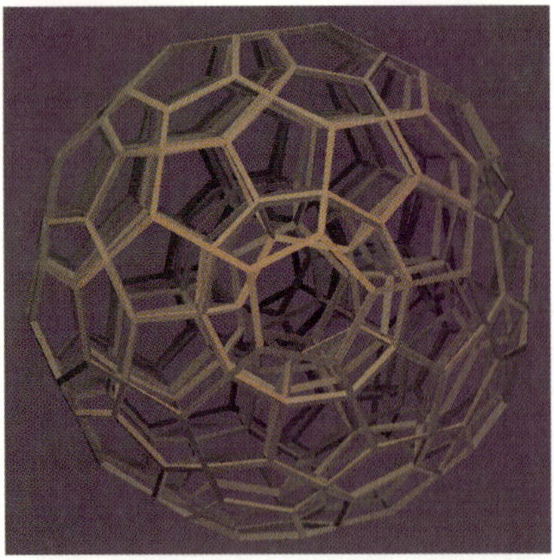

Plate 5. [GH]10: Edges of 120-cell orthogonally projected from 4D to 3D, along the direction to the center of a cell

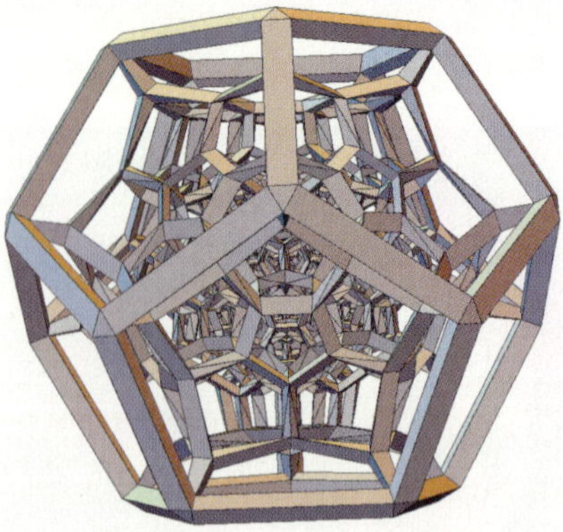

Plate 6. [GH]11: 120-cell projected in perspective, like Schlegel Diagram

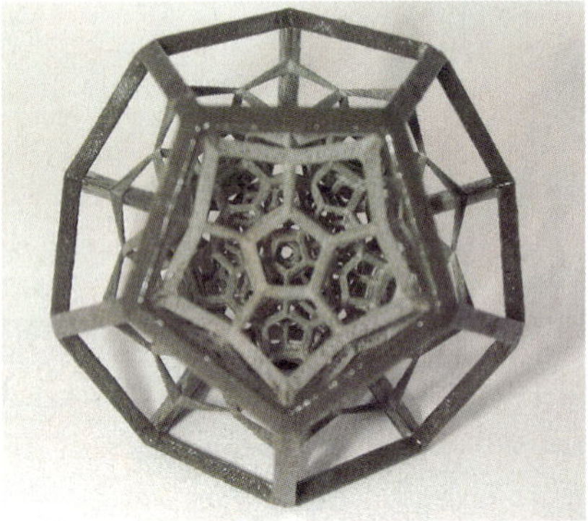

Plate 7. [GH]12: Wax realisation of Fig. 11, 5 cm diameter

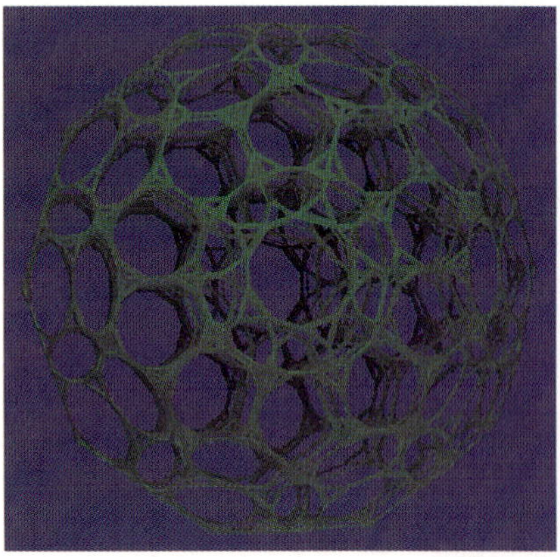

Plate 8. [GH]13: Projection of edges of truncated 120-cell

KONRAD POLTHIER

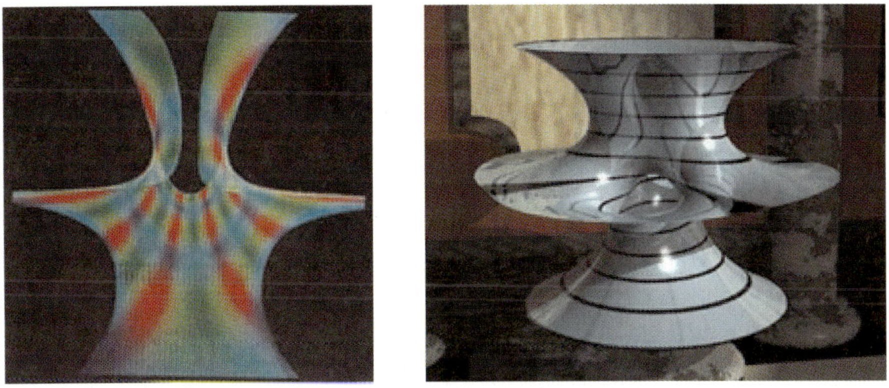

Plate 9. [KP]1: Costa-Hoffman-Meeks minimal surface (1985). The left image shows the very early computer pictures by Hoffman and Meeks (left). Years later, we see a raytraced image ot the Costa-Hoffman-Meeks minimal surface (right).

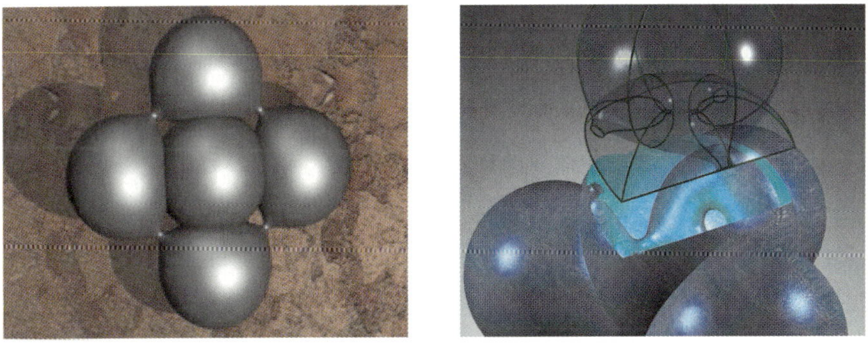

Plate 10. [KP]2: Compact Constant Mean Curvature Surfaces (1995, Große-Brauckmann - P., Oberknapp - P.). Free-boundary value problem for an elliptic partial differential equation. Numerical method based on notion of discrete differential geometry on triangulations. Compact soap bubble with genus four (left) and tetrahedral symmetry (right).

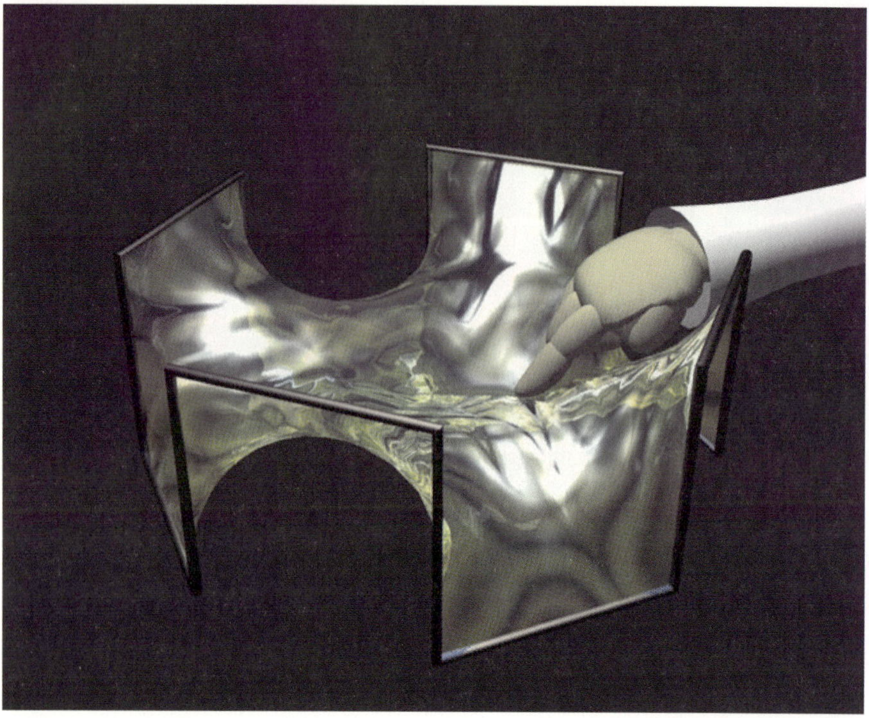

Plate 11. [KP]3: Snapshot of a minimal soap film taken from the video Touching Soap Films.

MIKE FIELD

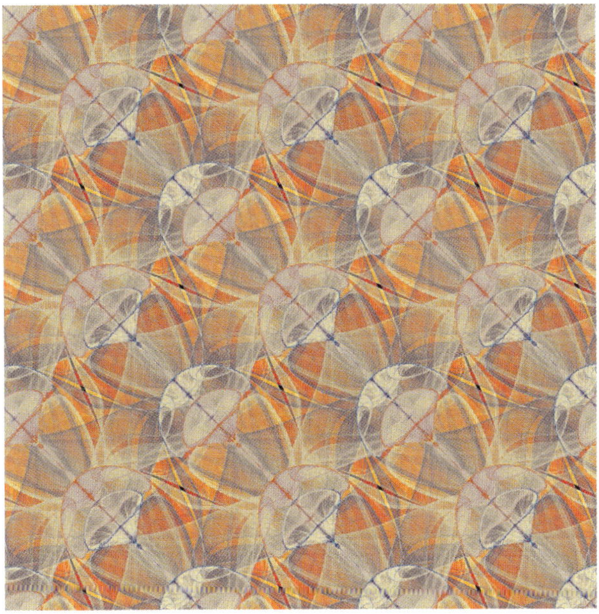

Plate 12. [MF]1: *Armies of the Night*: Two-colour pattern of type **c′m** (**pm/cm**)

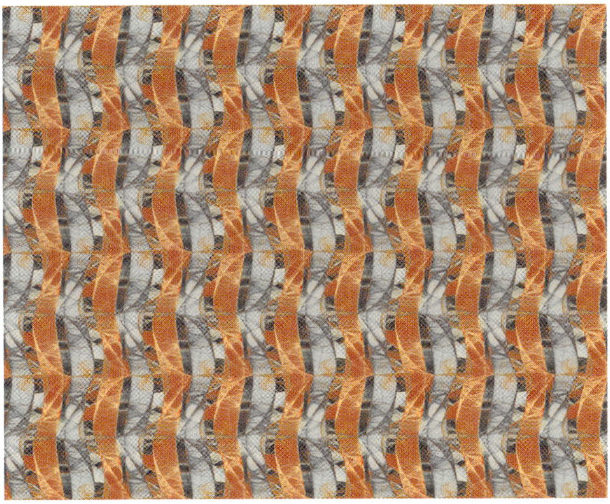

Plate 13. [MF]2: Two-colour pattern of type **pgg′** (**pgg/pg**)

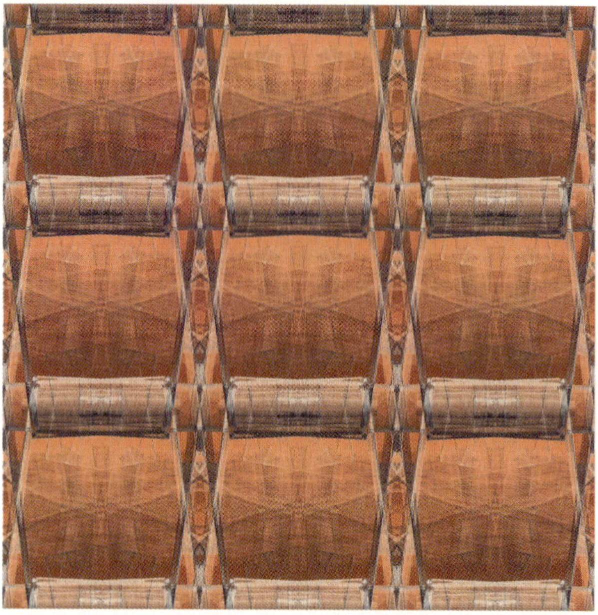

Plate 14. [MF]3: Two-colour pattern of type **pmm**′ (**pmm/pm**)

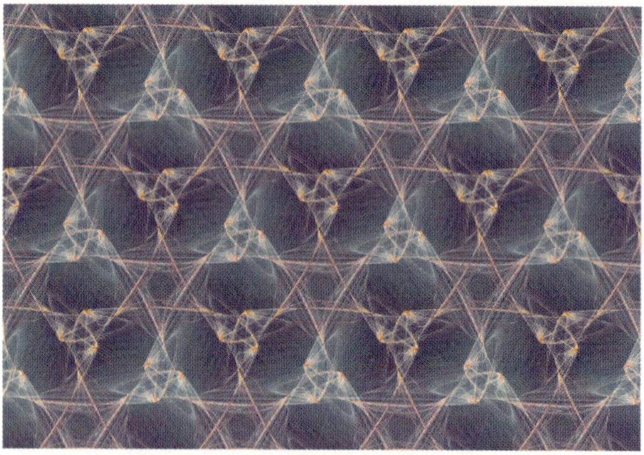

Plate 15. [MF]4: Two-colour pattern of type **p6**′ (**p6/p3**)

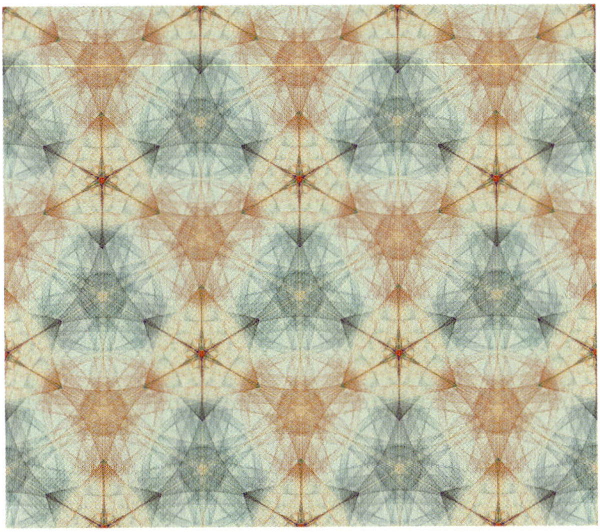

Plate 16. [MF]5

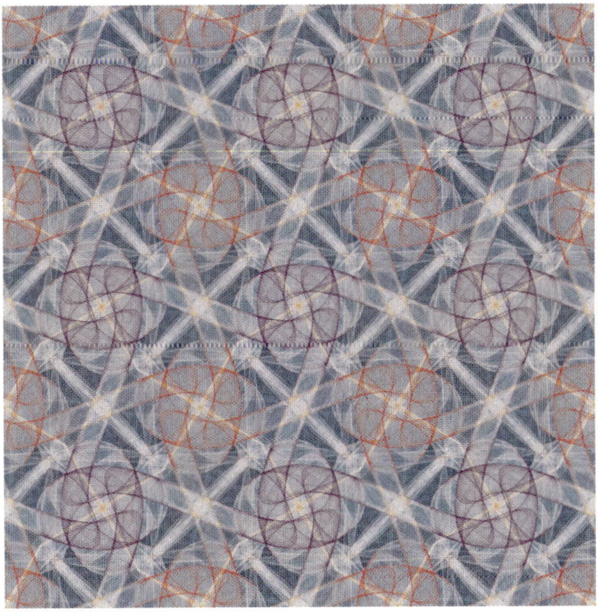

Plate 17. [MF]6

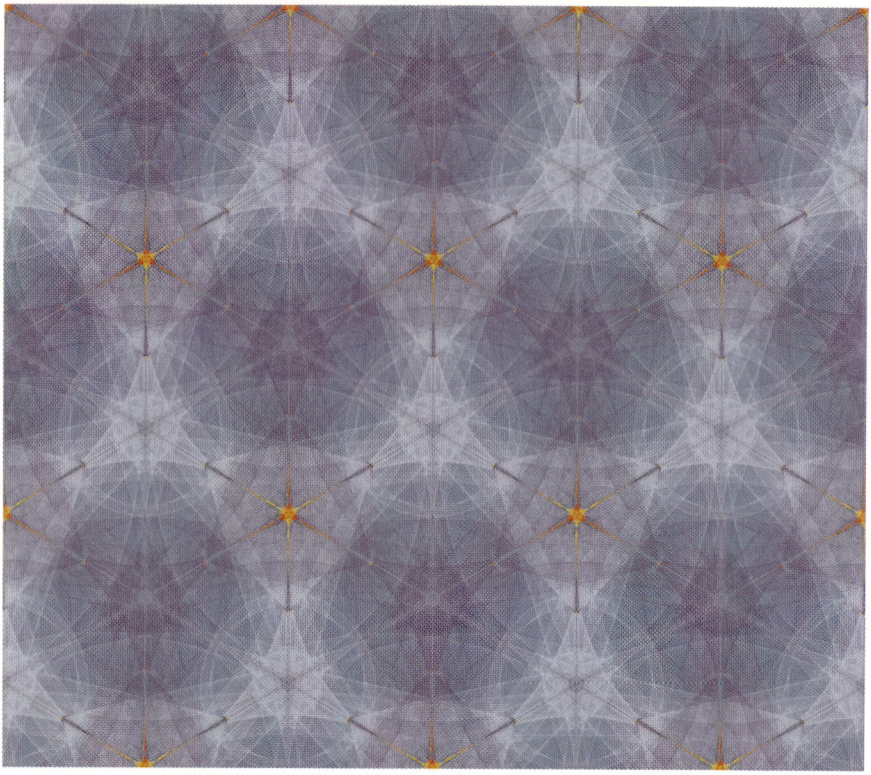

Plate 18. [MF]7

MARIA DEDO

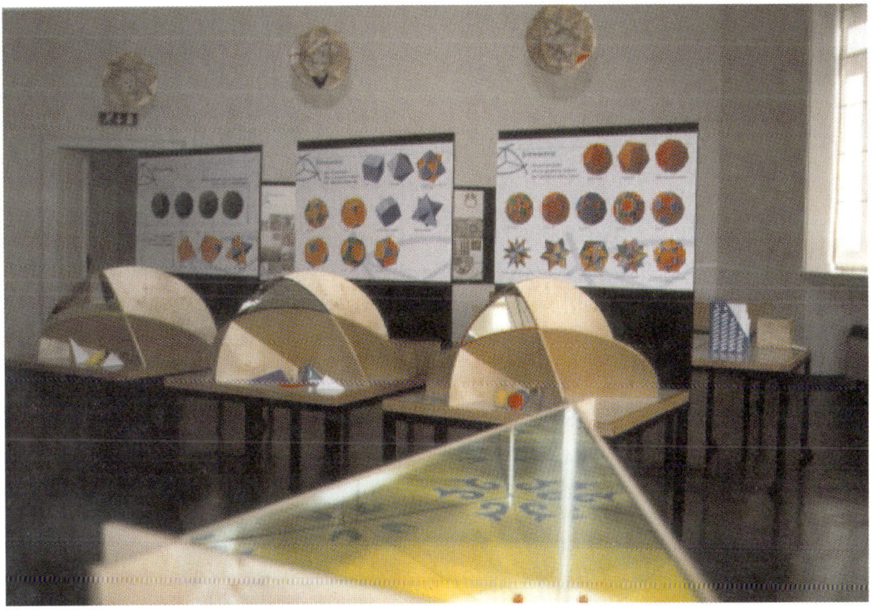

Plate 19. [MD]1: Exhibition "Simmetria, giochi di specchi", Dipartimento di Matematica "F. Enriques", Università di Milano

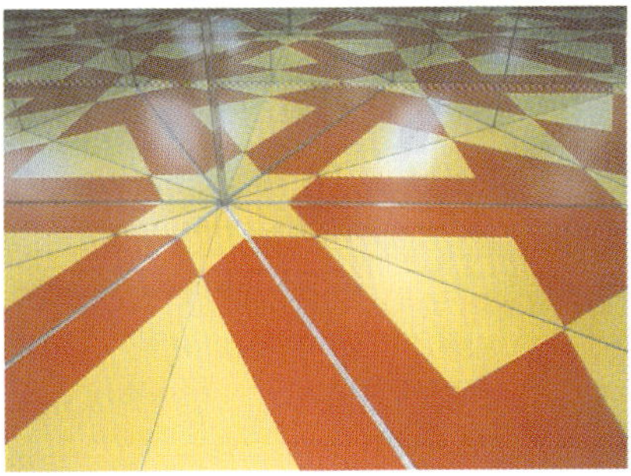

Plate 20. [MD]2: A plane tassellation seen in a mirror box (right triangle with angles $\frac{\pi}{6}$ and $\frac{\pi}{3}$)

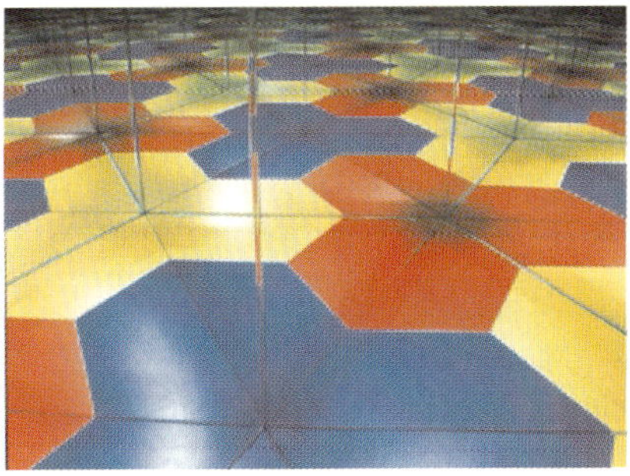

Plate 21. [MD]3: A plane tassellation seen in a mirror box (equilateral triangle)

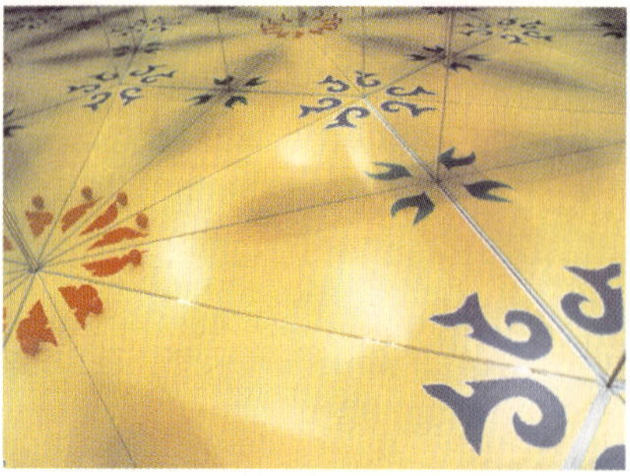

Plate 22. [MD]4: The group G with three generators A, B, C and relations $(AB)^2 = (BC)^3 = (AC)^6 = 1$ and $A^2 = B^2 = C^2 = 1$

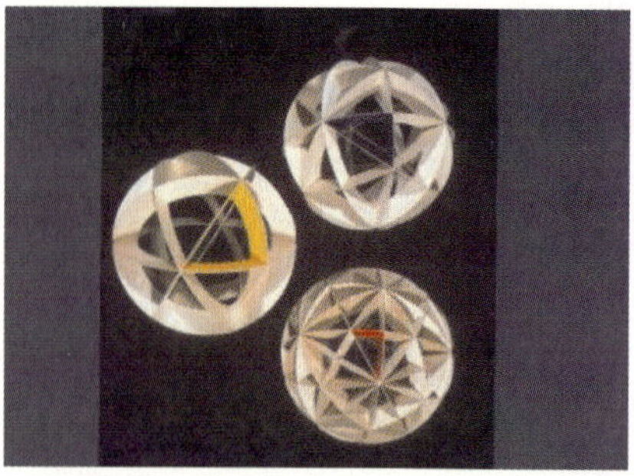

Plate 23. [MD]5: The three possible chambers for a finite irreducible Coxeter group in dimension three

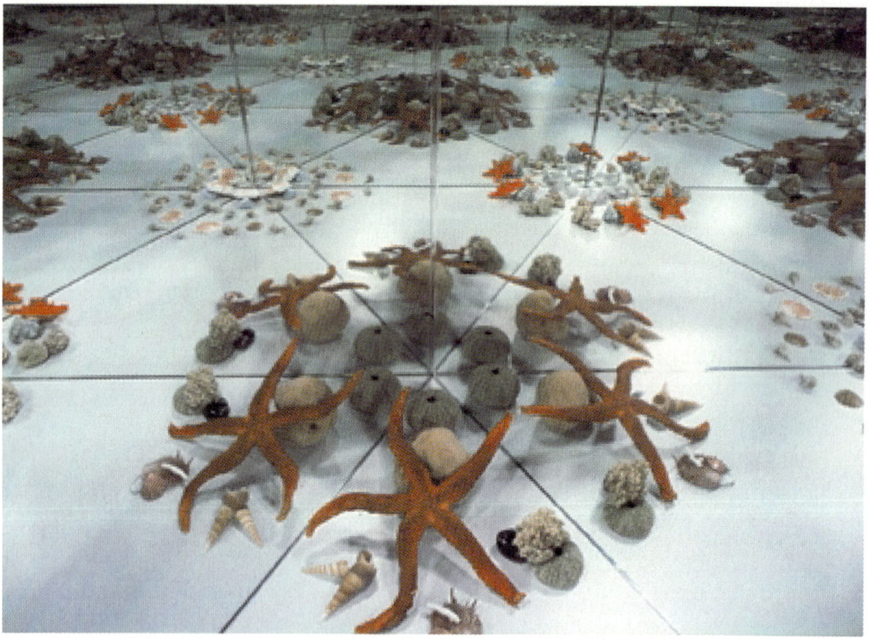

Plate 24. [MD]6: A landscape seen in a mirror box (equilateral triangle)

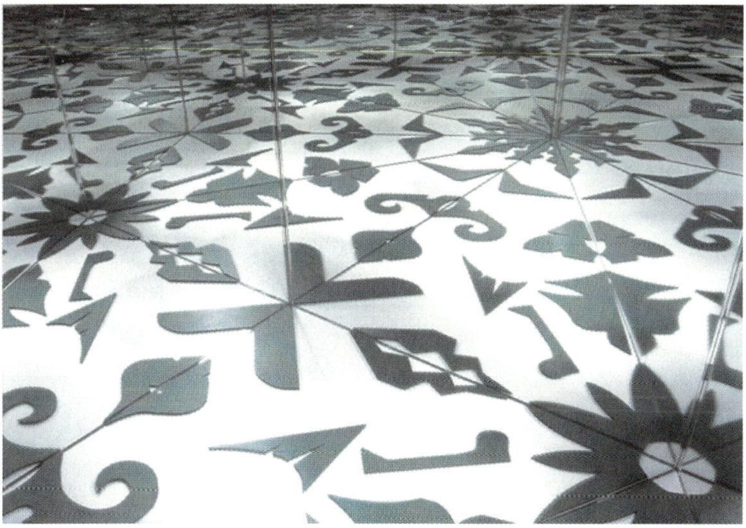

Plate 25. [MD]7: A fancy picture seen in a mirror box (right triangle with angles $\frac{\pi}{6}$ and $\frac{\pi}{3}$)

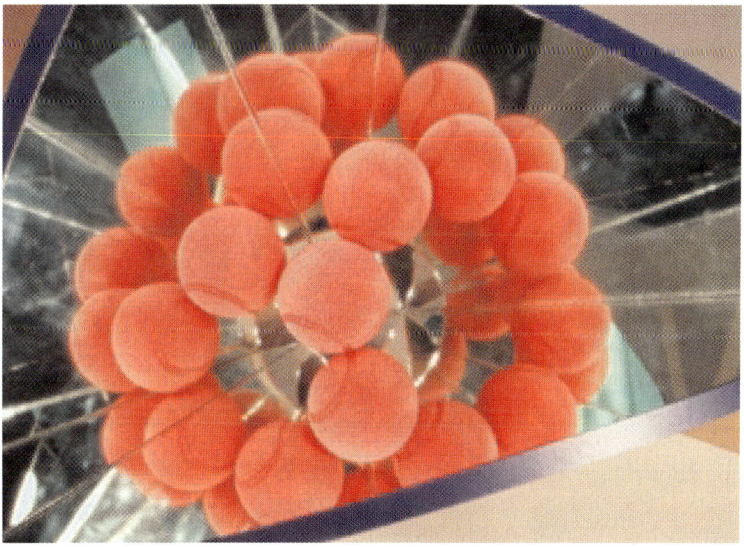

Plate 26. [MD]8: The effect of one ball in the kaleidoscope of the cube

CHARLES O. PERRY

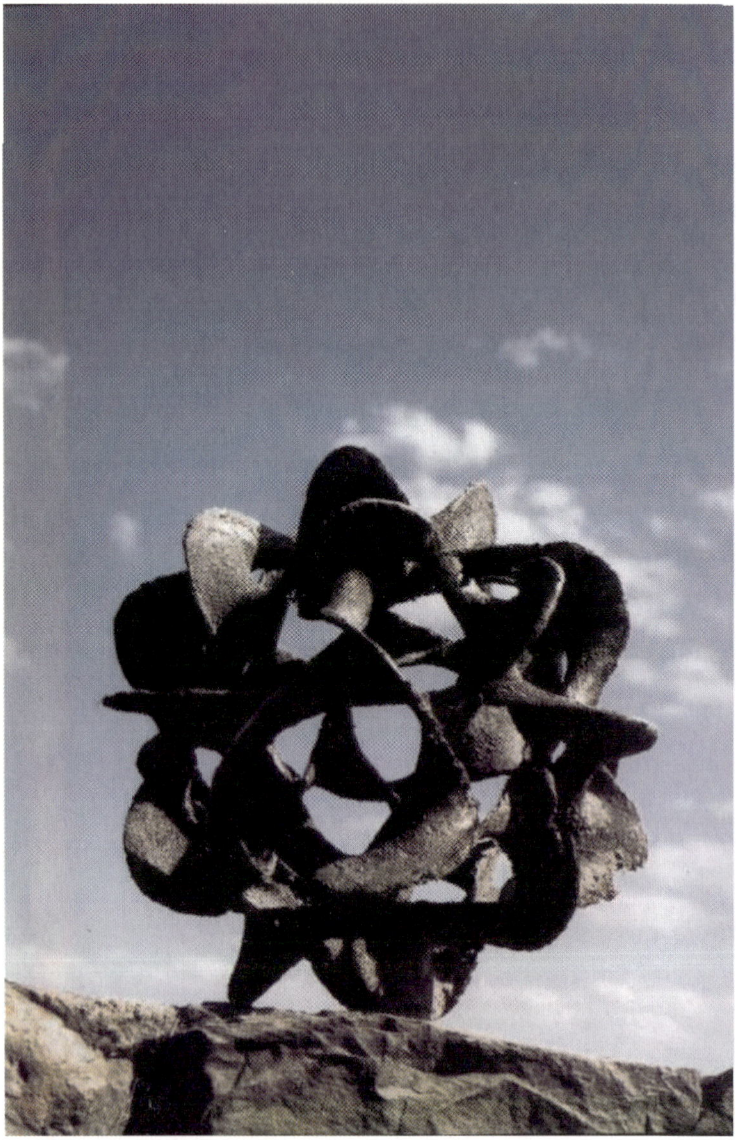

Plate 27. [CP]1: Star Cinder

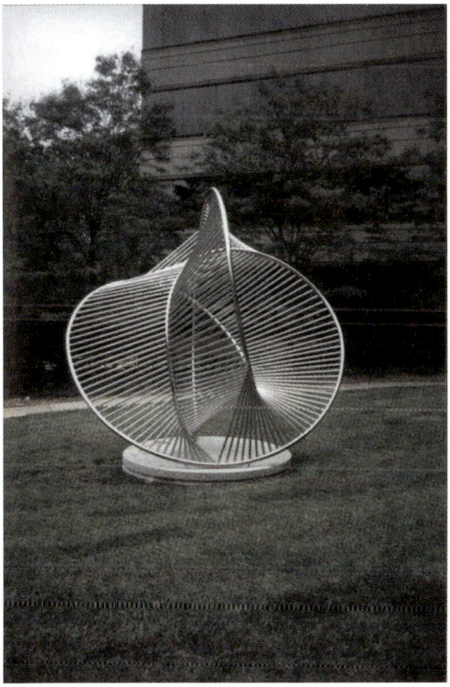

Plate 28. [CP]2: Ribbed Möbius Mace

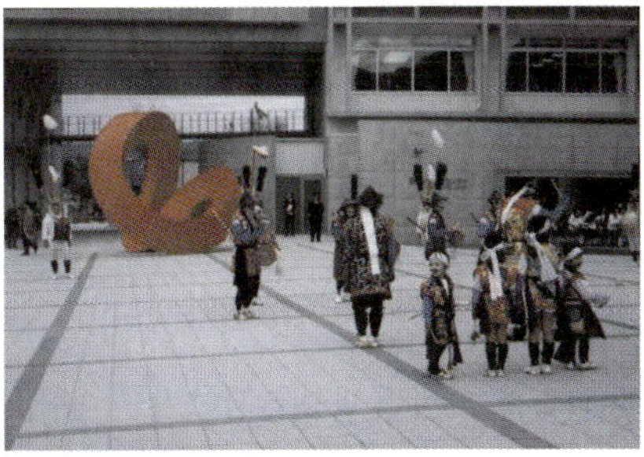

Plate 29. [CP]3: Regeneration

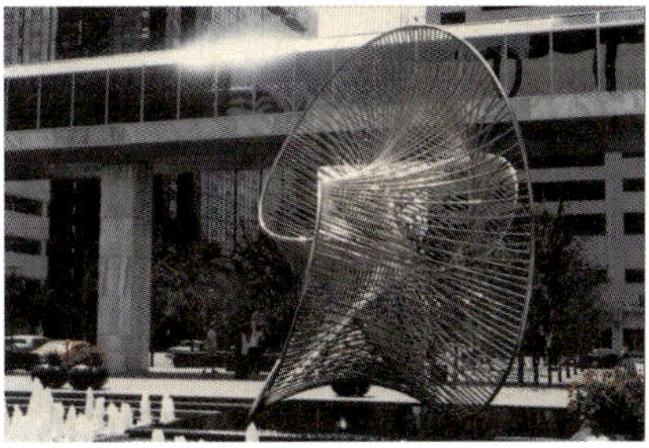

Plate 30. [CP]4: Solstice

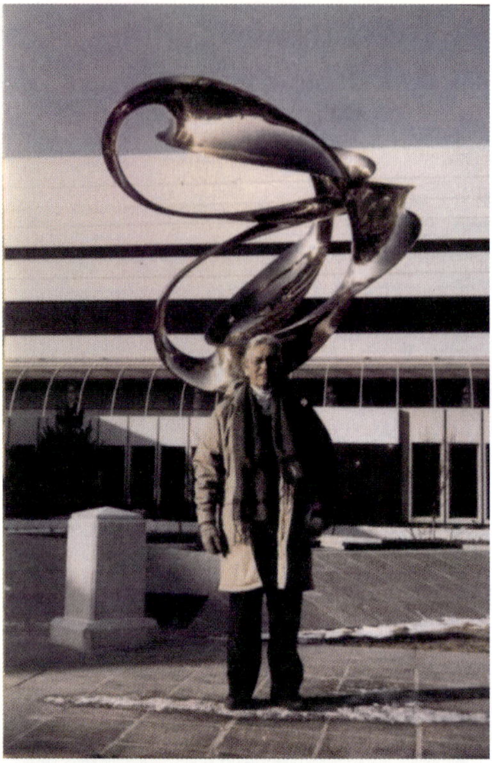

Plate 31. [CP]5: Double Knot

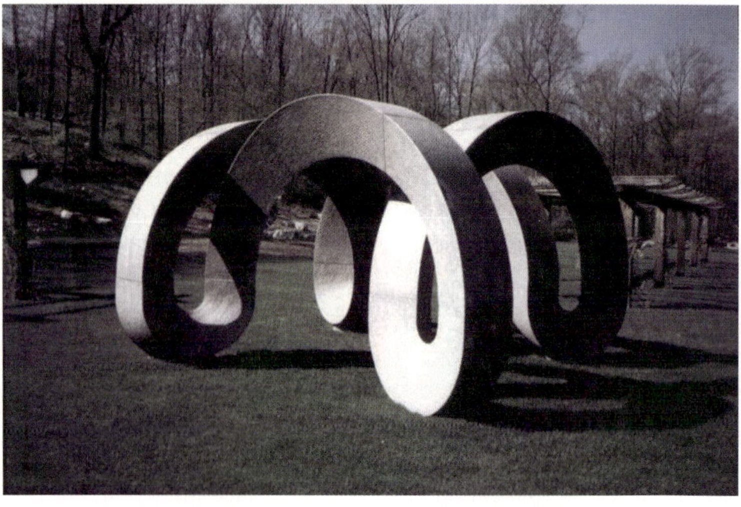

Plate 32. [CP]6: The Arch of Janus

JOHN HUBBARD

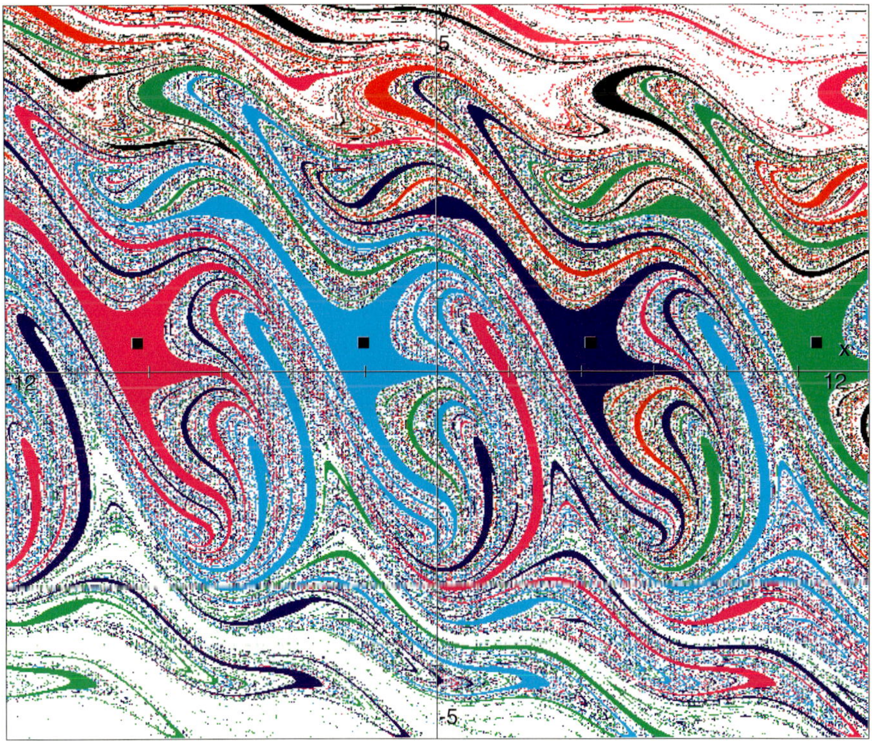

Plate 33. [JH]1: Lakes of Wada

SCOTT CRASS

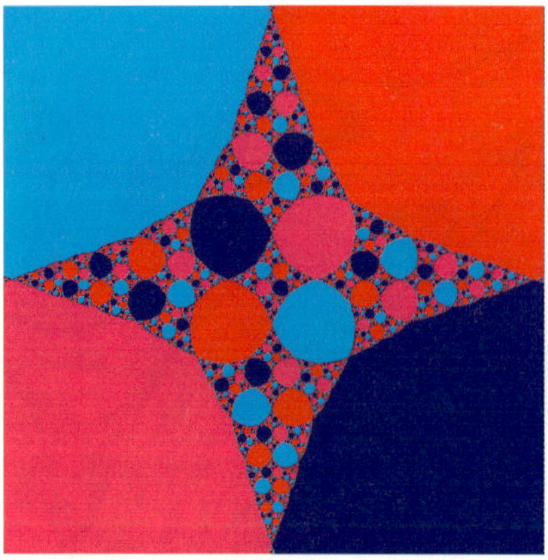

Plate 34. [SC]1: Four basins of attraction for the octahedral 5-map

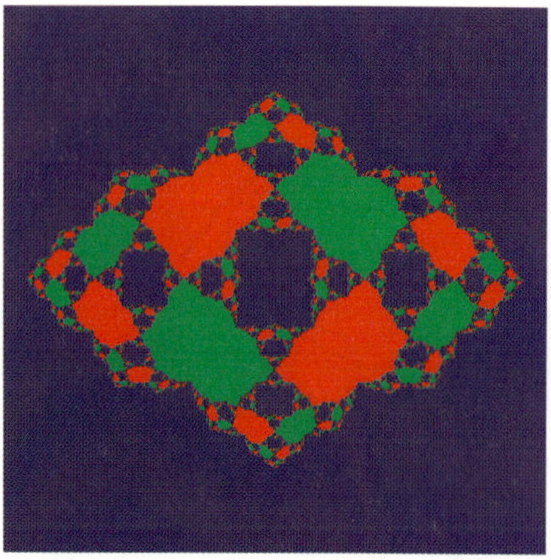

Plate 35. [SC]2: Three basins of attraction for h_{11} restricted to a plane

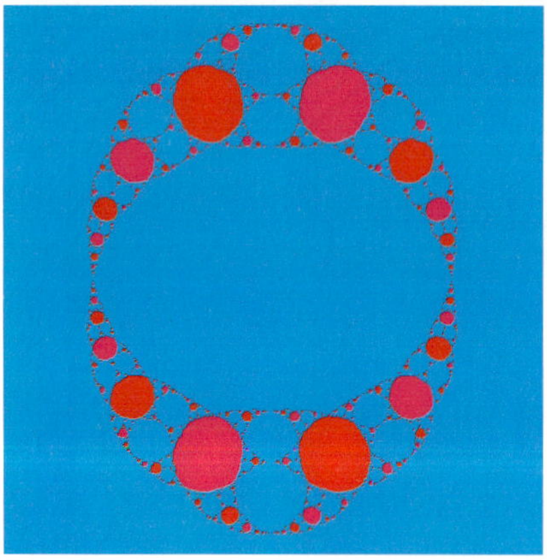

Plate 36. [SC]3: Three basins of attraction for h_{11} restricted to a plane

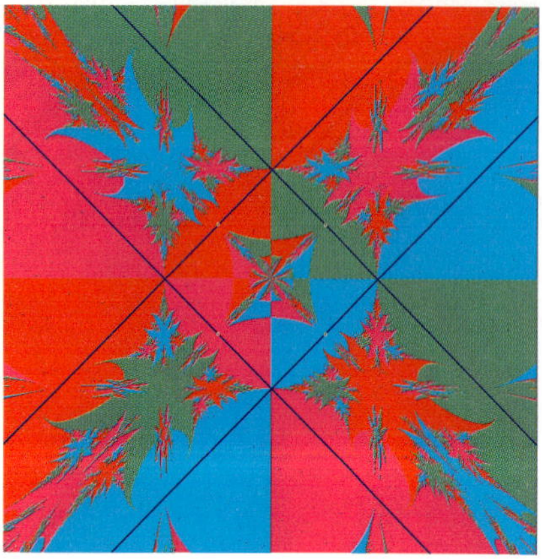

Plate 37. [SC]4: Chaotic attractors – the four lines – for h_{11} on a plane with octahedral symmetry

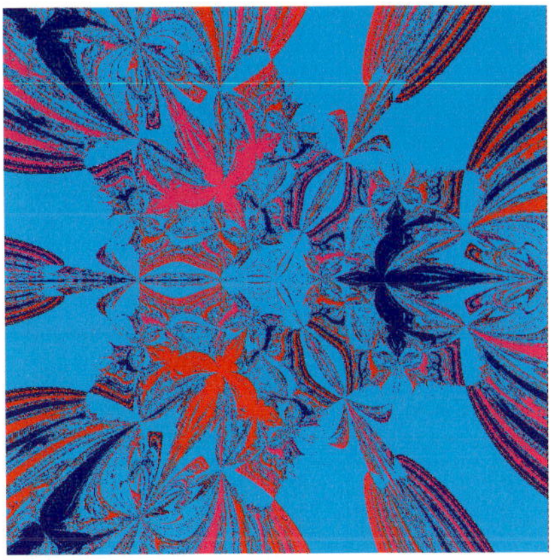

Plate 38. [SC]5: Chaotic attractor for h_{11} on a plane with the symmetry of a double triangular pyramid

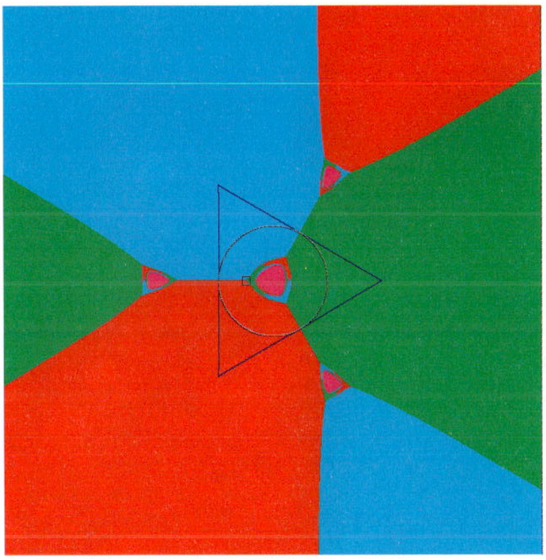

Plate 39. [SC]6: Four basins of attraction for f_6 restricted to a plane with the symmetry of a double triangular pyramid

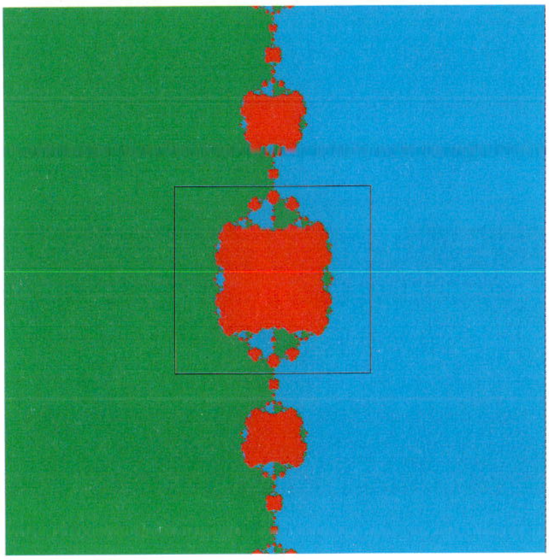

Plate 40. [SC]7: Three basins of attraction for f_6 restricted to a plane

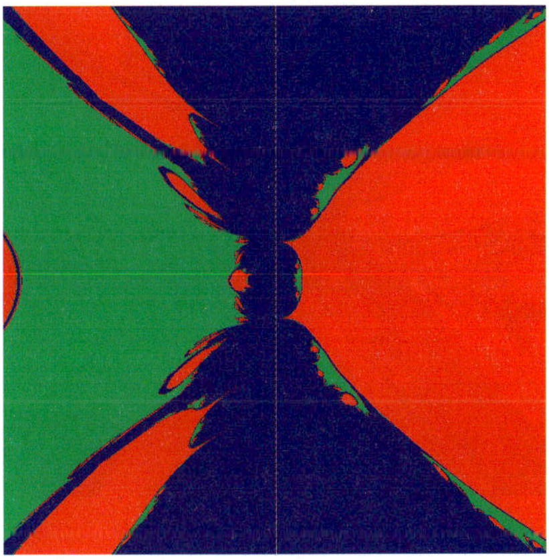

Plate 41. [SC]8: Chaotic attractor – the vertical line – for f_6 on a plane

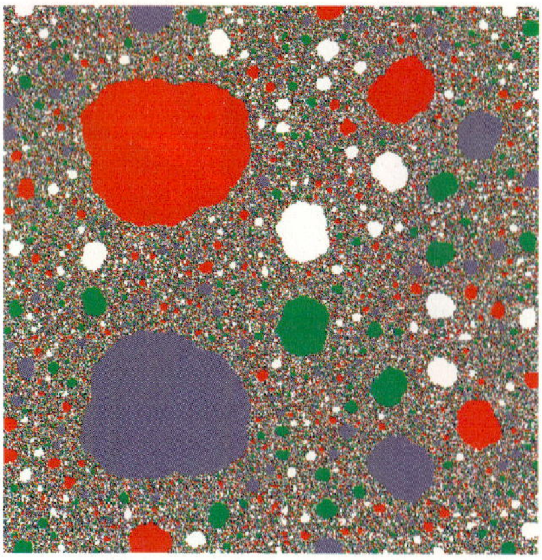

Plate 42. [SC]9: Dynamics of the degree-16 map.

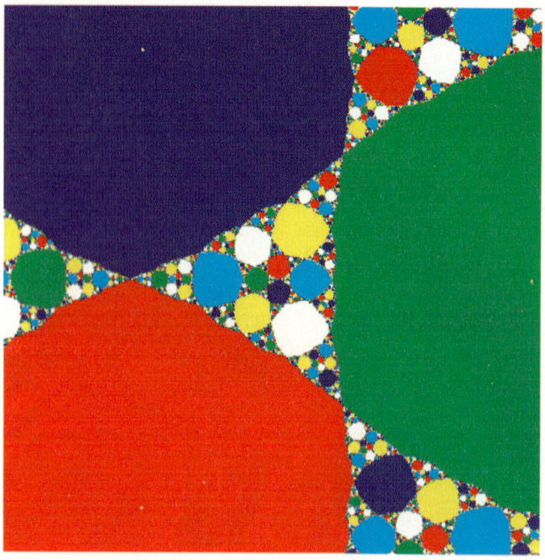

Plate 43. [SC]10: Icosahedral dynamics of the degree-19 map.

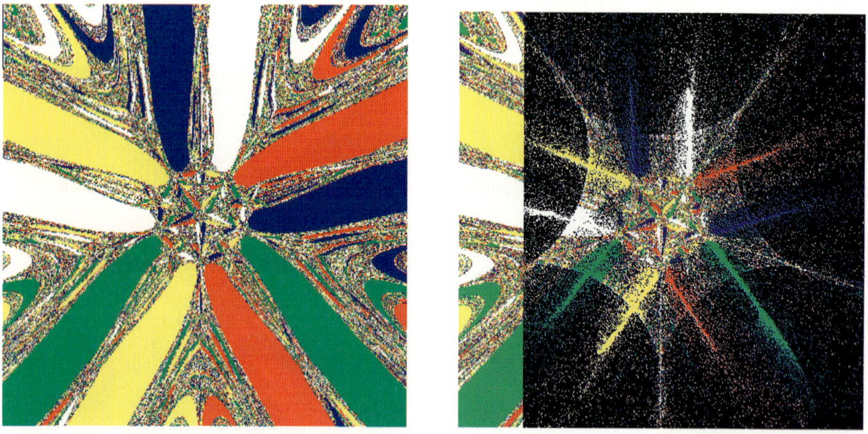

Plate 44. [SC]11, [SC]12: Dynamics of h_{19} on a special plane with the symmetry of a double pentagonal pyramid.

MICHELE EMMER

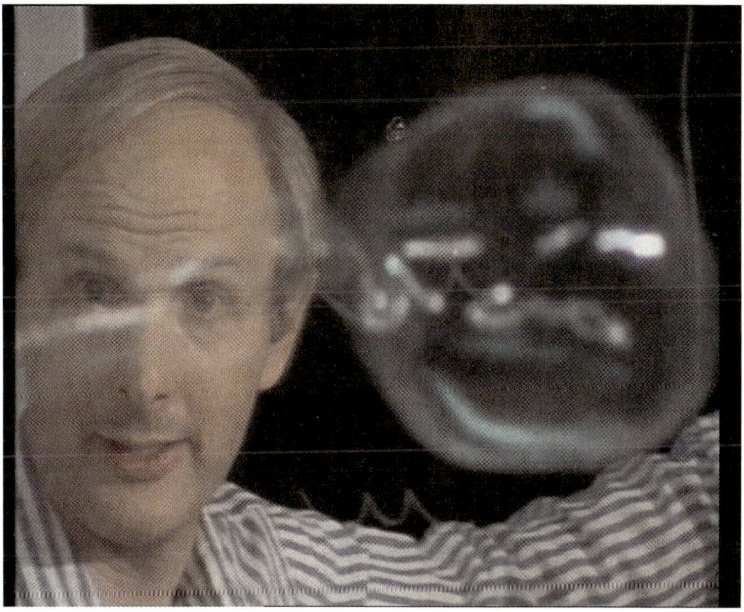

Plate 45. [ME]1: Fred Almgren in his studio at Princeton University

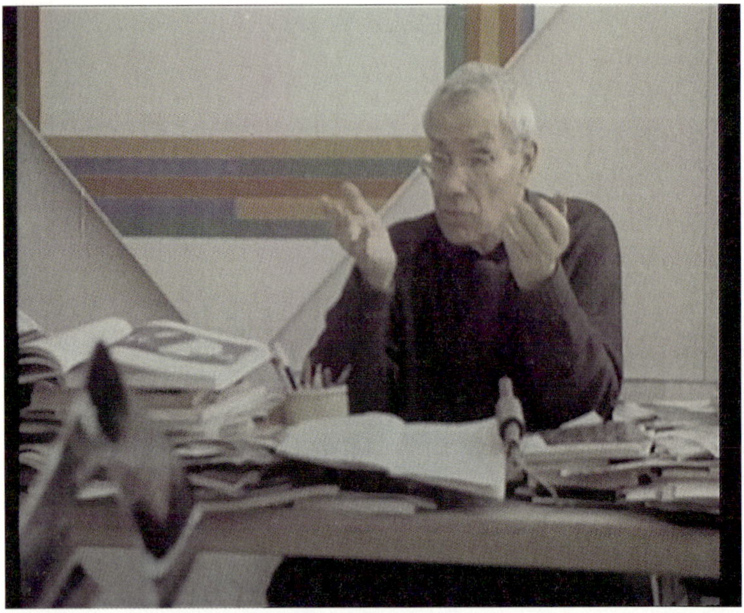

Plate 46. [ME]2: Max Bill in his Zurich studio

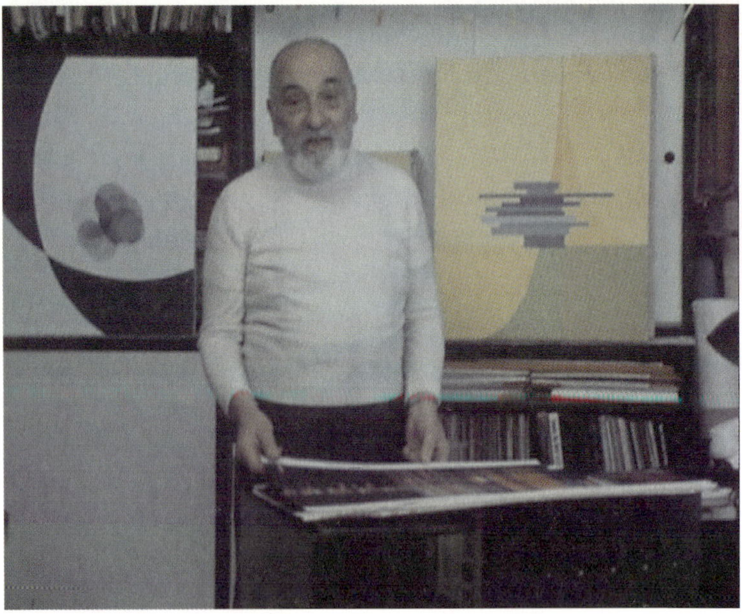

Plate 47. [ME]3: Luigi Veronesi in his Milan studio.

JEAN-FRANÇOIS COLONNA

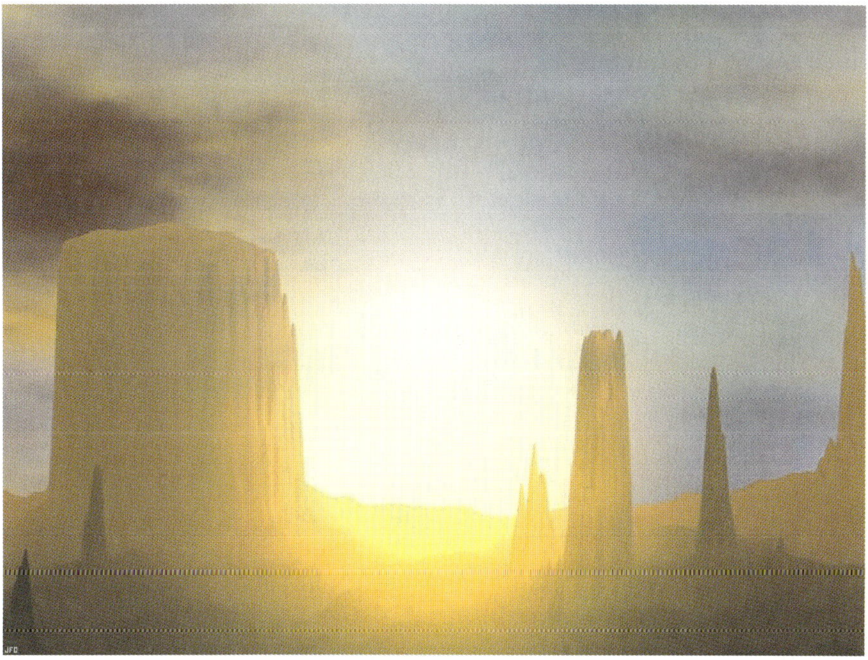

Plate 48. [JFC]1: Monument Valley at sunset

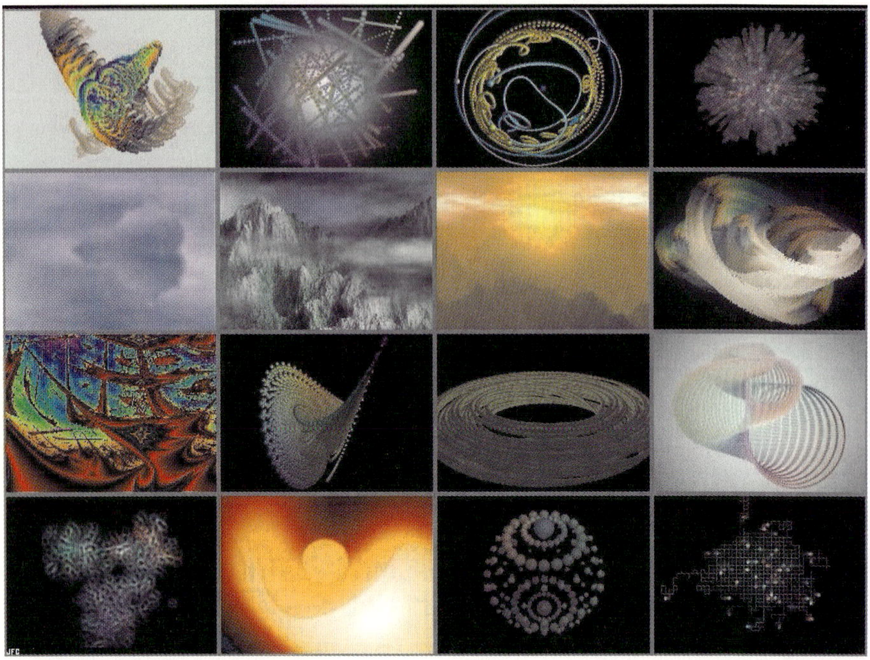

Plate 49. [JFC]4: From the quark and gluon structure of the nucleon (bottom left) to the Universe (top right)

JOHN ROBINSON

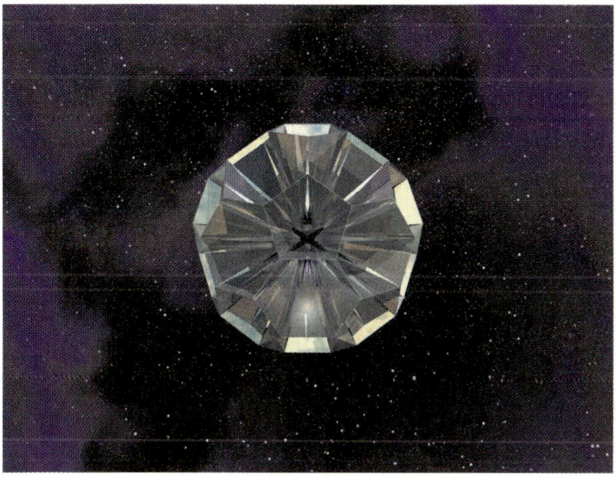

Plate 50. [JR]1: Pulse

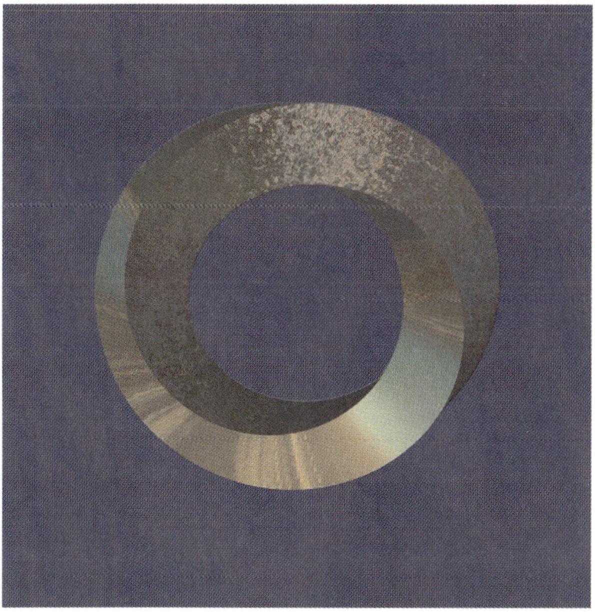

Plate 51. [JR]2: Dependent Beings

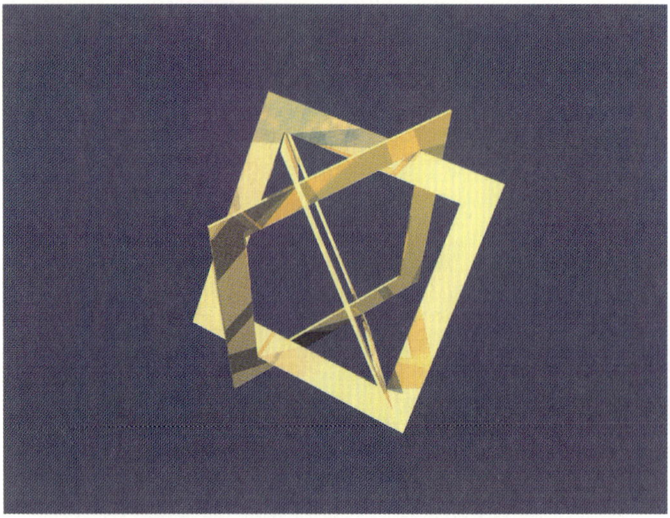

Plate 52. [JR]3: Creation

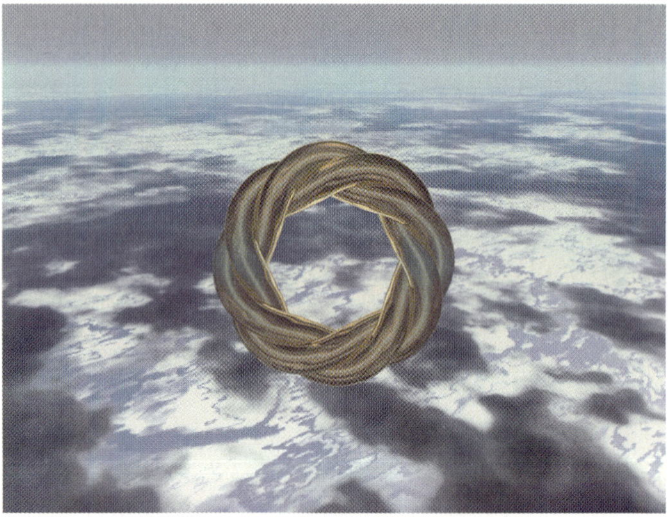

Plate 53. [JR]4: Gordian knot

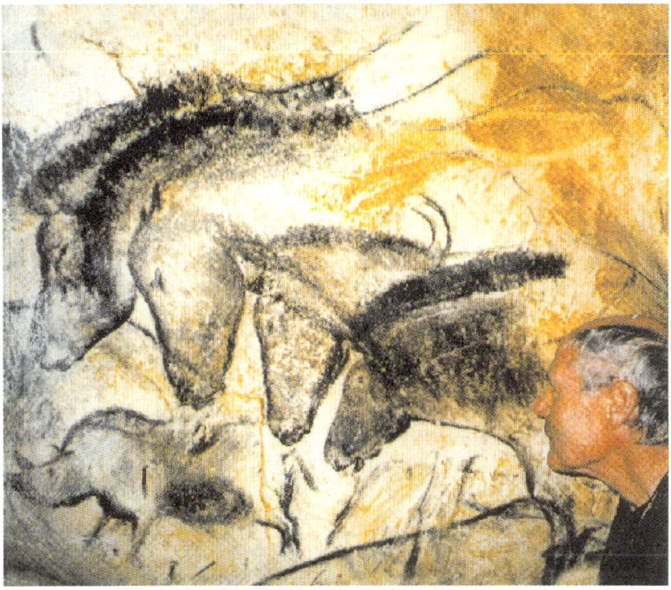

Plate 54. [JR]5: Chauvet Cave

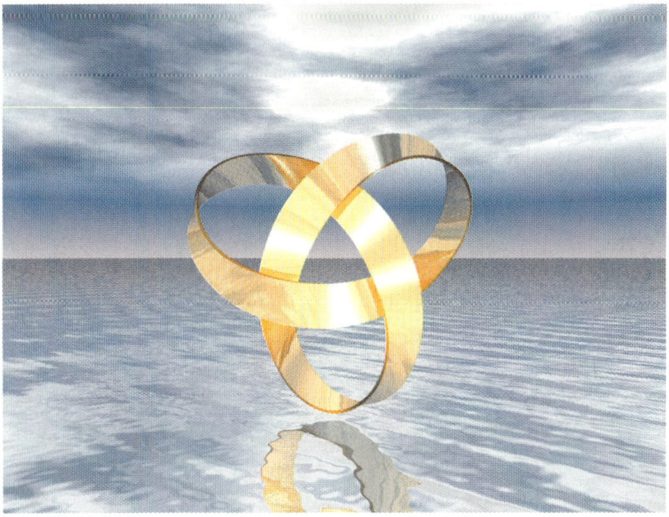

Plate 55. [JR]6: Immortality

MANUEL CHAVES

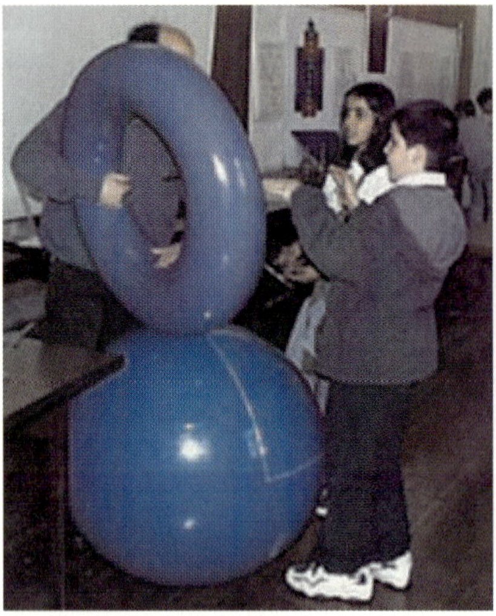

Plate 56. [MC]1

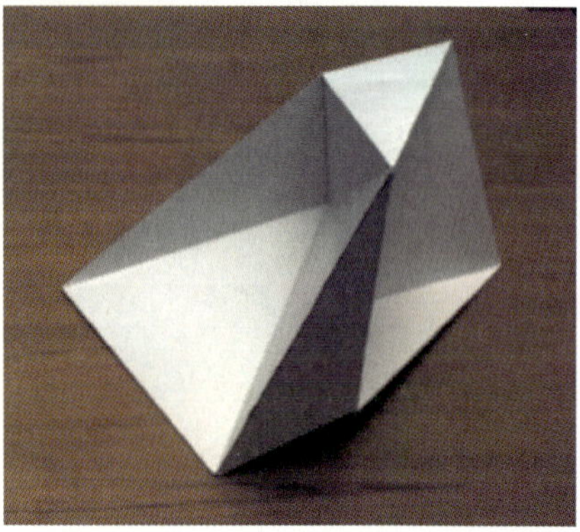

Plate 57. [MC]2

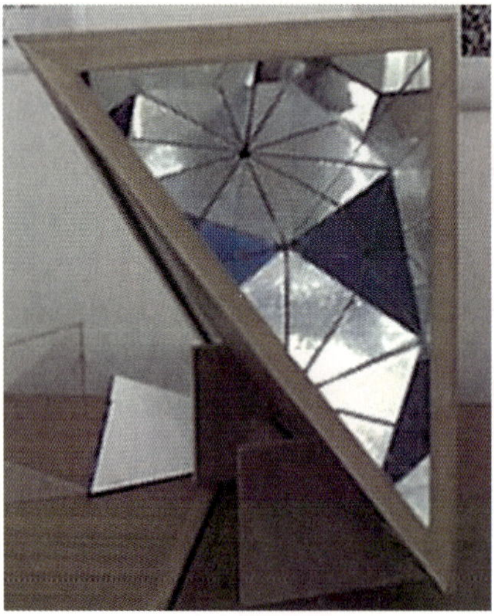

Plate 58. [MC]3

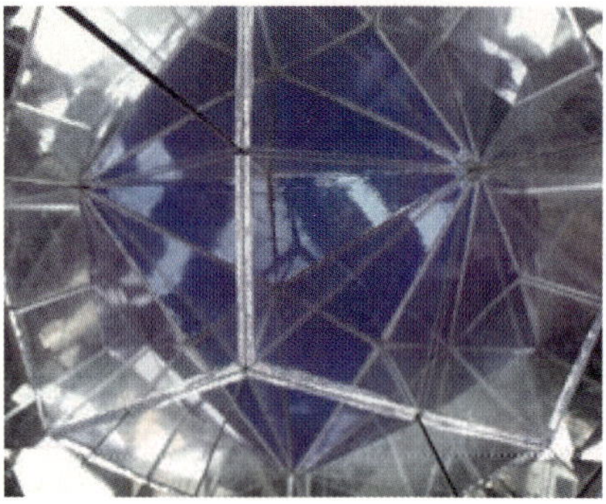

Plate 59. [MC]4

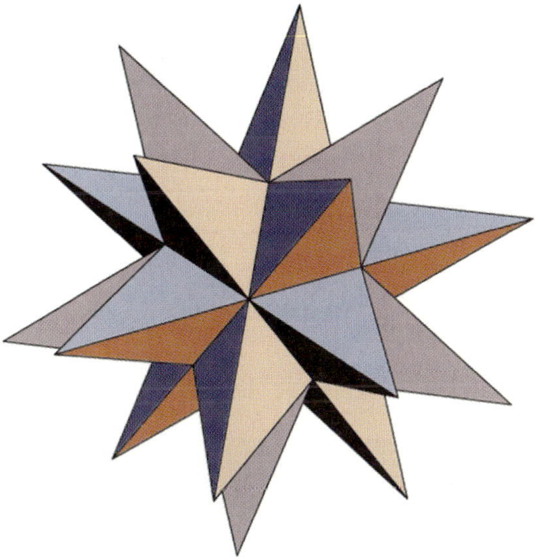

Plate 60. [MC]5

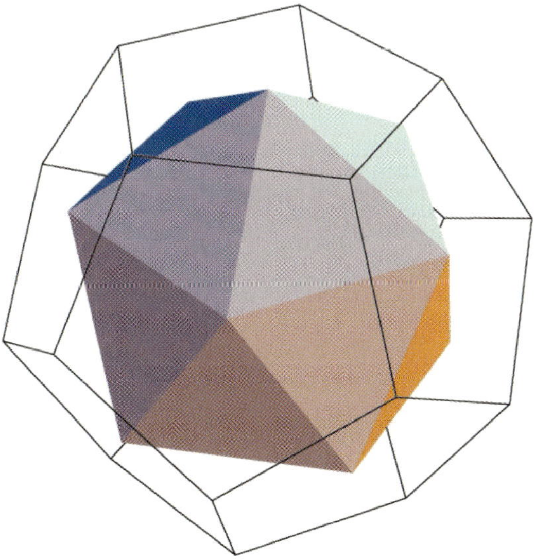

Plate 61. [MC]6

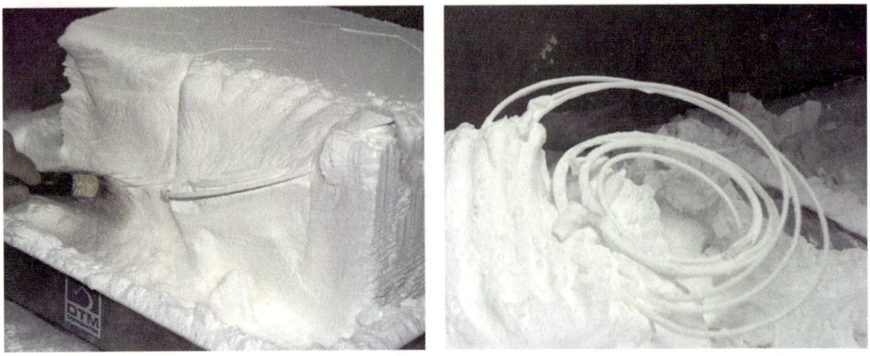

Plate 62. [MC]7: Lorenz Attractor

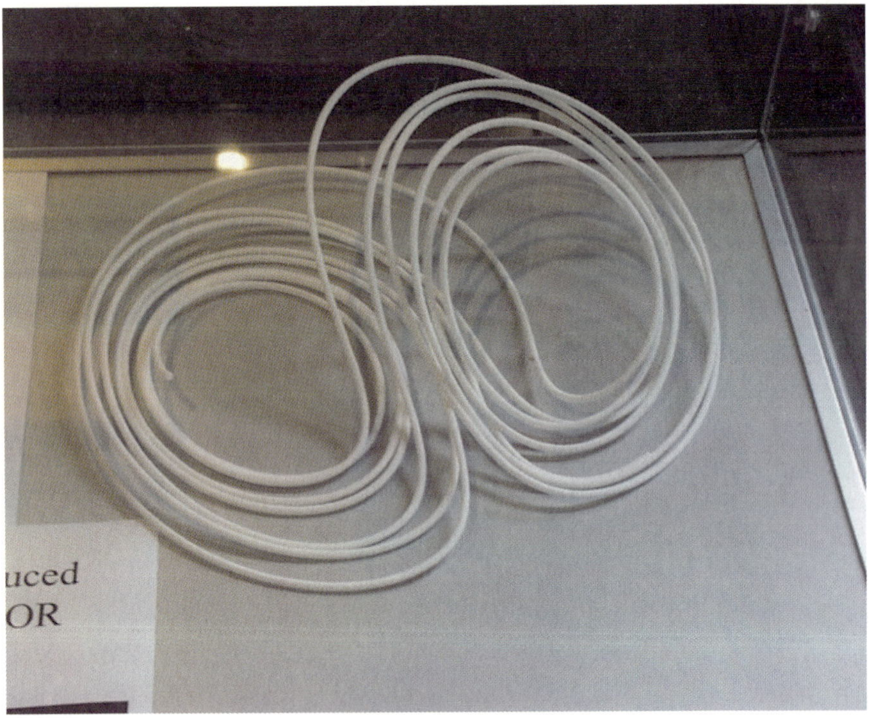

Plate 63. [MC]8: Lorenz Attractor

DICK TERMES

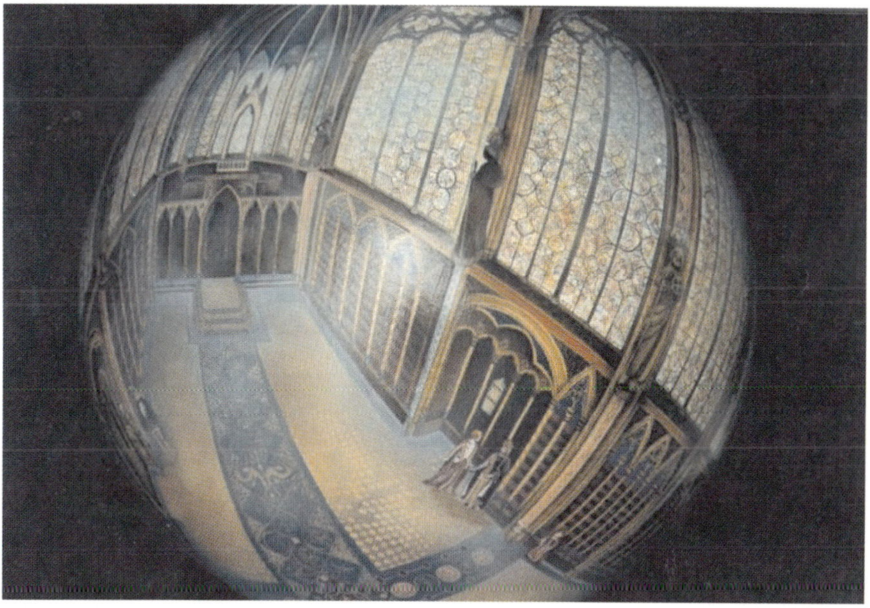

Plate 64. [DT]1: Sainte Chapelle in Paris

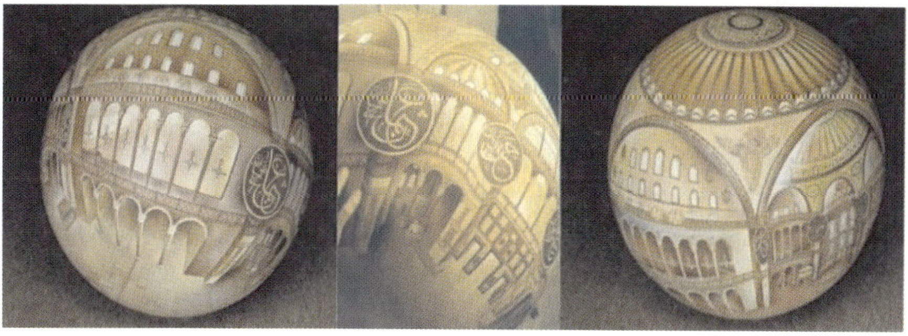

Plate 65. [DT]2: Hagia Sophia

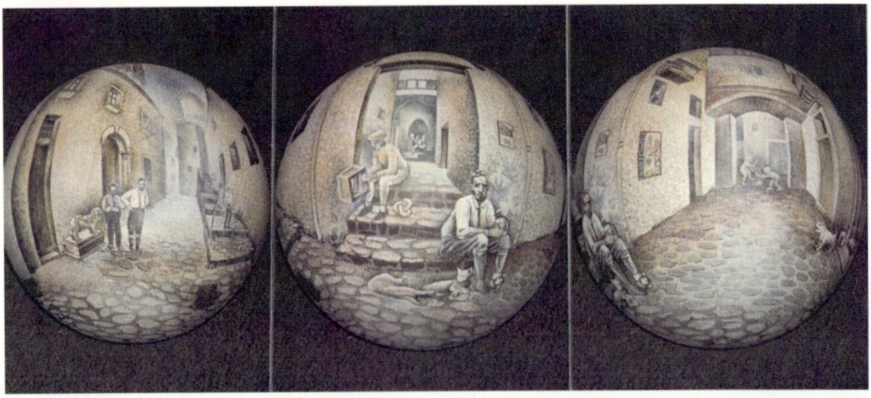

Plate 66. [DT]3: Inspirations for Escher

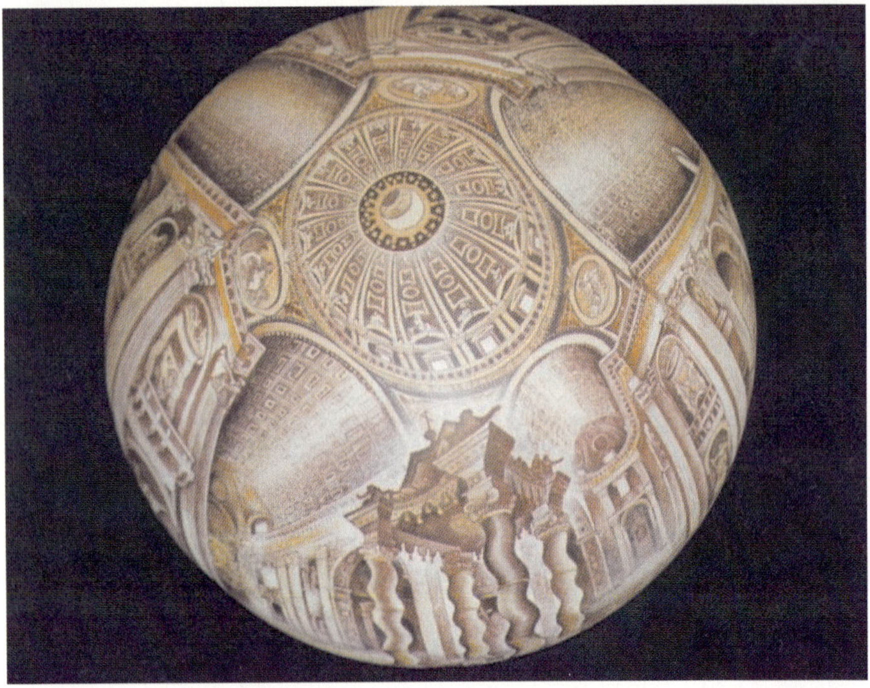

Plate 67. [DT]4

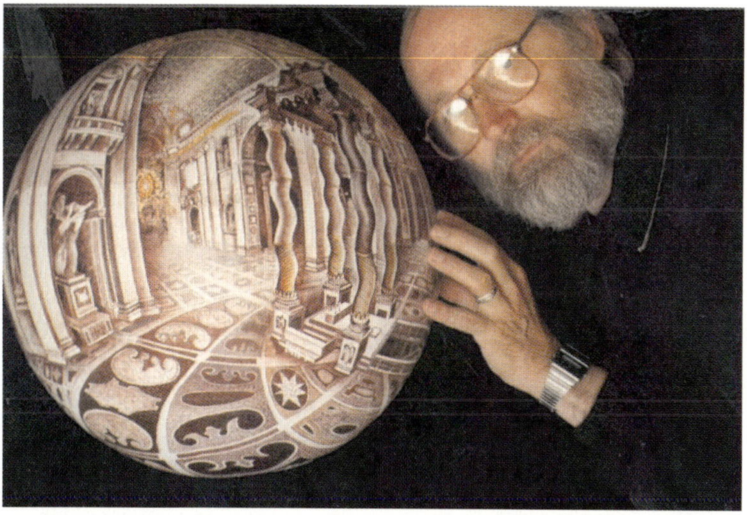

Plate 68. [DT]5: St Peters

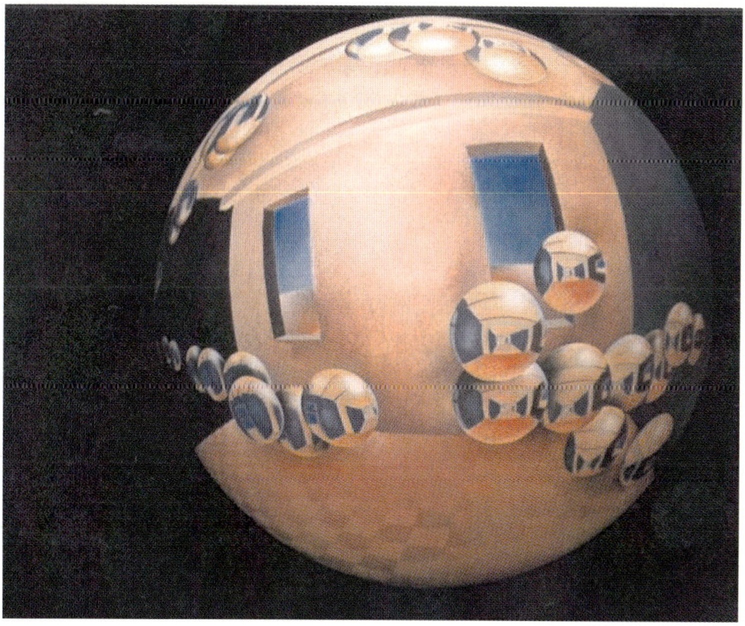

Plate 69. [DT]6: Concave Bubbles

FRANÇOIS APÉRY

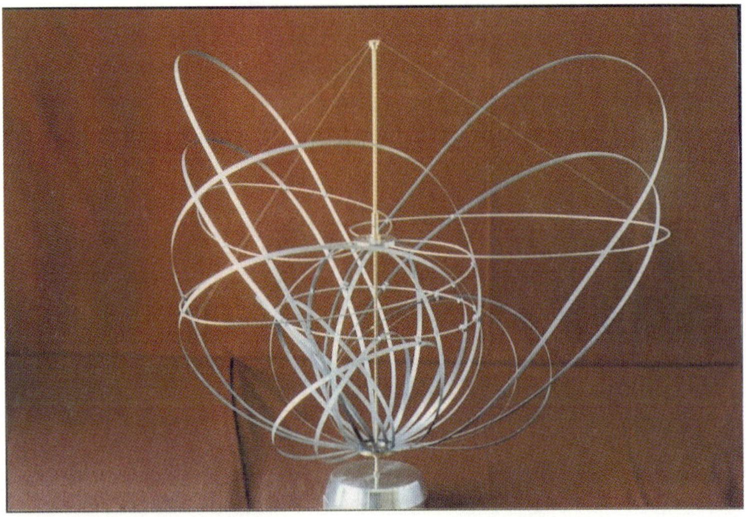

Plate 70. [FA]1: Boy surface in steel strips realized by F. Apéry

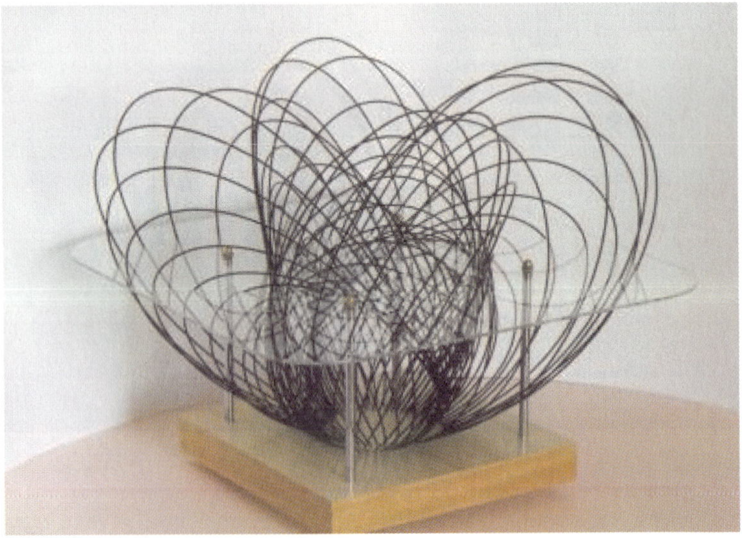

Plate 71. [FA]2: Halfway model in Teflon and Plexiglass realized by G. Franzoni

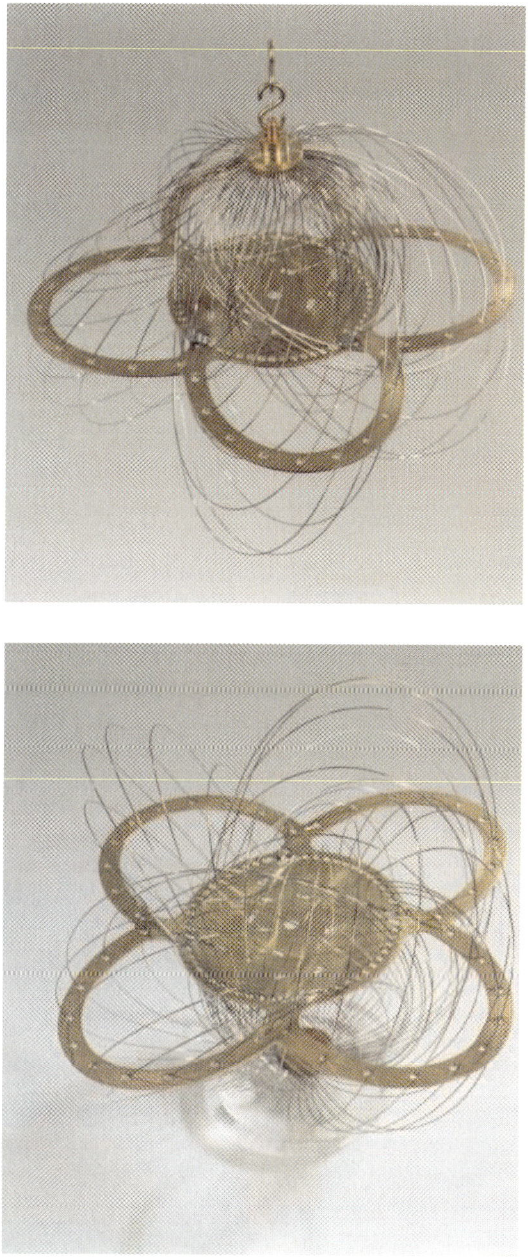

Plate 72. [FA]3: A view of the wire halfway model realized by F. Apéry

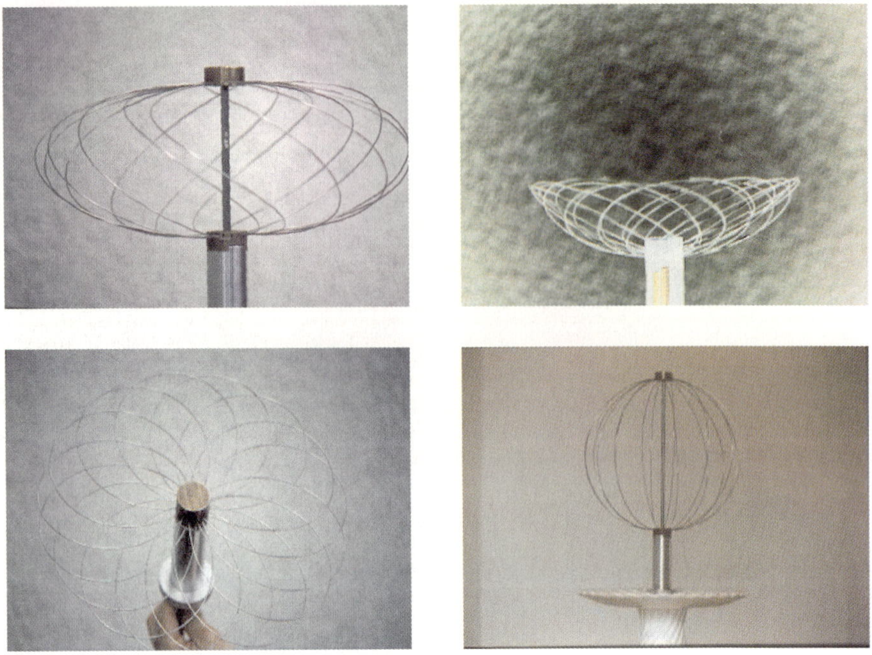

Plate 73. [FA]4: A view of the gastrulation model realized by F. Apéry

JOHN SULLIVAN

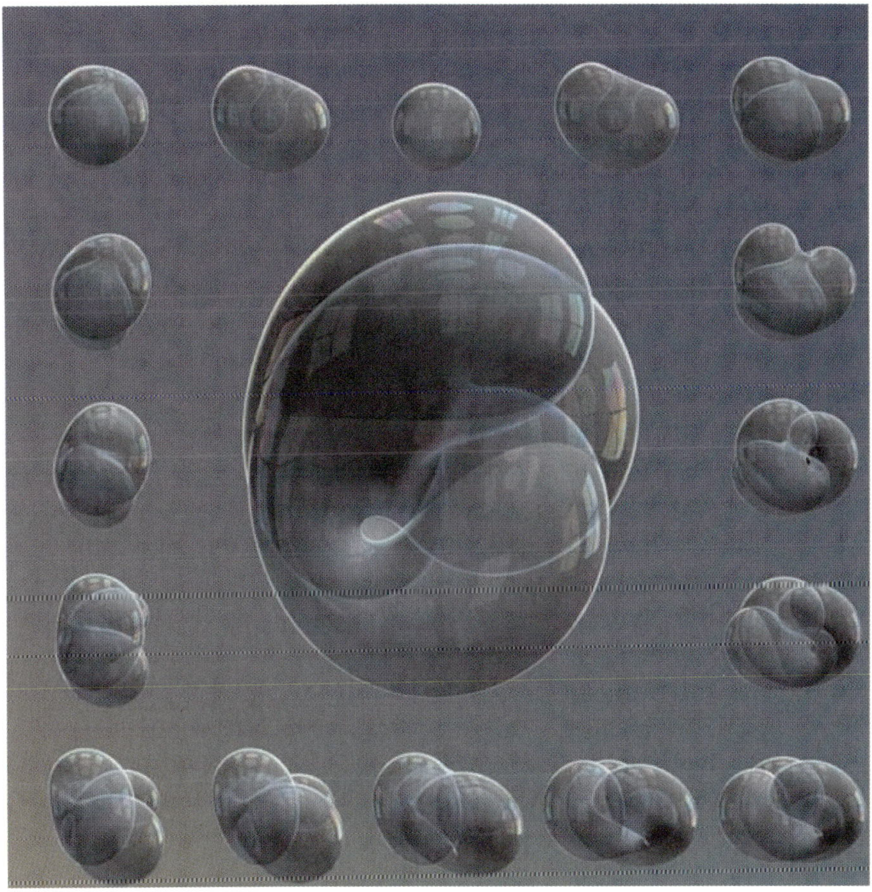

Plate 74. [JS]1: Transparent Eversion Montage

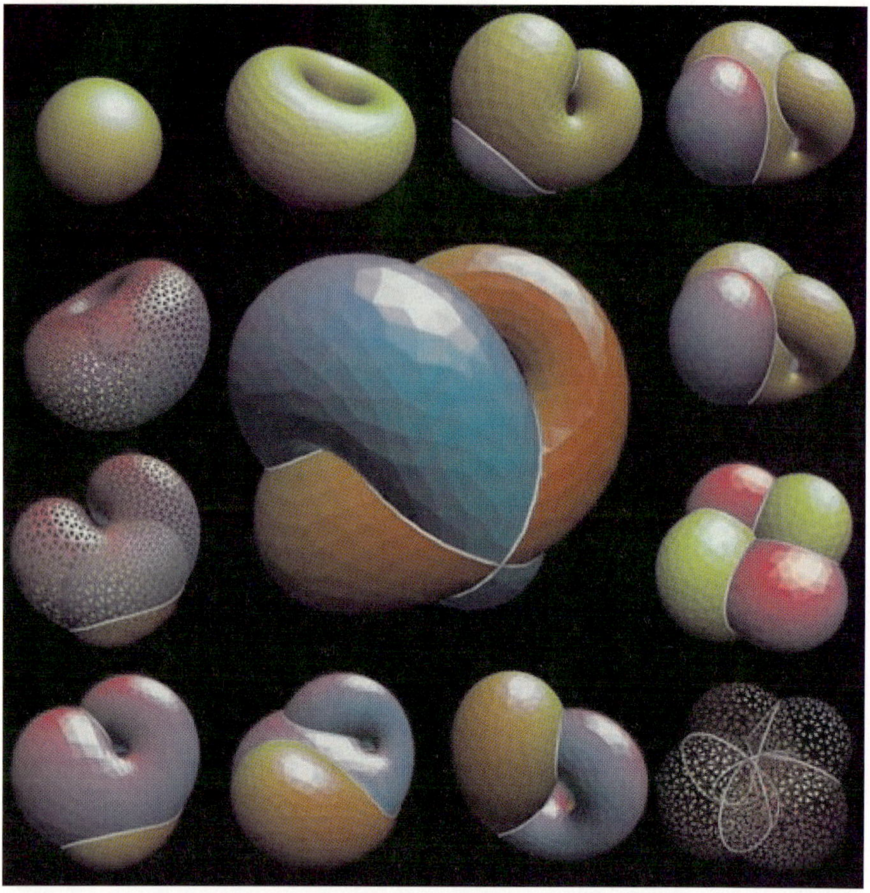

Plate 75. [JS]2: Two-fold Eversion Montage

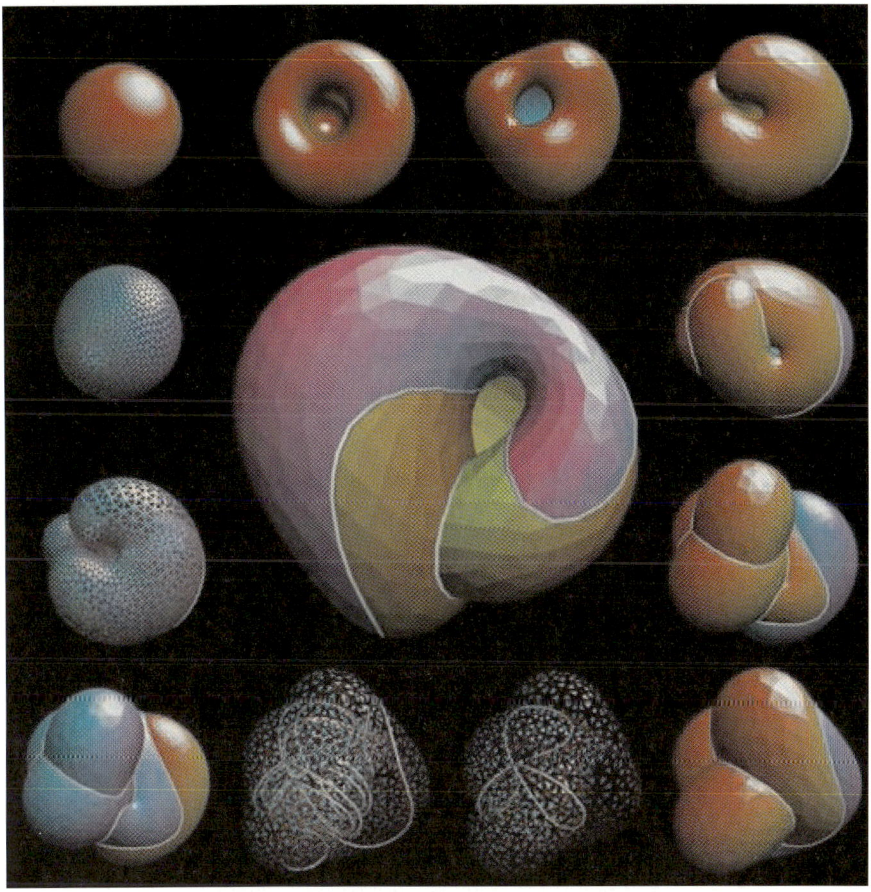

Plate 76. [JS]3: Three-fold eversion

STEWART DICKSON

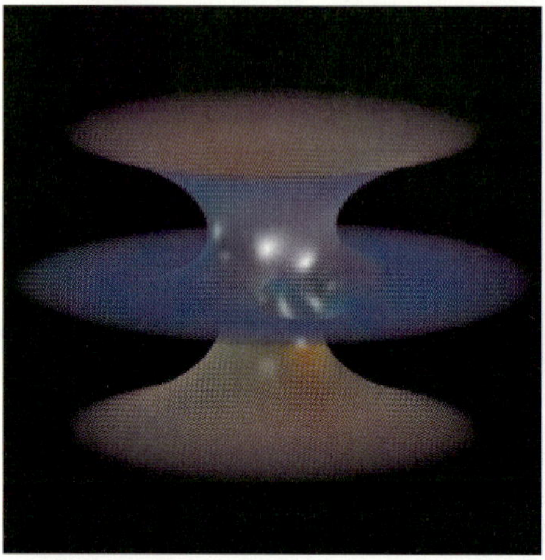

Plate 77. [SD]1: Costa's Three-Ended Minimal Surface, Image by David Hoffman and James Hoffman

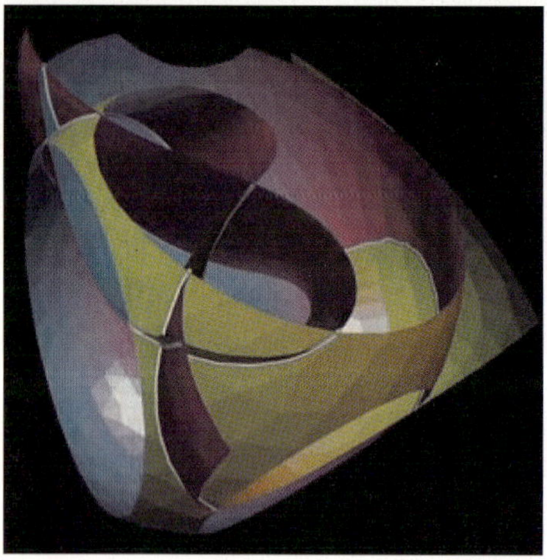

Plate 78. [SD]2: The Sphere as a Color Map during the Eversion Metamorphosis (from the Optiverse)

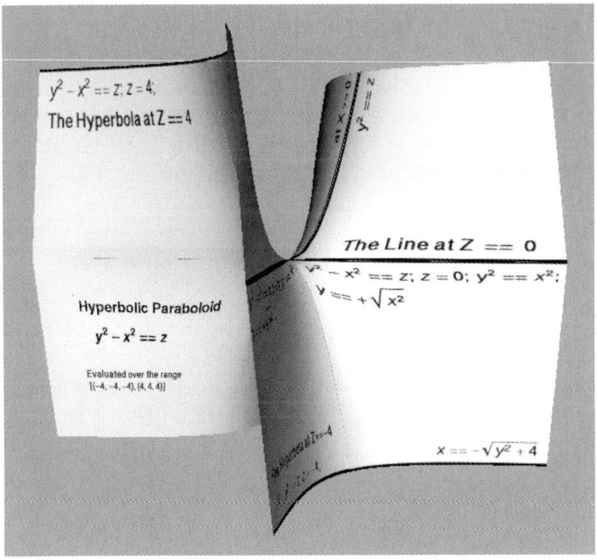

Plate 79. [SD]3: Annotated Hyperbolic Paraboloid – Computer Rendering by Stewart Dickson

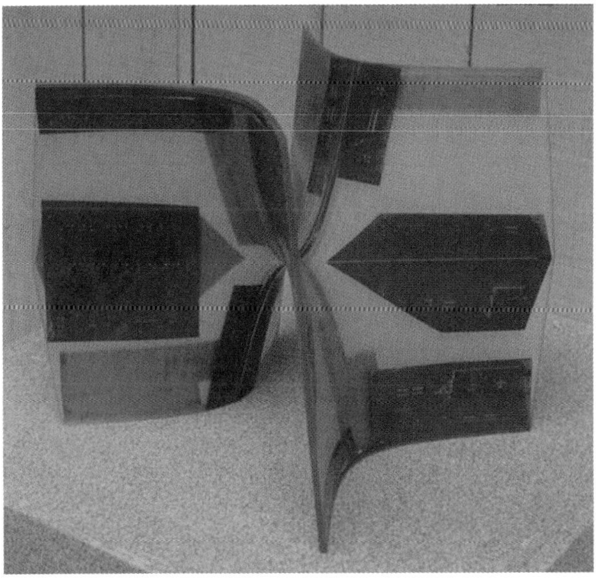

Plate 80. [SD]4: Hyperbolic Paraboloid 3-D Model (Stereolithograph) Annotated with Self-Adhesive Braille Captions

NATHANIEL FRIEDMAN

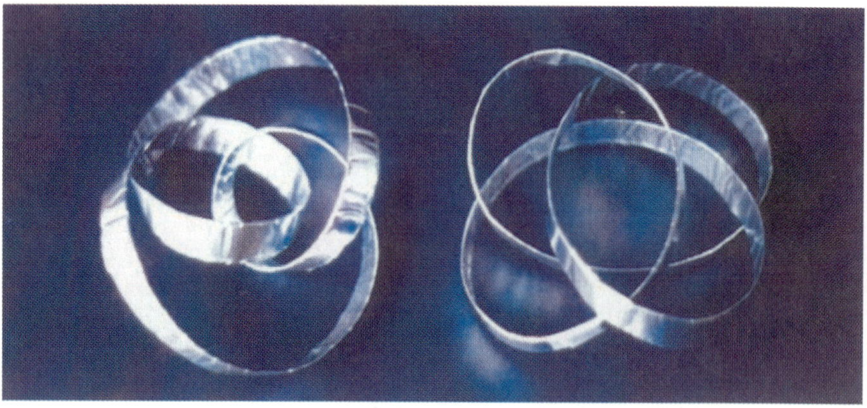

Plate 81. [NF]1: Aluminium Foil Knots

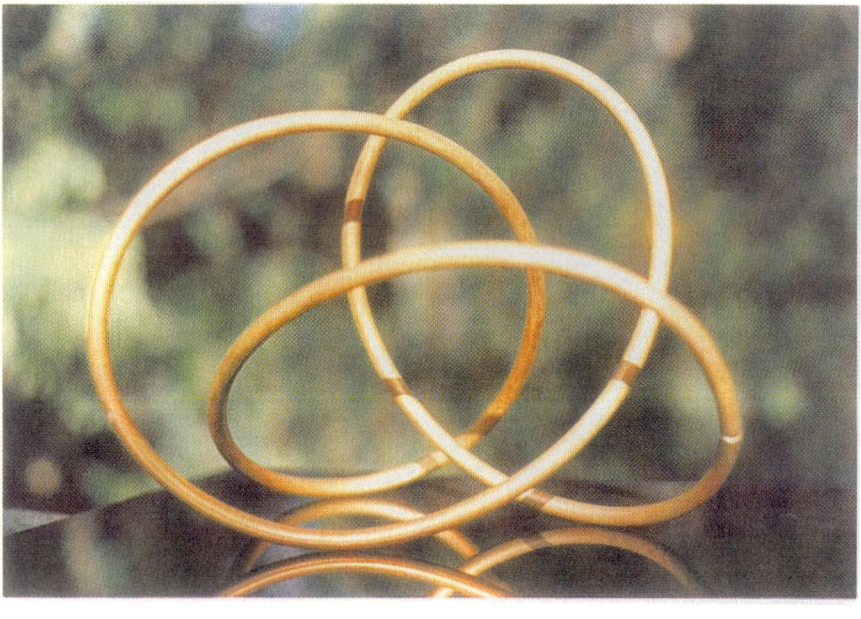

Plate 82. [NF]2: Copper Tubing Knot

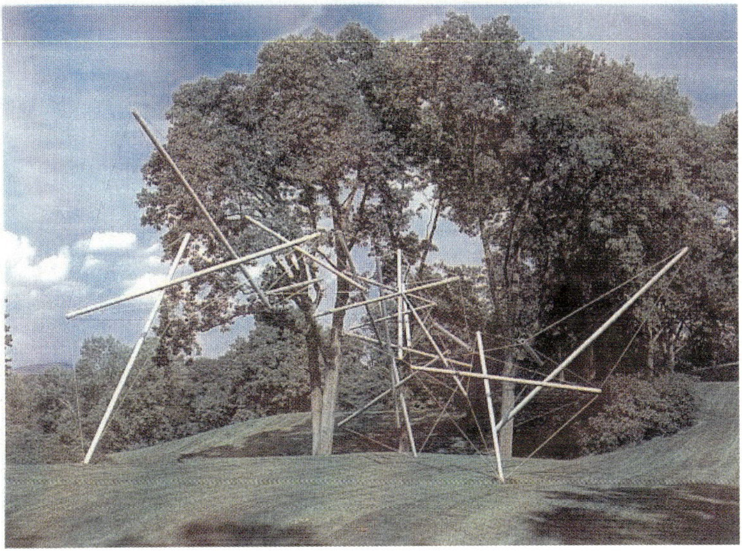

Plate 83. [NF]3: Free Ride Home

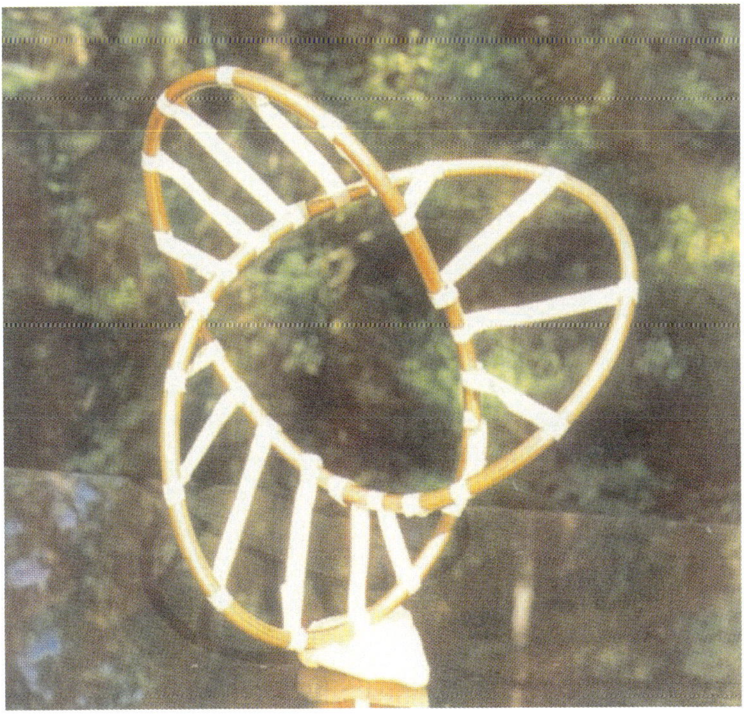

Plate 84. [NF]4: Trefoil Taped Surface

PHILIPPE CHARBONNEAU

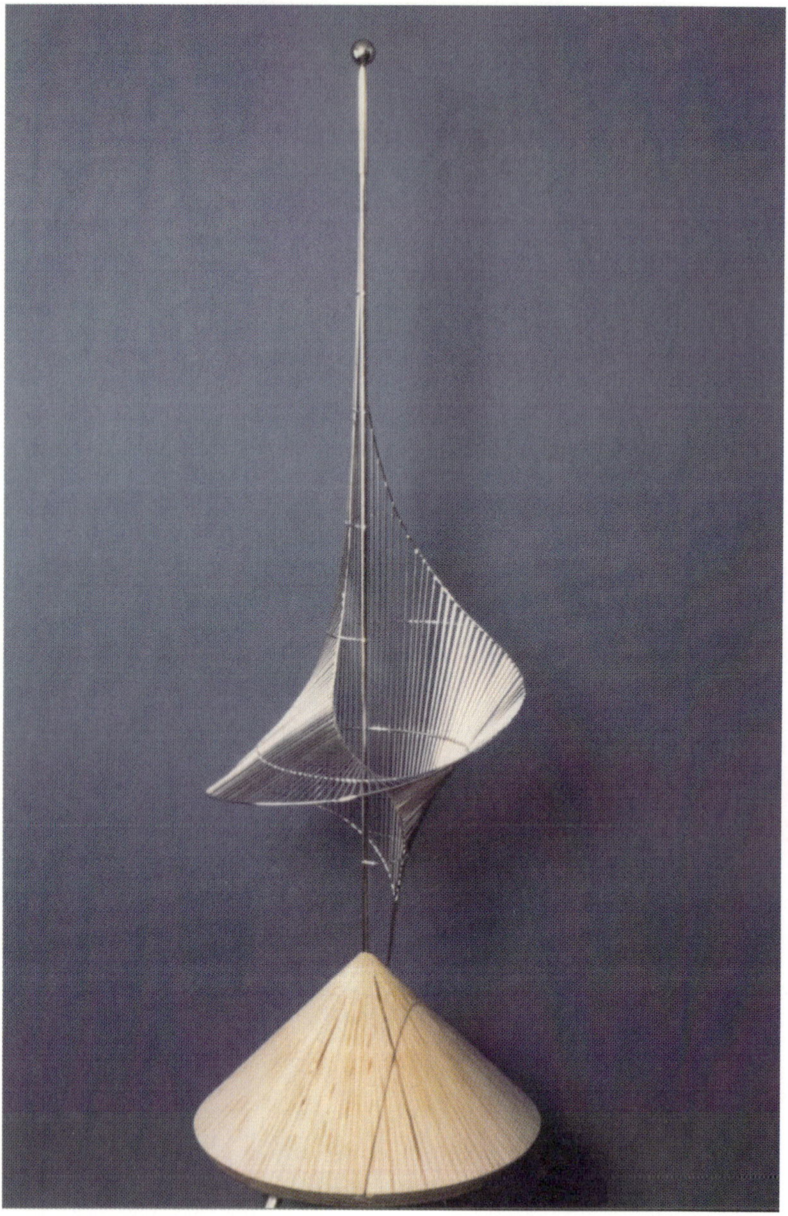

Plate 85. [PC]1: Biconique IV

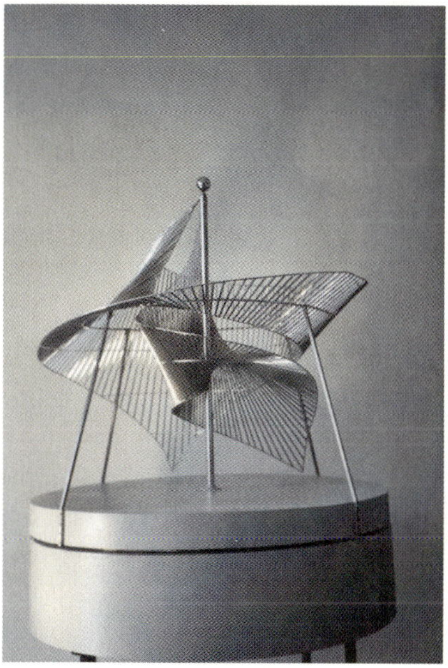

Plate 86. [PC]2: Biconique V

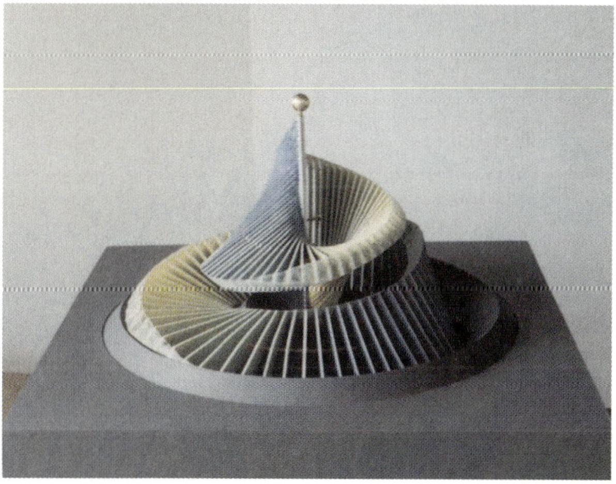

Plate 87. [PC]3: Biconique I

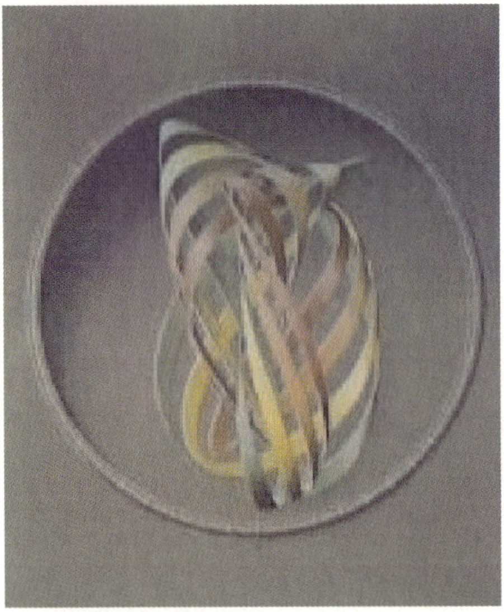

Plate 88. [PC]4: Entrelac quatre rubens

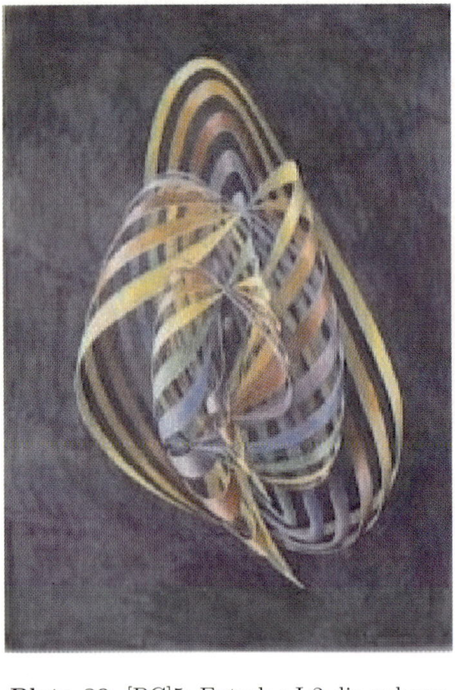

Plate 89. [PC]5: Entrelac I-2 dix rubans

BRUCE HUNT

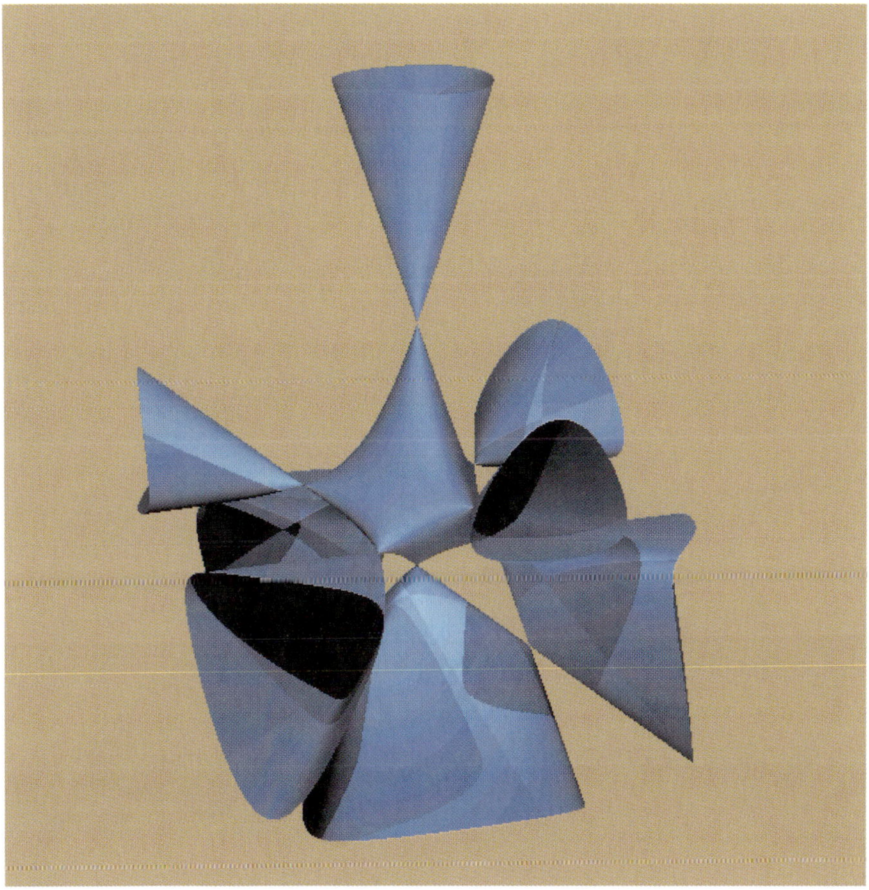

Plate 90. The Hessian of the Cayley Cubic

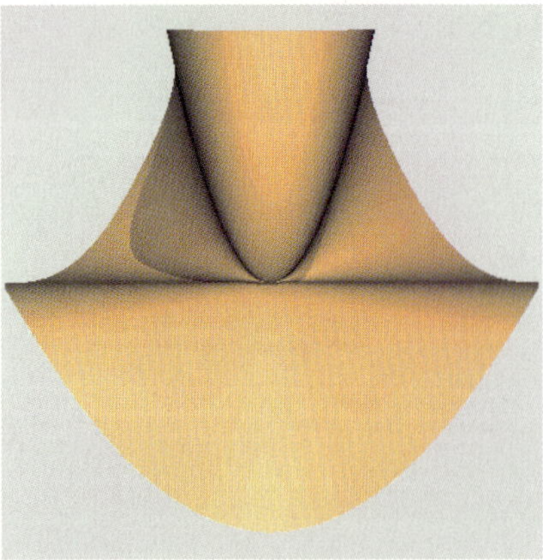

Plate 91. The E_6 cubic Surface

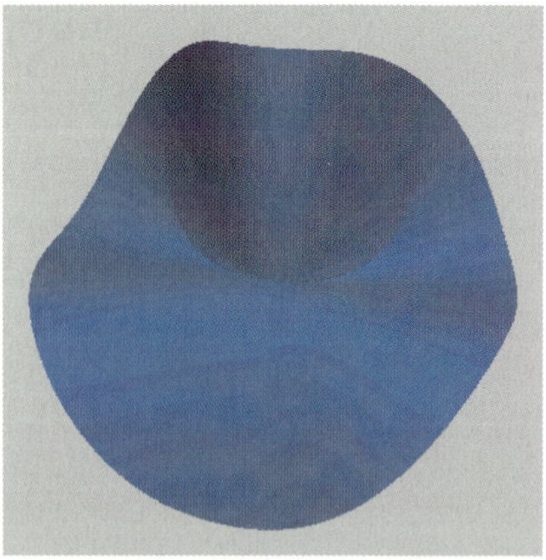

Plate 92. The E_6 cubic Surface, made of mirror glass

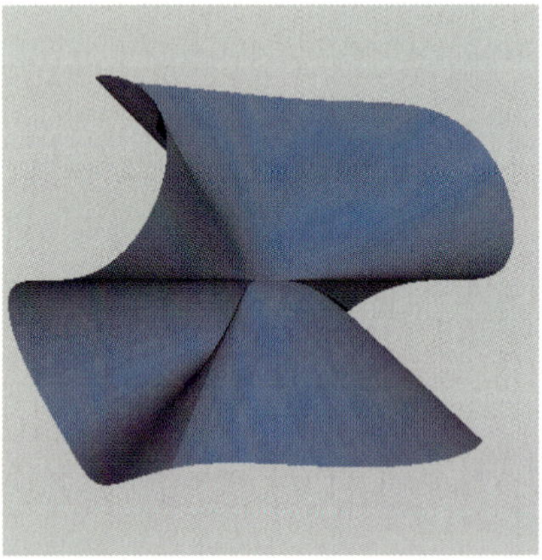

Plate 93. $xz + y^2z + x^3 - z^3 = 0$

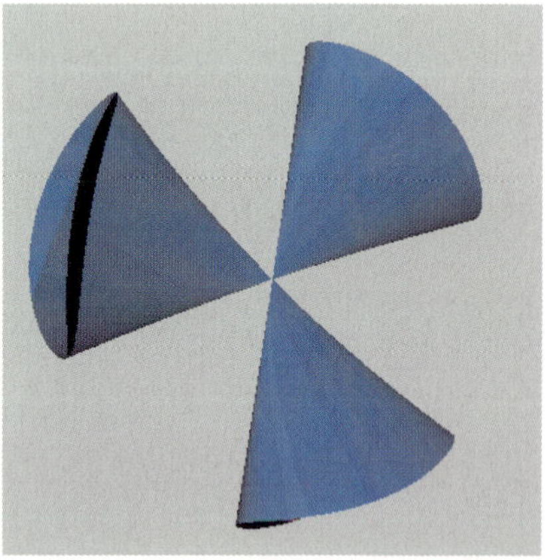

Plate 94. $(x + y + z)^2 + xyz = 0$

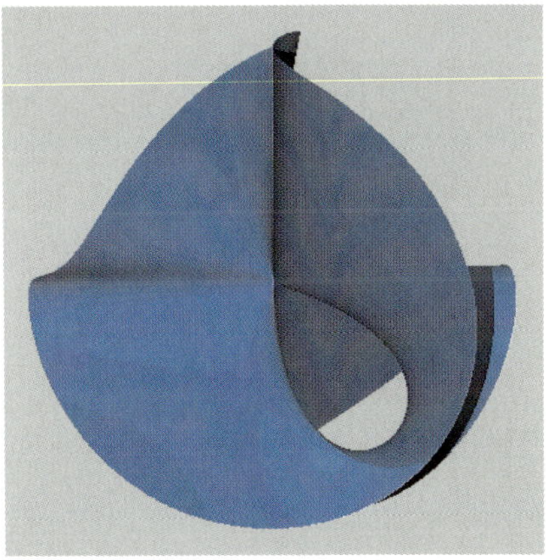

Plate 95. $xz + y^2(x + y + z) = 0$

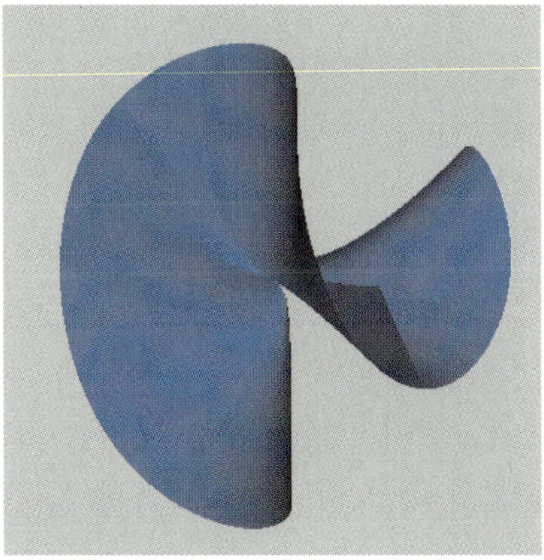

Plate 96. $xz + y^2z + yx^2 = 0$

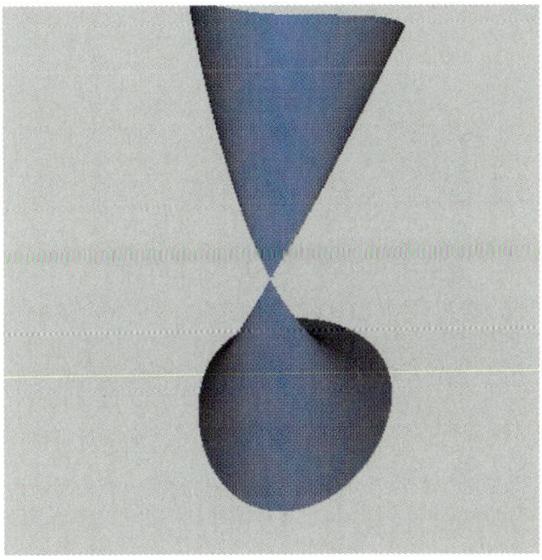

Plate 97. $x^2 + xz^2 + y^2z = 0$

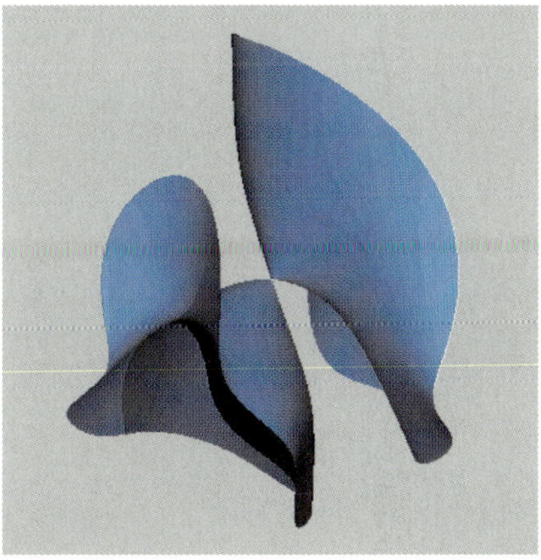

Plate 98. $(xy + xz + yz) + xyz = 0$

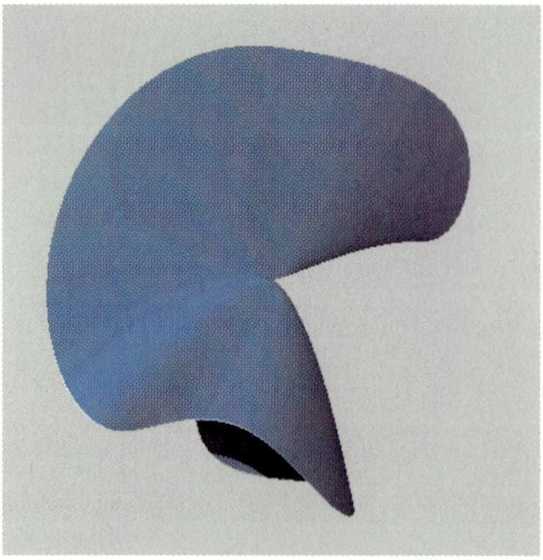

Plate 99. $xz + xy^2 + y^3 = 0$

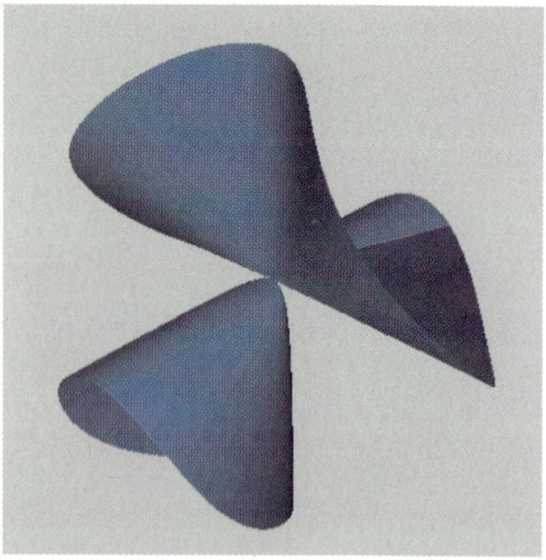

Plate 100. $xz + (x + z)y^2 = 0$

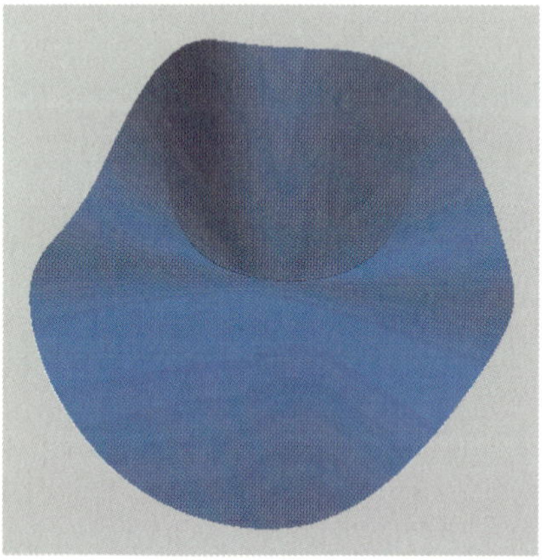

Plate 101. $xz + y^2z + x^3 = 0$

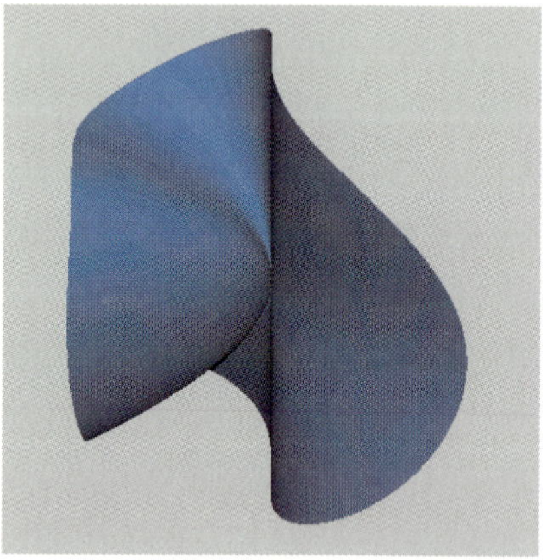

Plate 102. $x^2 + xz^2 + y^3 = 0$

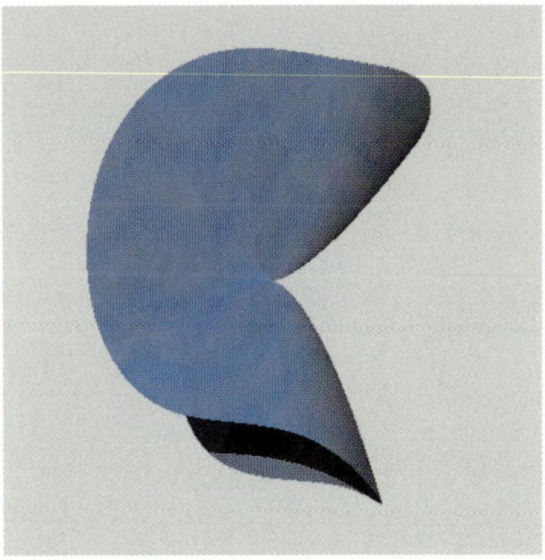

Plate 103. $xz + y^3 = 0$

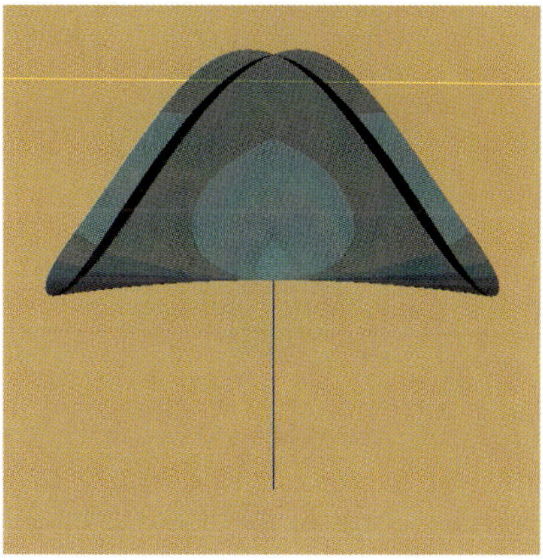

Plate 104. Whitney's Umbrella

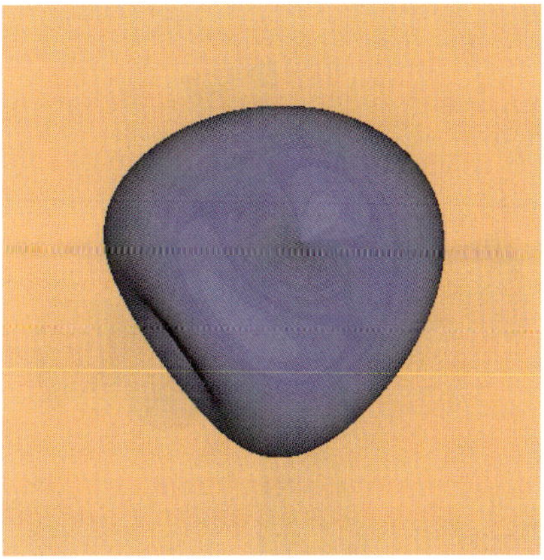

Plate 105. A degeneration of the Steiner surface, phase 1

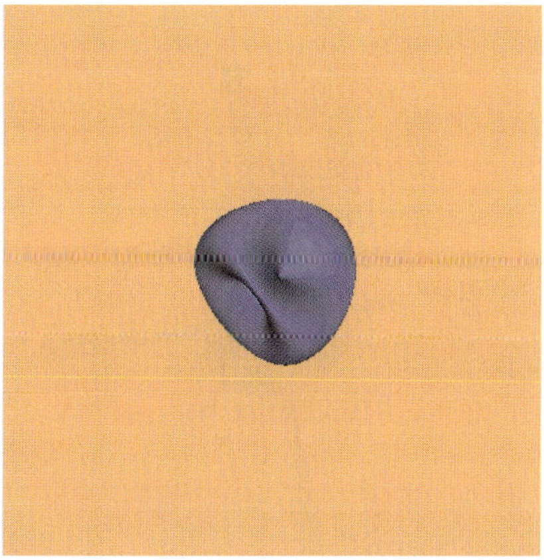

Plate 106. A degeneration of the Steiner surface, phase 2

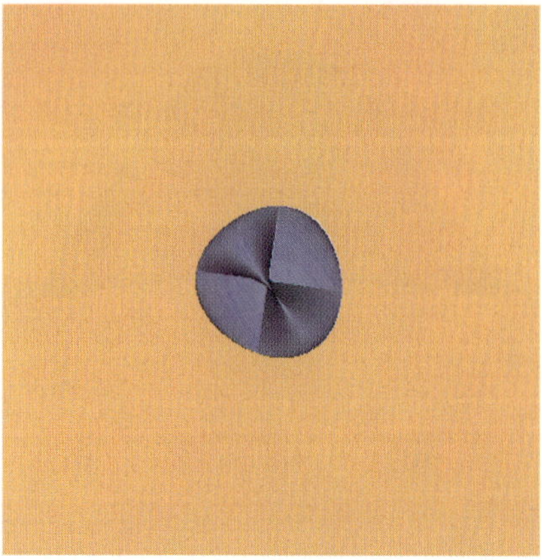

Plate 107. A degeneration of the Steiner surface, phase 3

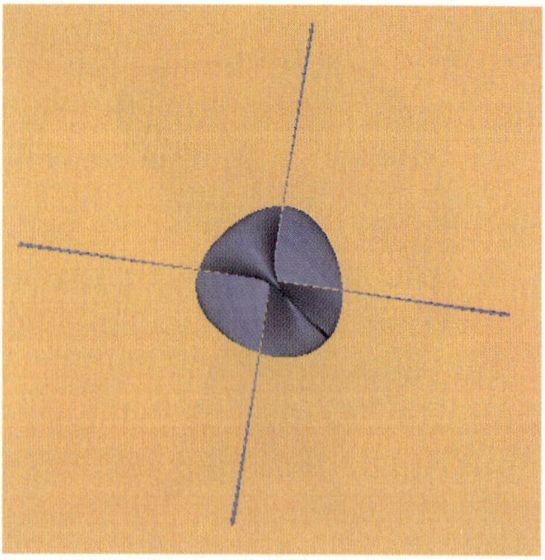

Plate 108. A degeneration of the Steiner surface, phase 4

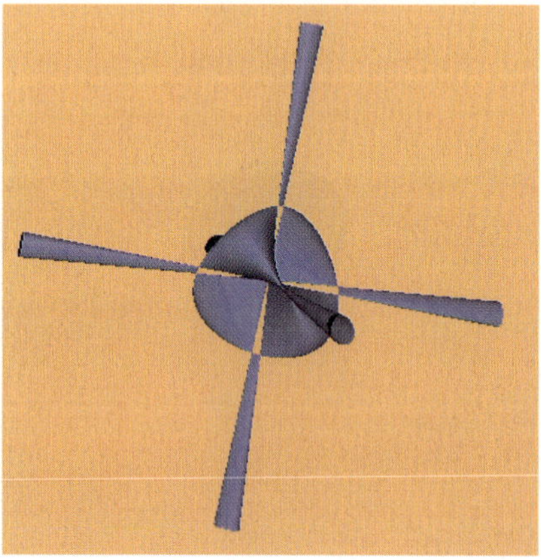

Plate 109. A degeneration of the Steiner surface, phase 5

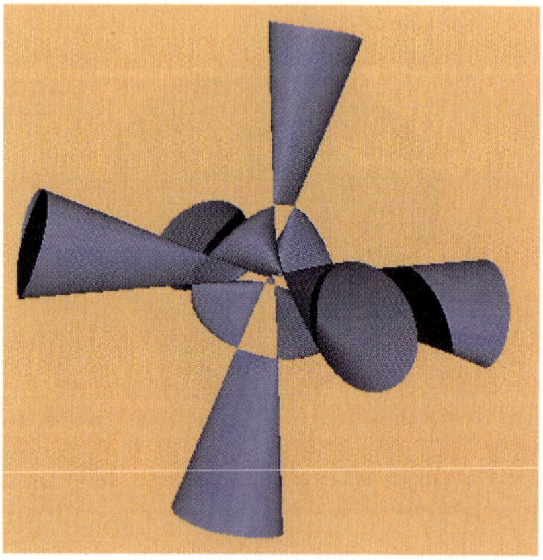

Plate 110. A degeneration of the Steiner surface, phase 6

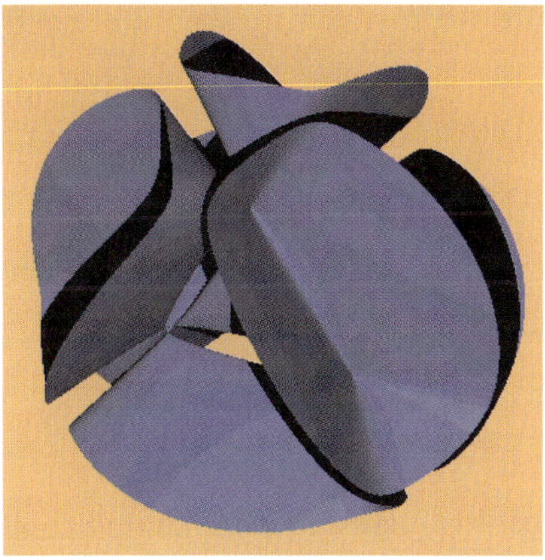

Plate 111. A degeneration of the Steiner surface, phase 7

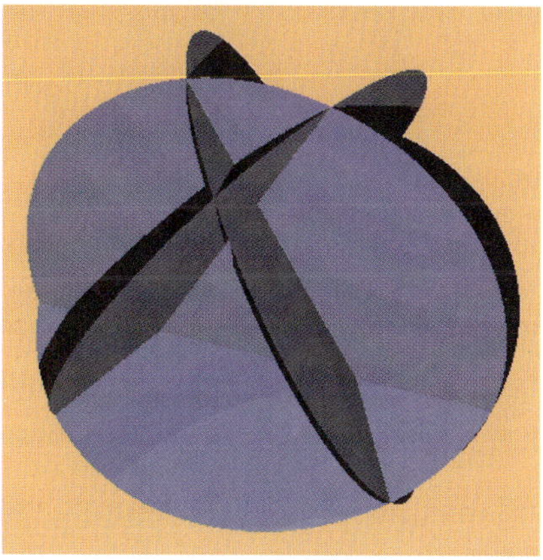

Plate 112. A degeneration of the Steiner surface, phase 8

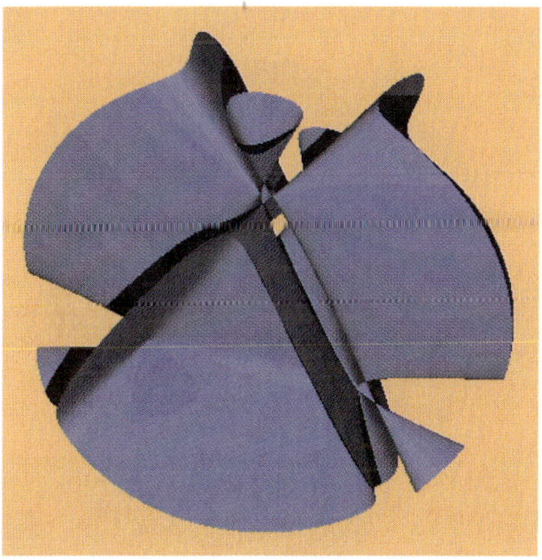

Plate 113. A degeneration of the Steiner surface, phase 9

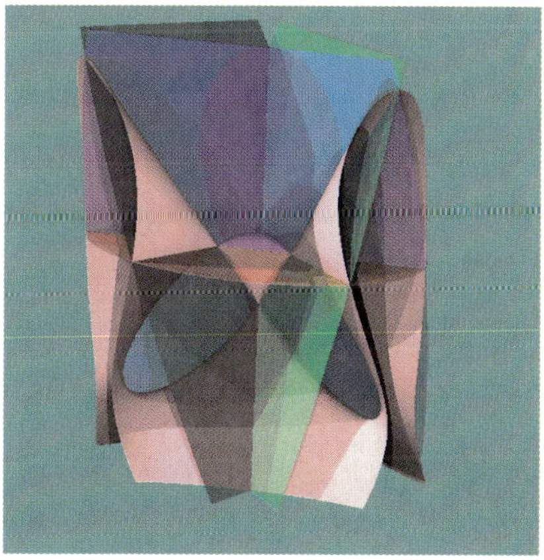

Plate 114. The Cayley cubic with some tritangents and lines

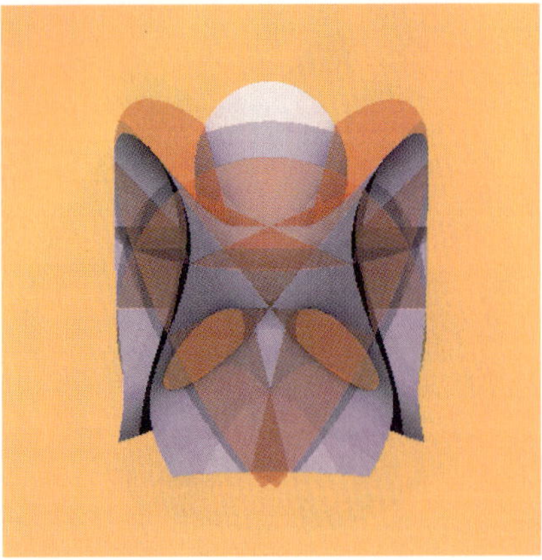

Plate 115. The Clebsch cubic with some tritangents and lines

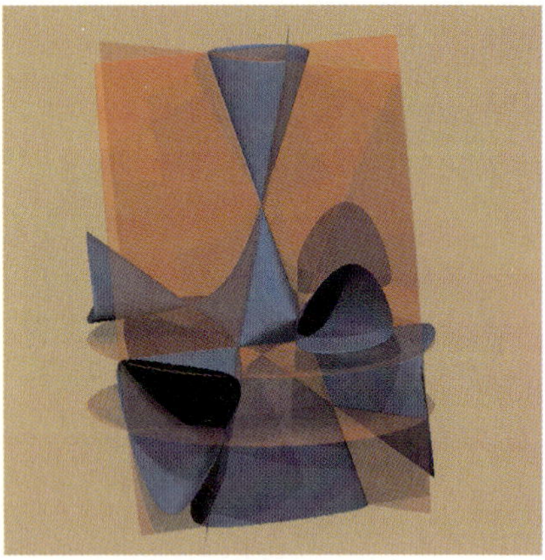

Plate 116. Lines on the Hessian of the Cayley Cubic

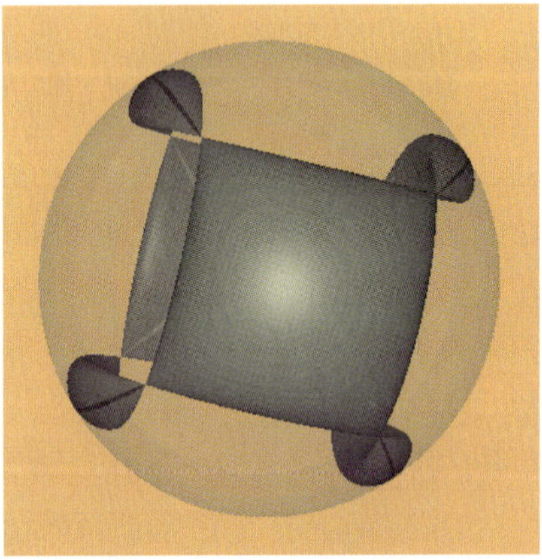

Plate 117. A quartic and quadric surface in contact

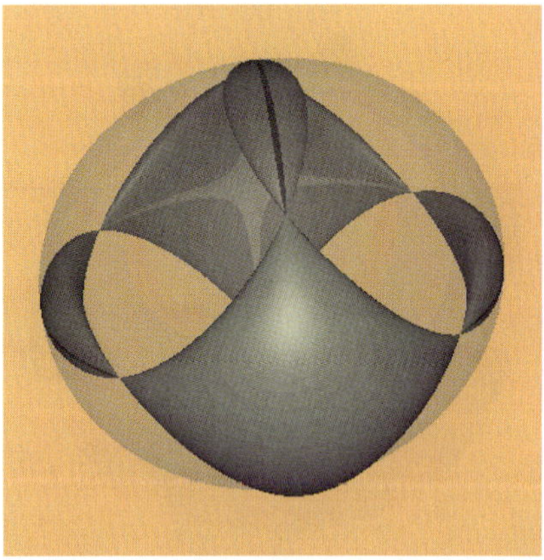

Plate 118. A quartic and quadric surface in contact

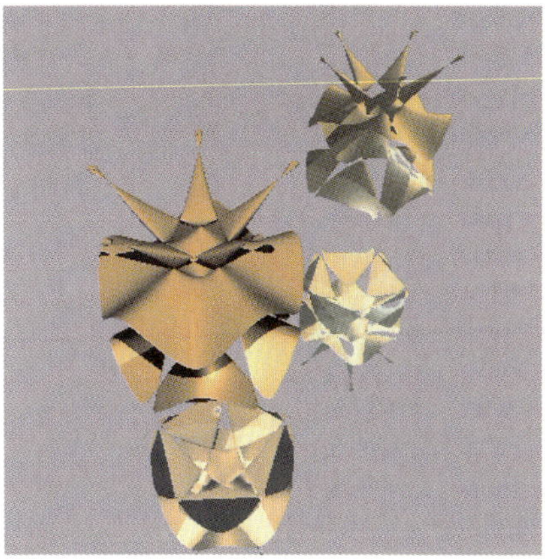

Plate 119. A Togliatti Quintic

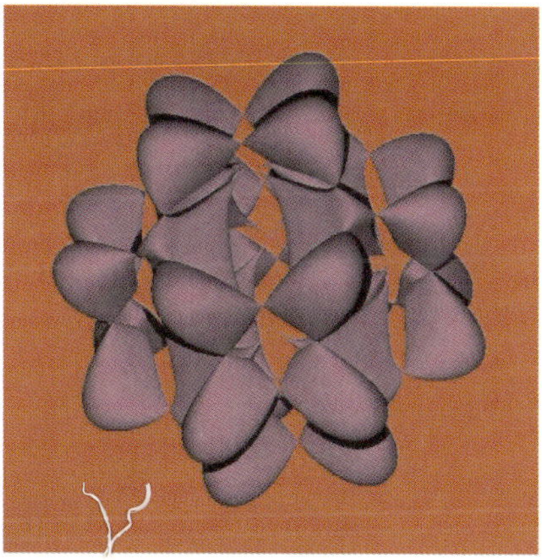

Plate 120. A Čmutov octic with 112 nodes

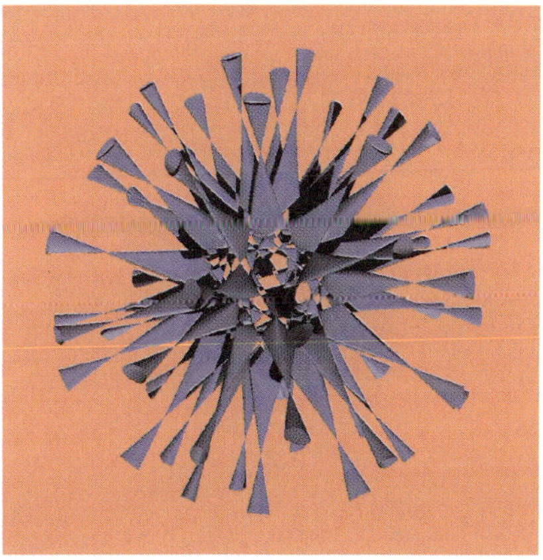

Plate 121. A Čmutov surface of degree ten with 345 nodes

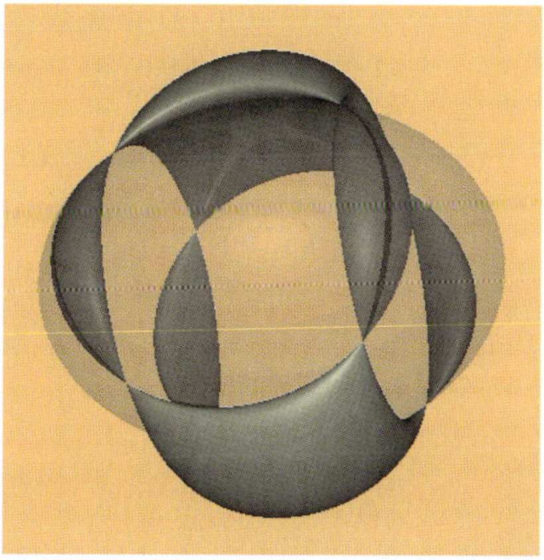

Plate 122. A quartic and quadric surface in contact

Plate 123.

PATRICE JEENER

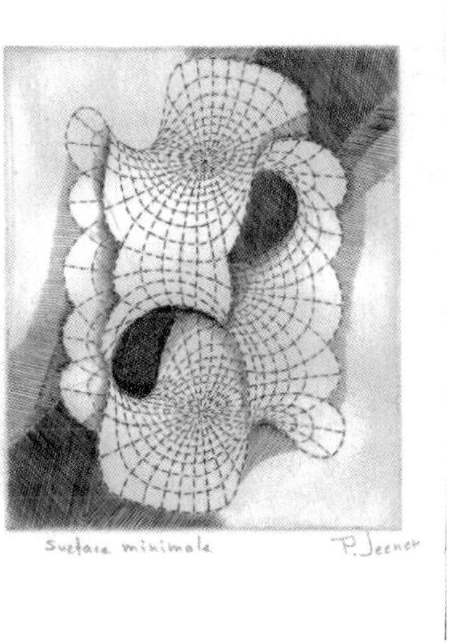

Plate 124.

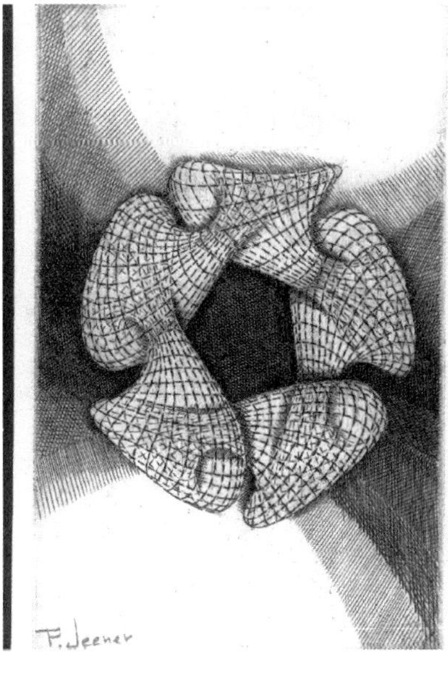

Plate 125.

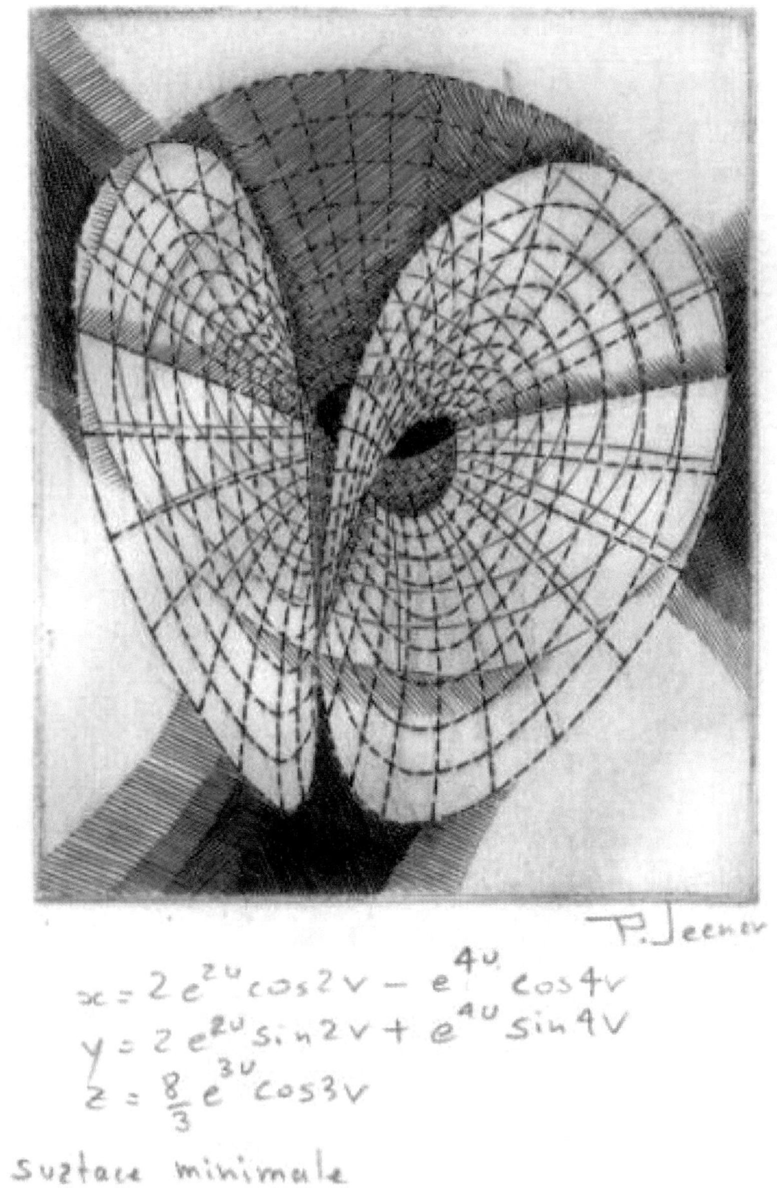
P. Jeener
$x = 2e^{2u}\cos 2v - e^{4u}\cos 4v$
$y = 2e^{2u}\sin 2v + e^{4u}\sin 4v$
$z = \frac{8}{3}e^{3u}\cos 3v$
surface minimale

Plate 126.

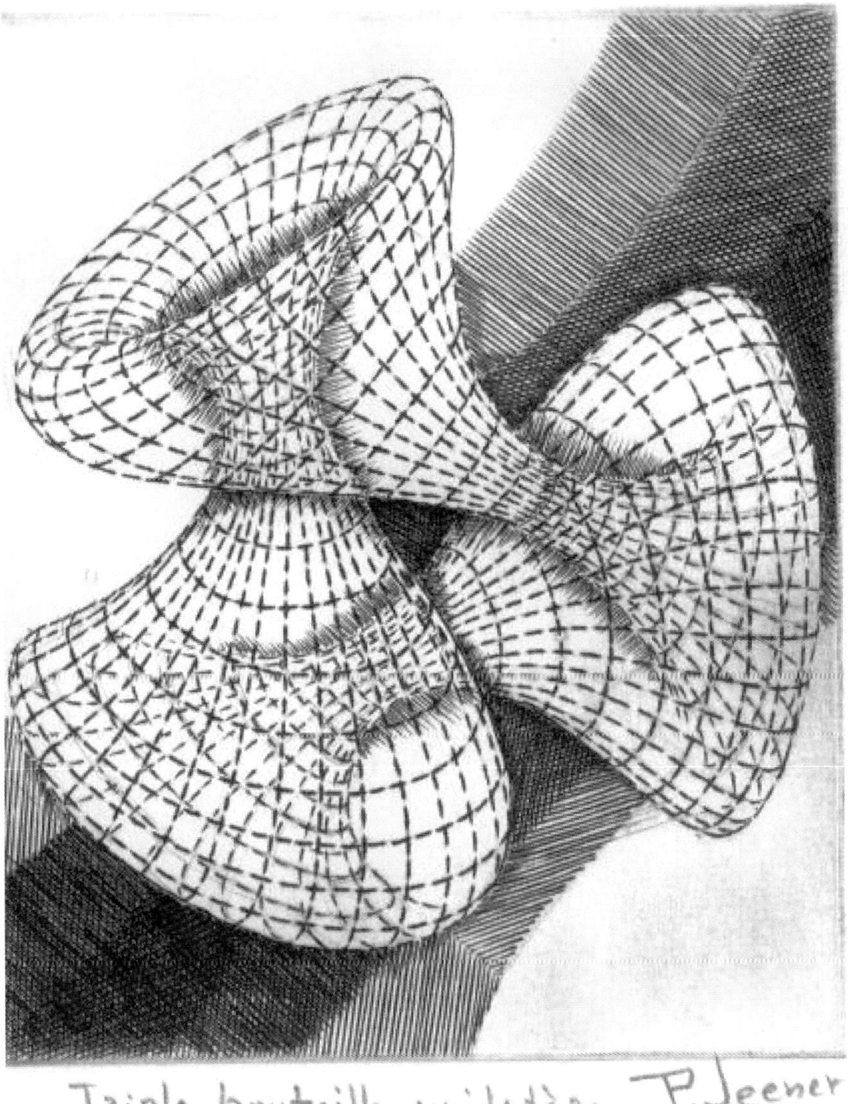

Plate 127.

Index

[FA] François Apéry, [RB] Ronnie Brown, [CB] Claude Bruter, [PC] Philippe Charbonneau, [MC] Manuel Chaves, [JCF] Jean-François Colonna, [SC] Scott Crass, [MD] Maria Dedo, [SD] Stewart Dickson, [ME] Michele Emmer, [MF] Michael Field, [NF] Nat Friedman, [GH] George Hart, [YH] Yves Hellegouarch, [JH] John Hubbard, [BH] Bruce Hunt, [PJ] Patrice Jeener, [EN] Erich Neuwirth, [RP] Richard Palais, [RP] Richard Perry, [KP] Konrad Polthier, [JR] John Robinson, [CS] Carlota Simoes, [JS] John Sullivan, [DT] Dick Termes, [AV] Alexandre Vitkine

Printing: Mercedes-Druck, Berlin
Binding: Stein + Lehmann, Berlin